# READINGS
## IN
## RISK

# READINGS
## IN
## RISK

Theodore S. Glickman
Michael Gough
editors

Resources for the Future
Washington, D.C.

© 1990 RESOURCES FOR THE FUTURE

Fourth printing, 1995

Printed in the United States of America

Published by Resources for the Future
1616 P Street, N.W., Washington, D.C. 20036

Books from Resources for the Future are distributed worldwide by
The Johns Hopkins University Press.


*Library of Congress Cataloging-in-Publication Data*

Readings in risk / Theodore S. Glickman and Michael Gough, editors.
   p. cm.
   Includes bibliographical references.
   ISBN 0-915707-55-1 (alk. paper)
   1. Technology—Risk assessment. 2. Health risk assessment.
I. Glickman, Theodore S. II. Gough, Michael, 1939–  .
T174.5.R38  1990
658.4–dc20                               90-35402
                                                   CIP


∞ The paper in this book meets the guidelines for permanence and
durability of the Committee on Production Guidelines for Book
Longevity of the Council on Library Resources.


*This book is the product of the Center for Risk Management at
Resources for the Future, Michael Gough, director. The project editor
was Dorothy Sawicki. The book was designed by Kenneth J. Sabol
and the cover by Peggy Friedlander.*

*The cover illustration "Acrobats and Jugglers," a woodcut by Joseph
W. Hart, Cincinnati, 1870, is used by courtesy of the Library of
Congress.*

RESOURCES FOR THE FUTURE (RFF) is an independent nonprofit organization that advances research and public education in the development, conservation, and use of natural resources and in the quality of the environment. Established in 1952 with the cooperation of the Ford Foundation, it is supported by an endowment and by grants from foundations, government agencies, and corporations. Grants are accepted on the condition that RFF is solely responsible for the conduct of its research and the dissemination of its work to the public. The organization does not perform proprietary research.

RFF research is primarily social scientific, especially economic. It is concerned with the relationship of people to the natural environmental resources of land, water, and air; with the products and services derived from these basic resources; and with the effects of production and consumption on environmental quality and on human health and well-being. Grouped into four units—the Energy and Natural Resources Division, the Quality of the Environment Division, the National Center for Food and Agricultural Policy, and the Center for Risk Management—staff members pursue a wide variety of interests, including forest economics, natural gas policy, multiple use of public lands, mineral economics, air and water pollution, energy and national security, hazardous wastes, the economics of outer space, climate resources, and quantitative risk assessment. Resident staff members conduct most of the organization's work; a few others carry out research elsewhere under grants from RFF.

Resources for the Future takes responsibility for the selection of subjects for study and for the appointment of fellows, as well as for their freedom of inquiry. The views of RFF staff members and the interpretation and conclusions of RFF publications should not be attributed to Resources for the Future, its directors, or its officers. As an organization, RFF does not take positions on laws, policies, or events, nor does it lobby.

# Contents ——————————————————————

# Foreword ————————————————————————

A few short months ago, in late 1989, devastation was wrought by Hurricane Hugo on the southeast coast of the United States and by the San Francisco earthquake 3,000 miles away. These natural catastrophes let us forget for a moment—but only that—that our seeming national preoccupation with risk is focused more on health and safety problems caused by man than on acts of God. Even a casual review of headlines over the past decade brings memories of our self-inflicted environmental and technological wounds flooding back: nuclear power accidents at Three Mile Island and Chernobyl; the release of lethal gas in Bhopal; the creation of the ozone hole in the Antarctic; the *Exxon Valdez* oil spill in Prince William Sound; the fire and subsequent chemical spill on the Rhine River in Basel, Switzerland; the Alar-in-apples scare.

Concerns about hazards such as these have spurred the recent growth in the field of risk analysis. Building upon the pathbreaking work on natural hazards by geographer Gilbert F. White and others, risk analysis expanded to address the sorts of health and safety problems recalled above. As it grew, the field began to be subdivided, first in the 1983 report by the National Academy of Sciences that made the distinction between risk assessment and risk management, and later in the landmark paper by William D. Ruckelshaus, first published in 1985 and reprinted in this volume, which drew attention to the importance of risk communication. Briefly, the distinctions are these: *risk assessment* is the process through which we attempt to determine the likelihood and extent of harm that may result from a health or safety hazard. In *risk management* we combine information about risk with economic, political, legal, ethical, and other considerations to make public or private decisions regarding protective policies. And *risk communication* is the two-way exchange of information, concerns, and preferences about risks between decision makers and the public.

Because risk analysis is an extraordinarily interdisciplinary field—drawing, at a minimum, from biostatistics, chemistry, toxicology, epidemiology, economics, operations research, psychology, engineering, and physics—important and helpful papers have appeared in a wide assortment of journals not easily accessible to all interested parties. This poses a special problem for those teaching undergraduate or graduate courses in risk analysis, as well as for other experts in the field and for readers seeking background information on this subject of growing interest.

When Resources for the Future created the Center for Risk Management in April of 1987, it did so with several purposes in mind. The first among those was that the Center sponsor and conduct original research designed to shed light on important policy problems in contemporary risk management, and the second, that it help organize existing research and policy analysis in such a way that nontechnical audiences can better understand the complexities of contemporary risk management. So it was that the idea arose of collecting papers from disparate sources and presenting them in a single volume with introductory remarks and thought-provoking questions. *Readings in Risk* is that volume. It is intended for college and university faculty and their students, for policymakers in the legislative and executive branches of government, for corporate risk managers and environmental advocates, and for ordinary citizens interested in knowing more about the intriguing new field of risk analysis.

Having helped RFF launch the Center for Risk Management as its first director, and having recruited Theodore S. Glickman and the Center's newly appointed director Michael Gough to its staff, it is with some considerable pride that I introduce this first book from the Center. It is appropriate that the book should build on the outstanding work done by those who helped initiate risk analysis, and even more appropriate that it be an effort to reach out to and help educate the first full generation of students in the field. This emphasis on education has been a hallmark of RFF for thirty-eight years, and *Readings in Risk* is a worthy contribution in that rich tradition.

April 1990

Paul R. Portney
Vice President and Senior Fellow
Resources for the Future
Washington, D.C.

# Preface ———————————————————————————

The past few decades have witnessed the emergence of a new field of research concerned with assessing and managing risks to health, safety, and the environment. The progress that has been made in expanding our understanding of environmental contaminants and technological hazards is well documented in the burgeoning research literature, but not in a way that is readily accessible to a broad audience. This volume bridges that gap by reprinting nineteen papers written by thirty-four prominent experts in the field, reflecting many of the major advances in the research that has been done and the related scientific and social issues that have arisen. These contributions, originally published in thirteen different journals, represent the broad range of disciplines that deal with risk—engineering, political science, the physical and biological sciences, and the law—as well as cross-disciplinary areas of inquiry such as risk analysis and environmental studies.

This book will be of interest to students, practicing professionals, and other interested readers seeking a convenient, objective, and stimulating introduction to risk assessment and risk management, who would understandably be daunted by the prospect of having to sift through the research literature on their own. Little if any reader familiarity with the subject matter is presupposed; the volume contains only papers that are largely self-contained and that do not require the reader to have a sophisticated technical background (although some mathematical formalism springs up here and there in the discussions of risk assessment, which is inherently a quantitative endeavor.)

The papers are organized in six parts, according to these topics: (1) basic concepts, (2) risk comparisons, (3) regulatory issues, (4) health risk assessment, (5) technological risk assessment, and (6) risk communication. The process of selecting the papers began with a survey of the reading lists for risk-related courses in colleges and universities across the United States. We filled the niches in our six-part organizational scheme as well as we could with the papers that showed up most frequently on these lists and at the same time satisfied our criteria of self-containment and technical approachability. An external peer review of our proposed choices then generated a number of candidates for addition, substitution, and deletion, along with two strong recommendations: that papers with real-world applications be made part of the collection and that the published correspondence relating to the most controversial papers also be included. Upon reviewing the candidates and considering the recommendations, we decided on the final set of selections.

For the benefit of course instructors, we have included with each paper a number of questions that can be used for classroom discussion, homework assignments, or examinations. We hope that readers who are not *required* to answer these questions will be motivated to ponder them nevertheless. We also urge those who are interested in learning more about risk assessment and risk management to pursue the references in the individual papers and to consider the list of books we suggest for further reading.

We are grateful to our colleagues Alyce Ujihara and Marilyn Voigt for their assistance, to Adam Finkel for his advice, and to Paul Portney for his encouragement and support. We also thank Dorothy Sawicki, the project editor.

April 1990

Theodore S. Glickman
*Senior Fellow*

Michael Gough
*Director and Senior Fellow*

Center for Risk Management
Resources for the Future
Washington, D.C.

# P A R T  1

# BASIC CONCEPTS

# Introduction ——————————————————————————

Mankind has always sought to eliminate unwanted risks to health and safety, or at least to bring them under control. Its successes are impressive, as evidenced by advances such as the discovery of vaccines that prevent polio and smallpox and the development of building methods that enable stadiums and skyscrapers to withstand powerful earthquakes. Despite our best efforts, however, some familiar risks persist while others that are less familiar are found to have escaped our attention far too long and new ones continue to arise.

Ironically, some of the risks that are most difficult to manage or accept arise from technologies that are intended to improve our standard of living. The invention of the automobile, the advent of air travel, the development of synthetic chemicals, and the introduction of nuclear power all illustrate this point.

The respects in which the world is or is becoming an unhealthy or unsafe place to live and the views on how to allocate resources to do something about it are the subjects of continuing debates. These debates underline the need for clear concepts and useful definitions to help solve the risk problems that we face. The four papers on basic concepts in part 1 reflect much of the current thinking on these matters. While they all embody the idea that risk means the possibility of harm, they also demonstrate that conceptualization and definition are only the beginning of understanding risk.

In "Probing the Question of Technology-Induced Risk," M. Granger Morgan describes a conceptual framework in which two phenomena—"exposure processes" and "effects processes"—can produce changes in the technologically related risks borne by us and our environment. Depending on our perception and evaluation of any given risk, he explains, we either accept it or not, and if not, then we attempt to reduce our exposure to it or act to mitigate its effects. The need to decide which risks to address draws attention to the intriguing question of why some risks are more worrisome than others, and he sheds light on that subject as well.

In "Choosing and Managing Technology-Induced Risk," a sequel to the first paper, Morgan draws our attention to risk assessment and risk management. These two activities are motivated, he tells us, by the realization that individually and collectively we must decide first how much risk we can live with and then how we will deal with the risks we choose to limit. In practicing risk assessment,

3

which he characterizes as part art and part science, the author warns against putting too much effort into the process of generating numerical values at the expense of exercising careful human judgment. When making decisions about which risks to abate and how much to spend to get the job done, risk managers must accept the fact that there are inevitable uncertainties and prepare to deal with the societal impacts of their decisions.

Authors Baruch Fischhoff, Stephen R. Watson, and Chris Hope develop the thesis, in "Defining Risk," that there is no single definition of risk. The particular situation determines, for instance, whether risks are assessed in terms of morbidity or mortality and whether they are limited to immediate effects or extended to include delayed effects. The appropriate measure of exposure to use when determining a fatality or injury rate also depends on the context, the authors say. Then they analyze the risks from alternative methods of generating electricity to demonstrate how the determination of risk can be influenced by the relative emphasis placed on various contextual factors.

Addressing the issue of acceptable risk in "Risk Analysis: Understanding 'How Safe Is Safe Enough?,' " Stephen L. Derby and Ralph L. Keeney suggest that acceptability depends on what it takes, financially or otherwise, to achieve reductions in risk. They argue that if one considers the advantages and disadvantages of all the available alternatives for diminishing a given risk, then by definition the *best* alternative is "safe enough." They dismiss as inappropriate the tenets that "only no risk is safe enough" and that "only the safest alternative is safe enough." The discussion then turns to the means by which technical, social, political, and ethical considerations enter into the decision process.

# Probing the Question of Technology-Induced Risk

## M. GRANGER MORGAN
*Carnegie Mellon University*
*Pittsburgh, Pennsylvania*

If there were a top 40 for the hot problems of our day, risk would be high on the charts. Many people are writing about it. Private organizations and government regulators are trying to assess and manage it. Lobbyists and interest groups would like the public either to ignore it or to worry more about it. And, whatever action is taken, lawyers are thriving because of it. Why this big concern with risks?

The statistical evidence shows that Americans live longer, healthier, and wealthier lives today than they did at any time in the past. Perhaps, some economists argue, we worry more about risk today precisely because we have more to lose and because we have more disposable income to spend on risk reduction. Such economic factors are undoubtedly important, but they probably do not tell the whole story. Why, for example, do risks from new technologies—like microwave ovens, which present a relatively low probability of death or injury—often get more attention than older, well-established risks like motor-

cycles, which routinely are involved in the deaths of or injuries to substantial numbers of people? Why do people who have slept under an electric blanket every night for years suddenly become terrified about possible adverse health effects when a new 765-kilovolt power line is built a few hundred meters from their home?

Our ancestors, more resigned to death, disease, and injury, understood that nothing in life is risk-free. But we have done so well in reducing risks that now many seem to forget that no activity or technology can be absolutely safe.

With increasing frequency, scientists, engineers, and others who have thought carefully about risk have been arguing that the real problem is not the unachievable task of making technologies risk-free, but the more subtle problem of determining how to make the technologies safe enough. Most engineers find it rather natural to think about risk questions in these terms—to think, for example, about trading off the benefits of improved medical diagnostic capabilities against the risks of exposure to diagnostic X-rays. But engineers often become disturbed when they begin to discover that the question "How safe is safe enough?" has no simple answer.

"Probing the Question of Technology-Induced Risk" by M. Granger Morgan is reprinted, with the permission of the publisher and the author, from *IEEE Spectrum*, vol. 18, no. 11, pp. 58–64 (November 1981). © 1981 IEEE.

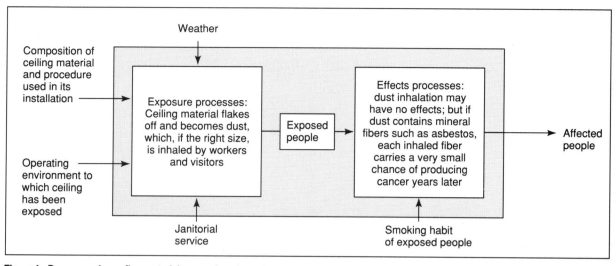

**Figure 1.** Does sound-proofing material sprayed on the ceiling of a room years ago in a manufacturing facility pose a health risk to visitors and workers? The answer depends upon the composition and history of the material and its application, a set of current environmental factors, and a set of exposure processes and effects processes.

## Typical Examples of Risk

Some examples may help to develop a framework for thinking about technological risk to health, safety, and the human environment.

Each year visitors tour the U.S. Mint in Denver, Colo. Last summer some noticed that the ceilings in some of the work rooms, sprayed years ago with a layer of sound-proofing material to deaden the noise from the machinery, had begun to flake and strands of fibers could be seen dangling here and there, gently wafting in the breeze from open windows and ventilating fans. In some places visitors could reach up from the tour balcony and touch the ceiling. Did this old, flaking ceiling represent a health risk? To answer that question, the processes by which people are exposed to the chance of undergoing some change (exposure processes) and the processes by which these changes occur (effects processes) must be examined.

Through age, vibration, air motion, and perhaps human contact, pieces of ceiling fiber were falling off. The physics and chemistry of the situation determined how many of these fibers fell off and whether they fell off only as large chunks or also as small submicrometer particles. Other factors that probably influenced the amount and size of dust particles in

the air included the way in which the janitors and maintenance people cleaned the machinery and floors, as well as the air-circulation patterns that were set up by the fans and open windows.

The human upper respiratory system is remarkably good at filtering out dust particles. But particles from a fraction of a micrometer to a few micrometers can enter deep into the lungs. Hence if the exposure process involved submicrometer ceiling particles, then some of these particles were probably ending up in the lungs of workers and a few in the lungs of visitors.

What is the possible effect of submicrometer ceiling particles lodged deep in the lungs? In the short run there probably are no effects at all. Dust is present in the air all the time, and in this case the amount of ceiling dust was fairly low. In the long run, the situation is less clear, depending strongly on the chemical and physical properties of the particles. There is a fair likelihood that the ceiling materials included mineral fibers, and asbestos in particular. If this were so, then there was a very small chance that 10 or 20 years after any given particle became lodged in a lung, cancer would develop. If the lung happened to belong to a smoker, the chance of cancer from exposure to asbestos might be as much as 15 times greater than the chance for a nonsmoker. These

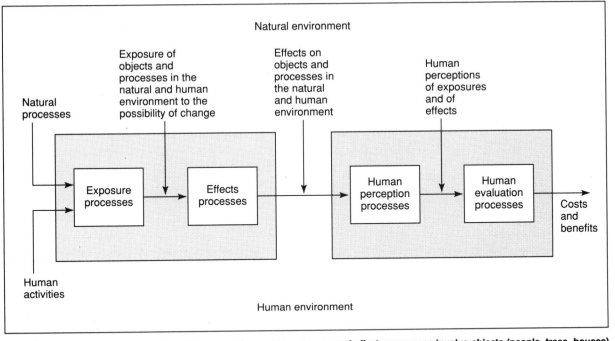

**Figure 2.** Interacting with the natural and human environments, exposure and effects processes involve objects (people, trees, houses) and natural events (weather). How these changes are perceived and evaluated by society and by people as "risks" depends in turn upon a set of perception and evaluation processes. All four processes shown—exposure, effects, perception, and evaluation—are also influenced by and can exert influence upon the natural and human environments in ways not shown here.

exposure processes and effects processes are summarized in a simple diagram [figure 1]. (The U.S. Treasury Department asserts that tests have been unable to detect the presence of airborne fibers at the Denver mint.)

The exposure and effects processes associated with other risks can be identified in a similar way. In considering the possible health effects on those who spend their lives working on or near high-frequency radar-transmitting antennas, the exposure processes involve anything that brings microwave radiation and antenna workers together. For example, workers may be required to lubricate and adjust the moving parts of the antenna mount periodically while the system continues to operate. The exposure processes may also involve protective shielding and intensification effects that may modify the incident field strengths to which workers are exposed. Effects involve the electrochemical and biological processes that may affect human health adversely, such as cataracts after years of high-level exposure.

An electric power-distribution system outage is an example of a risk that does not involve a direct threat to human health. The exposure processes may include the presence of thunderstorms, distribution lines in exposed places, inadequate automatic reclosure equipment, and the absence of an automated distribution system. The effects processes might involve a series of events when a lightning strike occurs in close proximity to an exposed portion of the line, resulting in the automatic opening of protection equipment and the loss of power to a neighborhood.

**Not All Exposure Is Bad**

The results of exposures and of effects can be beneficial as well as harmful. In a coal-burning factory or power plant, for instance, that produces the air pollutant sulfur dioxide ($SO_2$) as one of its combustion by-products, exposure processes disperse the pollutants and carry some of the $SO_2$ to a field of alfalfa that happens to be growing in sulfate-poor soil. The $SO_2$ is absorbed by the alfalfa plants, but

| | Probabilistic exposure | Deterministic exposure |
|---|---|---|
| **Probabilistic effect** | **Getting stung by a bee**<br><br>Those engaged in normal day-to-day activities do not usually get stung by bees, but there is always a small exposure probability. Once stung, most people simply have a painful bite, but a life-threatening allergic reaction is possible. | **Routine dental X-ray**<br><br>Exposure to dental X-rays is deterministic in that the patient chooses to have the procedure done. The possible health effects from exposure to low-level radiation are inherently probabilistic. |
| **Deterministic effect** | **Being on top of a large gas-main explosion**<br><br>Exposure to a big gas-main explosion is clearly probabilistic and very low for members of the general public. Those, however, who happen to be standing right above one will certainly be killed. | **Spending December on an arctic ice-floe in a tennis outfit**<br><br>The only way to get on an arctic ice-floe in a tennis outfit in December is intentionally. Once one has been there a month, the situation's effects are clearly deterministic. |

Figure 3. Risk involves the exposure to a chance of injury or loss. The chance or probabilistic aspect of risk can be introduced through exposure processes, effects processes, or both, yielding four possible combinations of exposures and effects.

because the concentration of $SO_2$ in the air is low, toxic levels do not build up in the plants. Instead, the $SO_2$ is metabolized by the plant into sulfates and other sulfur compounds, some of which are needed for healthy plant growth. The resulting effects processes give rise to an increase in the total yield from the alfalfa field.

Hence, in the framework of figure 1, "exposure" does not mean exposure to risk. Instead, it is the exposure of people, objects, and systems to the possibility of some change. Similarly "effects" does not mean costs or losses. It is simply a list of changes.

Many dictionaries define risk as "the exposure to a chance of injury or loss." Whether people perceive a particular change in the world as an injury or loss depends both on the processes by which they perceive the change and the processes by which they evaluate it [figure 2].

## The Chance Element

The chance, or probabilistic, element of risk may arise from the following:

1. The values of all the important variables involved are not or cannot be known, and precise projections cannot be made.

2. The physics, chemistry, and biology of the processes involved are not fully understood, and no one knows how to build adequate predictive models.

3. The processes involved are inherently probabilistic, or at least so complex that it is infeasible to construct and solve predictive models.

In addition, the extent to which a process is viewed as certain or uncertain often depends upon individual perspective. An individual driver is likely to view automobile accidents as highly unpredictable events. But an insurance company can usually predict with remarkable precision the number of accidents its policy holders will have during the year ahead.

The probabilistic element in risk can be introduced through exposure processes, or through effects processes, or both—yielding four possible combinations: a probabilistic exposure followed by a probabilistic effect, a deterministic exposure followed by a probabilistic effect, a probabilistic exposure followed by a deterministic effect, and a deterministic exposure followed by a deterministic effect [figure 3]. The last should probably not be included as an example of risk, because whereas the resulting injury or loss may be substantial, there is nothing probabilistic about its occurrence.

The four-way classification seems simple enough until it is applied. Then, depending upon the example, it can be seen why so many of the heated arguments about risk revolve around questions of where and how to classify things and draw the system boundaries. For example, some surgeons might argue that the risk of failure of a surgical implant constitutes an example of certain determin- istic exposure (inserting the device) with probabilistic effect (stemming from such unpredictable complica- tions as infection, rejection, and device failure). On the other hand, a malpractice lawyer might catego- rize this as a case of probabilistic exposure (the same injury treated by different doctors will get different treatment) with deterministic effects (the surgeon used poor technique so that infection occurred, or he damaged the device while inserting it, with the result that it ultimately failed).

The information about the uncertainty attached to risk processes usually fits into one or more of the following five categories:

1. Good direct statistical evidence on the process of interest is available. This is clearly the most desirable situation, but is rare for most categories of risk problems.

2. The process can be disaggregated with analyt- ical tools—such as fault trees, event trees, and vari- ous stochastic models—into subprocesses, for which good statistical evidence is available. Aggregate prob- abilities can then be constructed.

3. No good data are available for the process under consideration, but good data are available for a similar process and these data may be adapted or extended for use either directly or as part of a disaggregated model.

4. The direct and indirect evidence that is avail- able is poor or incomplete and it is necessary to rely to a very substantial extent on the physical intuition and subjective judgment of technical experts.

5. There is little or no available evidence, and even the experts have little basis on which to produce a subjective judgment.

Unfortunately a very substantial fraction of the risk problems that society must deal with fall into categories 3, 4, or 5.

## Other Factors Influence Judgments

Suppose that the following seven hypothetical exam- ples of risk all statistically produce the same expected number of deaths, D, when looked at on an annual basis. Despite this fact, perception and evaluation of risks vary considerably from case to case.

*Case 1:* People have begun to buy small, two- passenger electric automobiles to do their local shopping. These slow vehicles are occasionally in- volved in fatal accidents, often with larger conven- tional vehicles. At present, though they are much cheaper, these small cars are not as safe as larger cars. The incremental expected number of deaths for the current fleet of small cars is D deaths per year.

*Case 2:* Commuter airlines that fly twin-engine aircraft to small airports are becoming increasingly important around the world. Viewed on a fatalities- per-passenger-mile basis, they are generally quite safe. The commuter airline system in one country now operates with an annual expected rate of acci- dental deaths of D deaths per year. These accidents have a fatality rate between 0.5D and 2D per acci- dent.

*Case 3:* The large residential community of Wil- lowbend is three miles downstream from a large earth dam that is part of a regional irrigation and flood-control project. Should the dam fail abruptly, it has been estimated that approximately half of the N people living in Willowbend would die. The best available estimate on the annual probability of fail- ure for such dams is P. The product of $P \times N/2$ yields an annual expected statistical mortality rate of D deaths per year.

*Case 4:* With the ever-increasing price of oil, coal is once again playing an important part in the energy picture of the world. Underground coal mining is dangerous, but it usually does not require a very large fraction of a country's work force. In one country accidental coal mine deaths are now averaging D deaths per year.

*Case 5:* Twenty years ago workers in a major chemical industry were routinely exposed to a com- pound whose trade name was TZX. Nobody thought very much about the situation. It has since been learned that TZX causes a fatal degenerative disease of the central nervous system. Last year D deaths were attributed worldwide to occupational exposure to TZX in the early 1960s.

*Case 6:* On a per-passenger-mile basis the charter bus lines of most countries have an excellent record. In one country the mortality rate is averaging about D deaths per year. Indeed, there were precisely D deaths last year, and they all occurred in just one accident in which a charter tour bus, carrying the award-winning high-school concert band from a small, traditional, rural town, went out of control and over an embankment.

*Case 7:* Sulfur air pollution is produced when coal is burned to generate electric power. It is difficult to estimate the health impact of coal-fired power plants because neither the exposure processes nor the effects processes are well understood. For a given coal-fired power plant, it is estimated that the most likely value of the impact on excess mortality (that is, death occurring at least one year earlier than it would without the presence of the pollutant) is D deaths per year. But it is also estimated that there is a 20-percent probability of no increase in mortality and a 10-percent chance that the increase is as great as 4D deaths per year.

# Death: A Risk Faced Daily

Though no technology can be made absolutely safe, neither are everyday activities risk-free. Everyone faces "background" risks all the time. These can be compared with risks from various technologies, and they take a variety of forms: death, disease and injury, and undesired changes in environments. Of these, death is usually the easiest to identify and quantify.

The statistical rate at which people die depends strongly on a number of variables, with age and sex among the most important. Death rates start out fairly high at birth and then fall rapidly over the first several years of life as the various processes of infant mortality are passed by. Then there is a broad minimum followed by steadily increasing death rates as people grow older. The rate at which the death rate increases with age is higher for males than it is for females.

By analyzing a computer file of all the death certificates from 1900 to 1970, Salvador Bozzo of the Brookhaven National Laboratory in Upton, N.Y., has produced contours of constant age-, sex-, and race-specific mortality rates for the United States [figure A]. It is obvious from these plots

Deaths per 1000 per year among white males

Deaths per 1000 per year among white females

| Activities that increase chance of death by one in a million yearly | Cause of death |
| --- | --- |
| Smoking 1.4 cigarettes | Cancer, heart disease |
| Spending 1 hour in a coal mine | Black lung disease |
| Spending 3 hours in a coal mine | Accident |
| Traveling 10 miles by bicycle | Accident |
| Traveling 300 miles by car | Accident |
| Flying 1000 miles by jet | Accident |
| Flying 6000 miles by jet | Cancer from cosmic radiation |
| Living 2 months in Denver | Cancer from cosmic radiation |
| Living 2 months in average stone or brick building | Cancer from natural radioactivity |
| One chest X-ray taken in a good hospital | Cancer from radiation |
| Living 2 months with a cigarette smoker | Cancer, heart disease |
| Drinking 30 12-ounce cans of diet soda | Cancer from saccharin |
| Living five years at site boundary of a typical nuclear power plant in the open | Cancer from radiation |
| Drinking 1000 24-oz soft drinks from recently banned plastic bottles | Cancer from acrylontrile monomer |
| Eating 100 charcoal-broiled steaks | Cancer from benzopyrene |

**Figure A. Examining age-specific mortality rates from 1900 through 1970 shows how U.S. mortality rates have changed during this century.**
*Source:* Salvador Bozzo, Brookhaven National Laboratory, Upton, N.Y.

*Source:* "Analyzing the daily risks of life," by Richard Wilson, *Technology Review.* February 1979. [Reprinted with permission from *Technology Review,* copyright 1979.] Because of the nature of the data on which they are based, some of these examples are subject to considerable uncertainty, in a few cases involving probably as much as several factors of 10.

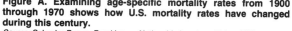

Despite the fact that all of these cases carry the same statistically expected mortality rate, most people do not react to each in the same way. Why? What are some of the factors that differentiate these hypothetical cases? The number of people killed in each accident is clearly one factor that varies. In the electric auto case D people are killed in D/2 to D

individual separate accidents. In cases 2 and 6 most or all of the D deaths occur in a single event. In the case of the dam failure, many more than D deaths occur, but the event has a low probability of occurring in any given year. The spatial distribution of the deaths also varies. Aircraft and dam accidents happen in one or a few localized places. Air pollution, by

---

that over this century mortality rates have decreased substantially. The effect of the 1918–19 influenza epidemic that killed 20 million people worldwide is clearly visible in both plots. The impact of World War II on the mortality rate of young men is also apparent. Women live longer than men, and from 1900 to about 1955 their expected lifetime was growing more rapidly than men's. This trend appears now to have stopped, perhaps partly because of the combined effects of smoking and the occupational hazards to which a growing number of working women have been exposed.

Over the century the causes of death have also shifted. There has been a dramatic decrease in deaths from infectious diseases, a rise in accidental deaths from automobiles, a modest decline in other accidental fatalities, and an increase in deaths from cancer and heart disease. The latter results in part from the fact that many people are now living long enough to die of these diseases, which primarily affect older people.

But while reasonably reliable direct evidence is available for the causes of accidental deaths, such data are not typically available for deaths from chronic exposure to such things as pollutants, ionizing radiation, or foods and drugs.

Estimates of the death rates produced by these kinds of risks are generally based on epidemiological studies or on various extrapolations from toxicological and other evidence. There is often very large uncertainty in the resulting values.

The risk of death from specific causes is most frequently stated in terms of the average probability of deaths per year per exposed person. Such numbers range from around $2 \times 10^{-4}$ for motor vehicle accidents to around 4 or $5 \times 10^{-7}$ for such events as tornados, hurricanes, and lightning. Since most people have trouble understanding what numbers this small mean, a useful approach is to compare how often a person must perform an activity each year to increase his or her chance of dying by one in a million per year [see table].

Finally, for accidental deaths that may kill varying numbers of people per event, one should ask about the distribution of such deaths. Figure B gives one such estimate.

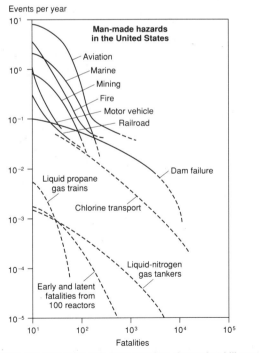

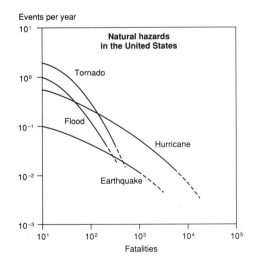

**Figure B. Natural and human-caused events that kill large numbers of people are generally less frequent than those that kill small numbers of people. Solid lines indicate estimates based on assessment studies and may carry considerable uncertainty.**
*Source:* Nuclear Regulatory Commission.

contrast, affects people over a very wide area.

The charter bus accident differs from most of the other accidents because all of the victims are socially related and their deaths will have a profound impact on the remaining residents of a small, traditional, rural town. The coal mine example differs because a few workers bear a high individual risk in order to produce a benefit that is shared by many. Death is immediate in the case of most risks involving accidents but occurs years after initial exposure in cases such as the TZX and sulfur air-pollution examples. Finally, the values of P and D may be known quite precisely for cases 1, 2, 4, and 6, but they may be extremely uncertain for cases 3, 5, and 7.

Studying many such cases shows that the important factors in determining how people perceive and

## Effects Processes Can Be Complex

It is tempting to think of an effects process in terms of a function that assigns a single effect for any given exposure in a particular environmental context. This concept can be useful but it has limitations.

In its simplest form, the magnitude of the effect depends on the total accumulated exposure. It might even be linear, like data on smoking and lung cancer in smokers who have been smoking for at least 10 years [figure A]. Frequently, however, dose-effect curves show substantial nonlinearities, as illustrated by the incidence of liver cancer in rats exposed to vinyl chloride monomer in the air they breathe [figure B].

Not all exposures have effects that most people would characterize as undesirable. For example, there is a decrease in decayed, missing, and filled teeth with increased fluoride concentration in the water that people drink as children [figure C].

Sometimes exposures have thresholds below which no effects are observed. In the case of living organisms, this may result for various reasons. For example, the organisms may be able to metabolize low levels of some materials, converting them into less harmful materials or excreting them. In the case of exposure to something like nonionizing radiation, a given power density may be necessary before certain effects begin to occur. For exposure to very low levels of ionizing radiation or to carcinogenic materials, biological processes may be able to repair or replace damaged cells.

The lack of solid experimental evidence frequently makes it difficult or impossible to know if a particular exposure does or does not display a lower threshold. This is the case, for example, with most air pollutants and ionizing radiation. Since

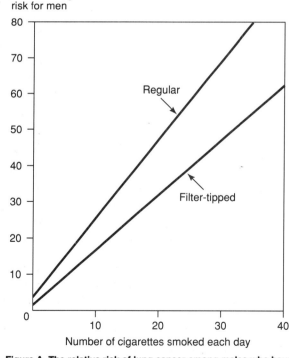

Relative lung cancer risk for men

Number of cigarettes smoked each day

**Figure A. The relative risk of lung cancer among males who have been smoking for 10 years or longer is reported to increase linearly as the number of cigarettes smoked increases.**
*Source:* [Adapted, by permission of] *The New England Journal of Medicine,* vol. 300 (1979) pp. 894–903.

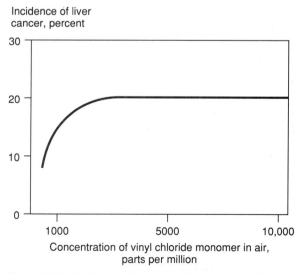

Incidence of liver cancer, percent

Concentration of vinyl chloride monomer in air, parts per million

**Figure B. The incidence of liver cancer in rats exposed to vinyl chloride monomer in the air they breathe reveals a nonlinear dose-effect curve.**
*Source:* [Adapted, with permission, from] *The Yale Journal of Biology and Medicine,* vol. 51 (1981), pp. 37–51.

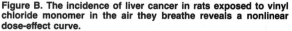

value a given set of exposure and effects processes can be captured by carefully answering two questions: "Who and what are exposed—when, where, how, and under what circumstances?" and "Who and what are affected—when, where, how, and under what circumstances?" Some of the factors that probably influence the way that people answer such

questions include the magnitude of the exposure and effects; the ways in which they are distributed in space and time; their characteristic time constants; whether effects are reversible; the physical, biological, and social circumstances of the exposures and effects; and the knowledge, resources, and capabilities of the population and the associated society.

## Insights into Human Perception

Until about five years ago, there were relatively few insights into how individual perception and evaluation processes work. Since then experimental psychologists have developed a number of hypotheses. Though still not universally accepted, they are being refined through experimental activity.

If a few hundred people are asked a series of such questions as "Considering the U.S. population as a whole, which is the more likely risk of death: emphysema or stroke? How many times more likely?", the answers allow some conclusions to be drawn about how various types of mortality risk are perceived. This is what three experimental psychologists, Sarah Lichtenstein, Paul Slovic, and Baruch Fischhoff and their colleagues at Decision Research in Eugene, Ore., did several years ago. Their questions involved 41 different causes of death. Their experimental subjects were college students and members of the League of Women Voters and their spouses.

The psychologists found that when the death rates between two causes of death differed by less than a factor of 2, their subjects could not pick out the more likely cause with any reliability. When the two death rates differed by more than a factor of 2, their subjects could usually get the order right. But they did not do very well on the ratios. For example, for a series of questions for which the true ratio was 1000 to 1, the average values of the answers to the individual questions ranged from less than 2 to 1 to roughly 5000 to 1.

When the study was repeated with a second group of subjects, high correlation was found between the answers from the two groups. Though the subjects were providing consistent, reproducible results, the results often did not correspond very well with the actual probabilities of dying.

In a subsequent series of experiments, new subjects were told that approximately 50,000 people die each year in the U.S. in motor-vehicle accidents, and they were asked to estimate the actual mortality

---

the presence or absence of a threshold can have important implications for public policy, this issue frequently becomes the subject of heated debate.

Moreover not all objects or living things that receive a given exposure will necessarily respond in the same way. In a report published by the National Academy of Sciences the risk of deaths from leukemia is reported to be about $10^{-6}$ deaths per year per rem for people who are 10 years of age or older at the time of exposure: about $2 \times 10^{-6}$ deaths per year per rem for children who are under 10 years old at the time of exposure, and about $25 \times 10^{-6}$ deaths per year per rem for children who receive exposure *in utero*.

Though the concept of a single effect for a given exposure can be useful, it often cannot adequately describe the effects of a pattern of exposures. This is particularly true when the object or organism exposed is a dynamic system whose response depends upon several factors: the environment, exposure history, and so on. In such cases the simple idea of a "single effects" function must be replaced with a more complex dynamic model of the effects process.

Incidence of decayed, missing, and filled teeth

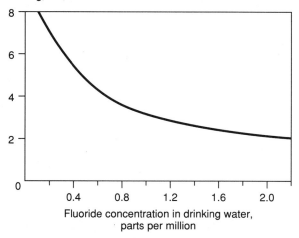

Fluoride concentration in drinking water, parts per million

**Figure C. At low concentrations, the incidence of dental problems decreases with increasing fluoride concentration in the water that those who were studied drank as children.**
*Source:* [Redrawn, with permission, from Hadge and Smith, "Some Public Health Aspects of Water Fluoridation," in J. H. Shaw, ed., *Fluoridation As a Public Health Measure* (Washington, D.C., AAAS, 1956). © 1956 AAAS.]

values for each of 40 other causes of death. The results revealed a clear tendency to underestimate the likelihood of high-probability causes of death and to overestimate the likelihood of low-probability causes of death.

For risks such as stroke that actually produce somewhat more than 100,000 deaths per year, the mean response of the subjects was around 10,000 deaths per year. The subjects' average response correctly estimated the number of deaths from risks such as asthma and tuberculosis at about 1000 deaths per year. Risks such as botulism that on average produce less than five deaths per year yielded an average response from subjects of roughly 500 deaths per year.

Why are people likely to underestimate the probability of dying from stroke and to overestimate the probability of dying from botulism? Two other psychologists, Amos Tversky of Stanford University and Daniel Kahneman of the University of British Columbia, have suggested that people assess the likelihood of an event by the ease with which they can recall or imagine examples of its occurrence. For common events, like the probability of encountering a highway patrol car during the drive to work in the morning, this heuristic—a simple-decision rule based on information available to the individual—works well. But psychologists have demonstrated that on problems like estimating the probability of dying from stroke or botulism, this "availability heuristic" can lead to serious biases.

Deaths from botulism are very rare but the population learns through the news media about nearly all botulism deaths that occur. Deaths from stroke are fairly common, yet unless a friend, relative, or famous person dies of a stroke, people rarely hear about such deaths. Some psychologists would argue that the availability heuristic causes people to overestimate the probability of the first and underestimate the probability of the second.

Another heuristic supported by experimental evidence is known as "anchoring and adjustment." Dr. Tversky and Dr. Kahneman hypothesized that when making judgments about an event's probability, people often start with an initial estimate that is then adjusted as they think of other factors in making their final judgments. But they generally do not adjust their answers enough, so that original values, regardless of how obtained, become an "anchor" that can bias their final judgments.

For example, the study just described was rerun with a second group of subjects who were not told that approximately 50,000 people die each year from motor-vehicle accidents, but instead were told that approximately 1000 people die each year from electrocution. The relative relation between average values of the subjects' responses for the various risks remained almost unchanged. But the absolute value of the responses fell by roughly a factor of two for the high-probability risks such as diabetes, stomach cancer, and stroke and by over twice that for the low probability risks such as botulism, tornado, and flood. The investigators ascribed this shift to the anchoring and adjustment heuristic.

## Overconfidence Is Troublesome

If, as hypothesized, people automatically use heuristics such as availability and anchoring and adjustment when they think about risk and uncertainty, various other problems can arise. Perhaps the most serious is overconfidence. Since people typically do not understand how unreliable their thought processes are when it comes to making probabilistic judgments, they generally have much more confidence in the correctness of their judgments than is justified.

For example, in the previously cited experiment with pairs of risks, subjects were asked to give odds on how sure they were that they had correctly chosen the more likely of the two causes of death. One out of eight of the answers was wrong in cases where subjects said that the odds that they were right were 1000 to 1 or greater—a sign of overconfidence in their choices.

On the other hand, someone may be asked to estimate the value of some poorly known quantity, such as the price of 90-octane, lead-free gasoline at a local gas station on Jan. 1, 1983, and asked not just for a best estimate but also for a confidence interval that will have a 98-percent chance of containing the correct answer. For one such study, Max Henrion of Carnegie-Mellon University recently reviewed 29 laboratory experiments that asked subjects questions like this, for which the answer could later be verified. If people were good at making such estimates, then the right answer should lie outside the confidence interval that was given only about 2 percent of the time. In the experiments reviewed, the actual frequency with which the correct answer lay outside the estimated confidence intervals ranged from a low of 5 percent to a high of 57. The average value was about 30 percent.

This poor performance may result partly from the operation of the anchoring and adjustment heuristic. People tend to anchor on their best estimate and then not adjust sufficiently when they try to estimate the associated uncertainty. Similar manifestations of overconfidence are clearly a serious problem in quantitative risk assessment.

## Assessing, Evaluating, and Managing Risk

Since risk cannot be eliminated, the problem people face, individually and collectively, is how much risk should they live with and how should they go about managing the risk?

To answer these questions, analytical tools must be built that will allow the processes in figure 2 to be understood and described for specific tasks. In parallel with this, the alternative social and technical implications of risk-management philosophies must be explored and philosophies selected that are compatible with the goals of society. Then a set of incentives and institutions must evolve that will implant them.

# Questions for Thought and Discussion

1. Morgan tells us that "the statistical evidence shows that Americans live longer, healthier, and wealthier lives today than they did at any time in the past." Yet, Americans appear to be more concerned about risk now than ever before. Are those concerns justified, given this statistical evidence? Why or why not? What accounts for the concerns—increased wealth or lack of familiarity with new technologies, as suggested by the author, or are there other reasons?

2. Describe each of the elements in figure 2 in terms of the health risk associated with radiation from the natural presence of radon gas in certain homes.

3. Morgan lists five categories of information about the uncertainty attached to risk processes. Based on your own knowledge and personal experience, provide a real-world example for each category.

4. The author presents seven hypothetical cases that all produce the same statistical expected number of deaths on an annual basis. Rank these cases in the order of their decreasing level of concern to society, and explain the basis for your choices.

5. Given the fact that people's perceptions of risk are not always accurate, what advice would you give to a regulatory official who is accountable to the public for his or her decisions? Should the official act solely on the basis of scientific estimates of risk, respond in strict accordance with the public's expressed priorities, or take an approach that is different from either of these?

# Choosing and Managing Technology-Induced Risk

## M. GRANGER MORGAN
*Carnegie Mellon University*
*Pittsburgh, Pennsylvania*

It is not hard to think of counterexamples to that old saying, "There's no such thing as a free lunch." What about the time you found a coupon that got you a free burger with a shake and fries? But add one more word to make the aphorism "There's no such thing as a risk-free lunch," and suddenly the counterexamples dry up. You might, for example, have tripped on your shoelaces and hit your head on the tile floor as you walked into the burger place. While you were there, you might have been injured in a shoot-out between the police and an escaped convict who had stopped for a fish sandwich. A food additive, or the carbon on the charcoal-grilled burger, might contribute to a future case of digestive system cancer. You might have choked to death on a piece of food because nobody in the restaurant knew how to perform the Heimlich maneuver to clear your airway. Life is full of such risks.

People don't worry very much about most of them. They try to take basic precautions, such as avoiding restaurants in notorious crime areas. They are vaguely aware that some risks, like those from certain food additives, are under study by private and government researchers and may be regulated in the future. They do what they can do to help reduce some general risks—for example, by contributing to the local paramedic team that teaches people how to perform the Heimlich maneuver and cardiopulmonary resuscitation. The simple example of the risks in having lunch raises important questions, however. Among these are:

1. How can the magnitude of the risks people face be known?
2. What techniques can be used to regulate or otherwise limit the levels of these risks?
3. To the extent that these risks can be controlled, what level of risk should be chosen?
4. Who should be responsible for managing the risks and what institutional arrangements should be used?

Though the nomenclature is not yet standard, the first of these problems is usually called risk assessment. The second goes by the name of risk abatement. The third and fourth are often referred to together as risk management. These problems are very much interrelated.

"Choosing and Managing Technology-Induced Risk" by M. Granger Morgan is reprinted, with the permission of the publisher and the author, from *IEEE Spectrum*, vol. 18, no. 12, pp. 53–60 (December 1981). © 1981 IEEE.

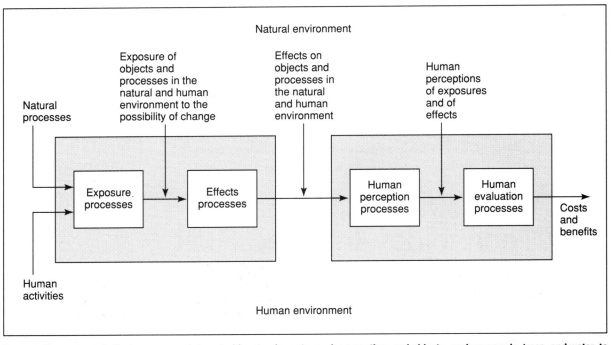

**Figure 1. Exposure and effects processes interact with natural events, such as weather, and objects, such as people, trees, and water, to produce some effect or change (left-hand side). How these changes are perceived and evaluated as risks depends in turn on a set of human perception and evaluation processes (right-hand side).**

## Approaches to Risk Abatement

Several years ago houses were built that contained aluminum wiring instead of the traditional copper. Aluminum was substituted because it was less costly. In those days people were not aware of the risks with aluminum house wiring. Unsuspecting homeowners occasionally smelled smoke whenever equipment that drew large loads of electricity, like toasters or food-warming ovens, was turned on. Some homeowners lost their homes and even their lives in fires caused by aluminum wiring.

Aluminum develops a surface oxide layer that has high resistivity, so connections may become warm whenever the circuit is used. As the circuit is cycled on and off through daily household use, the resulting heating and cooling of the connection does two things: it contributes to the loosening of the mechanical contacts, and it aids the further growth of the high-resistance oxide layer. Sooner or later the high-resistance connection may become so hot that insulation melts off the wires, plastic connection caps char and fall away, and, if there is other combustible material nearby, a fire may start.

The left-hand side of figure 1 suggests the following conceptual approaches to controlling risk in general:

1. The natural and human environment can be modified.
2. The exposure processes can be modified or avoided.
3. The effects processes can be modified or avoided.
4. The effects, once they occur, can be mitigated or compensated for.

For the aluminum-wiring example, some practical applications of these approaches are shown in figure 2. Modifying the natural and human environment on the grand scale is probably not useful. True, risks associated with electricity would vanish if families stopped using it, but most are unprepared to do so. However, some problems on a smaller scale may be easily dealt with this way—in the 1940s shoe stores had fluoroscope machines that let customers check how well shoes fit. Kids had great fun seeing their foot bones as they wiggled their toes. But by the early

**Abating the risk of house fire from faulty aluminum wiring**

| Modify environment | Modify exposure processes | Modify effects processes | Mitigate or compensate for effects |
|---|---|---|---|
| Do not use electricity | Replace all aluminum wiring | Install smoke detectors | Carry adequate insurance |

Figure 2. Four general strategies for risk abatement are modifying the background environment, modifying or avoiding exposure processes, modifying or avoiding effects processes, and mitigating or compensating for effects once they have occurred. For the risk of a house fire because of faulty aluminum wiring, examples of the four approaches are shown.

1950s people began to realize that these machines produced dangerous and unnecessary exposure to X-rays. The machines were junked.

The second approach—modifying or avoiding exposure processes—does make sense in the aluminum wire example [figure 2]. It would include such strategies as not using aluminum wire in the first place; replacing existing aluminum with copper; cleaning and treating the aluminum with ointment that inhibits oxide formation and inspecting the contacts periodically; and not using large loads on circuits suspected of containing aluminum wire. It would also include the strategy of pulling the fuse on circuits suspected of containing a problem until they had been fixed.

The third approach, modifying or avoiding effects processes, is also workable. It includes installing smoke detectors, placing flashlights at everybody's bedside for use in an emergency, and holding family fire drills occasionally.

The final strategy is mitigating or compensating for effects once they have occurred. In this case, they might include having a fast and capable local fire department or carrying adequate fire insurance. In general, such after-the-fact strategies are less satisfactory than those that avoid or prevent the problem.

Examples of the four approaches to risk abatement can be given for other risks [table I]. Not all the possible risk-abatement strategies that can be imagined for these cases will unambiguously produce the

**Table I. Risk-Abatement Strategies Fall into Four Major Categories**

| Risk | Risk-abatement strategy | | | |
|---|---|---|---|---|
| | Modify the natural or human environment | Avoid or modify exposure processes | Avoid or modify effects processes | Mitigate or compensate for effects |
| Risk of vehicle occupant injury in auto accidents | Live close to work and walk<br>Build rapid-transit systems | Change speed limits<br>Suspend licenses of drunk drivers<br>Train people to drive defensively | Wear seat belts<br>Install collapsible steering columns<br>Strengthen side panels | Carry auto insurance<br>Operate good emergency medical systems<br>Sue the other driver |
| Risk of developing cataracts from radiation from microwave ovens | Use gas or electric ovens<br>Don't cook food | Design ovens with good shielding<br>Design ovens with door interlocks | Provide users with Faraday shielding through goggles or helmets | Carry health insurance<br>Provide cornea transplants<br>Sue manufacturer and/or supplier |
| Risk of getting shot by someone with a handgun | Eliminate poverty, inequity, anger, mental illness, and so on | Ban handguns<br>Impose harsh penalties on crimes committed with handguns<br>Stay away from high-crime areas | Wear protective bulletproof clothing<br>Duck | Carry health insurance<br>Operate good emergency medical systems<br>Sue person who shot you |

desired risk reduction. For example, "carry own handgun for self-defense" could be added to the strategies for the risk of getting shot by someone with a handgun, but it is unclear whether this would increase or decrease the chance of getting shot. Similarly, compulsory high-school driver education might be thought to combat the risk of vehicle occupant injury in auto accidents, but because such classes inject large numbers of high-risk, teenage drivers into the driving population, they may not have the desired effect on the overall risk.

The two remaining processes in the conceptual framework for thinking about risk in figure 1 are human perception and human evaluation. These processes do not offer direct approaches to managing risk because regardless of my perception of the probability that acid rain from power plants will kill all the trout of certain lakes in the Adirondack Mountains, and regardless of how I value this potential event, the perception and evaluation will not change the rate of acid precipitation or the chemical and biological processes that go on in the lakes. But changes in people's perceptions and evaluation clearly do affect whether they consider a problem worth worrying about and the kinds of risk-management actions they consider appropriate.

Changing the perception and evaluation of risk suggests that one way government agencies, for example, could handle the problem of acid rain would be to persuade the people of upstate New York that there really are not many effects, and anyway there are better kinds of fish for their lakes than trout. This sort of thing certainly goes on. But working to change the perception of risks need not be pernicious. Helping develop a risk perception based on an accurate understanding of objective evidence may lead to more rational risk management.

## Acceptable and Optimal Risk

If a set of strategies has been chosen that will allow the abatement of a particular risk, the question of what level of risk should be chosen arises. If abating the risk costs nothing, the obvious answer is zero—get rid of the risk. But risk abatement almost always does cost money and time. When it is impossible to eliminate the risk completely with finite resources and time, the question becomes, "How much risk abatement should we buy?"

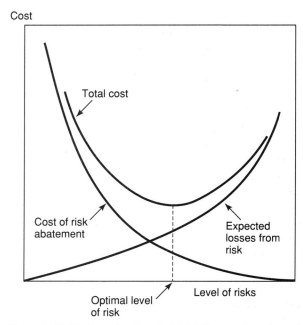

Figure 3. What is the optimal level of risk? The classic answer of economics is that the optimal level is where the sum of the cost of risk abatement and the expected losses from the risk is at a minimum.

But if the risk arises from some human activity or technological system, one should first ask whether, given the expected levels of risk, the activity or technology is necessary. In the case of medical diagnostic X-rays, the answer is yes. In the case of shoe-fitting fluoroscope, the answer is no.

Various criteria can be used to decide how much risk abatement to buy. Two of the most common are "acceptable risk" and "optimal risk." "Acceptable" implies a threshold level below which risk will be tolerated. "Optimal" implies a trade-off that minimizes the sum of all undesirable consequences. This notion of optimality is a classic idea of engineering-economic analysis, and it deserves careful attention.

For example, a technological development such as a computer information system produces benefits to banks in the form of convenient and efficient financial transactions. But it also imposes risks of social costs such as information theft, financial fraud, and privacy invasion. Assuming there is a way to quantify the benefits and risks of this system and that the system is worth building, if more is spent on risk abatement such as better computer security, the expected social costs can be reduced. How much risk abatement should be bought? The classic answer is that the optimal level is that at which the incremental or marginal cost of risk reduction just equals the marginal reduction achieved in societal cost [figure 3].

Some have argued that if society wants to be rational about risk abatement, it should operate so that a dollar spent in one place will buy about the same risk reduction as a dollar spent in another. This is not what people want to do if one measures risk simply in terms of expected mortality. On a per-death-averted basis, for example, most people want to spend more to prevent deaths in airliners—which typically kill many human beings at a time—than they wish to spend to prevent isolated deaths in automobiles. . . . Yet even if we accept variations of this sort, most people would probably not approve any great difference in spending.

Computing the marginal investment per life saved for risk abatement programs is extremely difficult. Recently John Graham and James Vaupel of Duke University did the next best thing for 57 programs of five Federal regulatory agencies in computing the average investment of dollars per life saved in 1978 [table II]. Their results made it clear that the investment rates of the Federal agencies varied by more than a factor of 20. This variation probably reflects the different political and historic environments of the National Highway Transportation Safety Administration and the Occupational Safety and Health Administration more than the reasoned judgment of our society.

Though the marginal cost approach may strike some as quite reasonable, it is only a normative solution. It answers the question of how people *should* choose the level of risk, but says nothing about how they *do* choose it. It offers no guarantees that if a careful marginal cost calculation establishes the optimum risk for a technological system, society will accept the answer. Society may balk if it does not understand the problem very well, if it doesn't believe the analysis can capture all the important costs, or if it rejects the idea of cost trade-offs in favor of an approach based on rights and obligations.

To some extent, public acceptability of risk can be studied empirically. We can examine the revealed preferences of people, as reflected in the levels of risk

**Table II. Breakdown of 57 U.S. Regulatory Agency Risk-Abatement Programs by Dollars Spent to Save One Life in 1978**

| U.S. regulatory agency | Programs spending less than $170,000 | Programs spending $170,000 to $3 million | Programs spending more than $3 million |
|---|---|---|---|
| National Highway Traffic Safety Administration | 22 | 7 | 0 |
| Department of Health and Human Services | 4 | 1 | 0 |
| Consumer Product Safety Commission | 4 | 2 | 0 |
| Environmental Protection Agency | 4 | 1 | 5 |
| Occupational Safety and Health Administration | 0 | 0 | 7 |
| Total number of programs | 34 | 11 | 12 |

*Source:* [From] J. Graham and J. Vaupel, "The Value of Life: What Difference Does It Make?" [*Risk Analysis*, vol. 1, no. 1 (1980), pp. 89–95].

they have accepted in the past. Or, we can seek their expressed preferences by asking them what levels of risk they find acceptable.

One of the first to analyze revealed preference was Chauncey Starr of the Electric Power Research Institute in Palo Alto, Calif. In 1969 he hypothesized that the historical levels of risk in a number of areas—such as motor vehicles, aviation, railroads, skiing, hunting, and smoking—were socially acceptable, since they were in wide use. He made rough estimates of the economic benefits associated with each technology, and from the resulting plot of risk versus benefit, as well as many other considerations, drew several tentative conclusions. Among these was that for a given level of benefit, people will voluntarily expose themselves to risks roughly 1000 times greater than those they are exposed to with no choice in the matter.

Although recent experimental work by psychologists casts doubt on the validity of this and some of Dr. Starr's other conclusions, his work has had an enormous impact over the last decade by stimulating critical thinking and promoting the development of risk-related research.

Using expressed-preference techniques, Baruch Fischhoff and colleagues at Decision Research Inc. in Eugene, Ore., have examined both the relationship between perceived risk and perceived benefit and the importance of the voluntary or involuntary nature of the risk assumed. For a set of 30 activities or technologies, they have found that people perceive, on the average, that current risks decrease as perceived benefits increase. But when the investigators adjusted the risks to levels their subjects told them would be acceptable, they found that these acceptable levels of risk actually increased with increasing perceived benefit.

## The Tools of Quantitative Risk Analysis

Quantitative analytical tools are important in risk assessment and management. When the level of the risks in a technological system are not known, such tools may help develop an estimate. When the efficacy of a proposed risk abatement strategy is unknown, the tools may help explore the question. Both the specific tools and the general form of quantitative risk analysis change, depending on the nature and quality of information available. In general, there are three forms of analysis—single-valued best estimate; probabilistic; and bounding.

When the science of a particular risk problem is well understood and good information is available, the obvious thing to do is to build deterministic models and use best-estimate values for the model's coefficients. This is the way an electrical engineer would compute the field strength to which a small object is exposed in the far field of a microwave antenna. Here Maxwell's equations accurately describe the science, and the coefficients, such as $\lambda$, $\epsilon$, and $\mu$, and the dimensions of the antenna are all known accurately. However, this same form of "single-valued best-estimate analysis" is often used on risk problems that are far less well understood. Such an approach then produces a "best guess" answer, often with little or no indication of the certainty that should be attached to the answers.

If the science of the problem is reasonably well understood but only limited information is available on some important coefficients, probabilistic analysis is more appropriate. It gives an explicit indication of the level of uncertainty in the answers. In such analysis the single-valued best estimates of coefficients are replaced by a probability distribution over the range of values that an expert believes the coefficient might take. If there is also uncertainty about the functional form of the model that should be used, this uncertainty may also be incorporated into the

analysis. Though it is rapidly growing in importance, probabilistic analysis is still somewhat rare. Many analysts continue to make single-valued, best-estimate analyses on problems that should be solved with probabilistic analyses.

Probably the most widely known example of probabilistic analysis is the U.S. reactor safety study called WASH-1400—an enormously complex undertaking because of the many possible failures that could lead to an accident. But this does not mean that all probabilistic analysis must be complex or difficult. At Carnegie-Mellon University, Max Henrion has developed an interactive software system called DEMOS, which allows performance of simple probabilistic analyses with about the same effort required for simple, single-valued, best-estimate analysis. A number of similar systems have been built.

Computational problems are not the only problems in such analysis. Experts in that field must also provide or construct a variety of quantitative, subjective judgments, usually in the form of probability distributions. There are a number of heuristics—such as "availability" and "anchoring and adjustment"—that are used in making judgments about uncertainty. These can lead to a variety of biases, including overconfidence. . . . Hence the process of "eliciting" subjective probabilistic judgments from experts must be approached with considerable care.

There are some problems about which so little is known that even probabilistic analysis is not appropriate. However, in some of these cases it may still be possible to use what little is known at least to bound the answer—to say, for example, that it couldn't be greater than this or less than that.

## Assessment Combines Art and Science

While it draws on many quantitative analytical tools, risk assessment is at least as much an art as a science. Success requires analytical teams that can combine an excellent understanding of the science and technology of the problem; a good command of a range of appropriate analytical tools; imagination, good taste, and judgment in setting up and working such problems; a sensitivity to the potential pitfalls of eliciting expert judgment; and an ability to describe and interpret the analysis to semitechnical and nontechnical people in a clear, jargon-free manner.

Yet despite the difficulties, government agencies concerned with regulating risks are beginning to turn to analytical techniques to help them to do a better, more consistent job.

But the use of such tools carries potential problems. Here is a look at five of them.

The first is a great temptation to use quantitative techniques to get "the answer." This is not an appropriate use. Quantitative risk assessment can provide understanding and insight, but it can never capture all the factors, such as quality of life, that are important in a problem, and it should not become a substitute for careful human judgment.

A second problem involves how to treat experts who disagree. If individuals or a private company are doing a risk analysis for their own use, they can, if they want, weigh and combine the various expert opinions to get a single answer. But in analysis for public decision making, one must take care before combining expert opinions or trying to produce consensus estimates. Though this may sometimes be useful, more time should be spent developing and using analytical techniques that tell how the range of expert opinions affects conclusions, as well as whether this range is really important in the risk-management policy at issue.

A third concern involves the inclination of many agencies to produce a single, definitive, analytical tool for dealing with their problems. This shying away from multiple approaches is unfortunate, since much can often be learned by studying and comparing the results of alternative formulations.

A fourth concern is that in some government regulatory agencies, quantitative risk-assessment tools may become a substitute for science. After all, it is generally much cheaper and easier to ask 8 or 10 experts what they believe the value of some variable is than to mount an experimental program to measure it. The results of most quantitative risk assessments look pretty technical, so it is sometimes easy to forget that the technical understanding they are based on is often quite incomplete. This risk is compounded by the fact that most federal regulatory agencies are dominated by people trained in law, who are often more concerned with getting answers than they are with developing complete understanding.

There are some simple institutional ways in which this danger can be overcome. For example, it would help considerably if agencies required that one of the outputs of their quantitative risk assessments be a critical examination of how much has been learned since the last time the problem was examined, whether that has improved the current analysis, and what kinds of new things need to be learned before the next analysis is done.

A final serious problem is that industry and public-interest groups will adopt a passive attitude when governmental agencies attempt to develop and use quantitative risk-assessment tools. They tend to adopt a wait-and-see attitude. This is unfortunate for two reasons: (1) the private sector may be more capable than government of building high-quality risk-assessment and management tools; and (2) the private building of such tools may be beneficial in shifting risk regulation from an adversarial approach to a more collaborative approach based on consensus.

## Managing Risks

A number of institutional structures have evolved to implement risk-management procedures. Of these, the four most important are:

1. Tort and other common law—particularly those laws related to negligence, liability, nuisance, and trespass.

2. Insurance—offered either by private companies, through joint private/government arrangements, or directly by government programs.

3. Voluntary standard-setting organizations—for example, the Underwriters Laboratory, the National Fire Protection Association (which produces the national electric code), the American National Standards Institute, the American Society for Testing and Materials, and engineering societies such as the IEEE.

4. Mandatory government standards or regulations.

One-hundred and fifty years ago only the first two of these mechanisms were important. The first technological risk to be regulated by the U.S. government was that of steamboat boiler explosions, and it took over 50 years, several hundred accidents, and more than 3000 deaths before Congress finally overcame its reluctance to meddle in private enterprise and passed an effective regulatory law. Regulation of technologically based risks did not become a major activity until after the innovations of the New Deal in

# Toy Airplane "Risk" Exemplifies Probabilistic Assessment Methods

I have a friend who is a bit of a braggart. At an office party a few months ago he started boasting that when he was a kid they used to have contests to see who could keep a rubber-band airplane flying for the longest period of time—and that he was the neighborhood champion. Unfortunately, he chose to brag to someone who won the state championship for rubber-band airplane flight duration as a kid, with a time of 22.3 seconds. My braggart friend bet $20 that he could better that time. The entire office witnessed the bet, so despite second thoughts, there is no way he can back down now.

A copy of the state championship rules says that the contestant will receive three balsa-wood, rubber-band airplanes still in their original cellophane wrappers. Without opening the package, the contestant must select the one he will fly and return the other two. The contestant opens the package and assembles and flies the plane once. No test flights are allowed. If he breaks this plane during assembly or while winding the propeller, no repairs will be allowed. For a while my friend thought he was off the hook because the brand of airplane specified in the rules did not seem to be available any longer. But that hope died when one of the secretaries gleefully announced that she'd located a hobby store that still had four in stock. She bought the three that would be needed. My friend promptly snuck out and bought the remaining one.

My friend faces two sets of problems. The first involves how to optimally adjust the wing and launch the plane. He knows from his experience as a youth that there is virtually no variation among individual planes in this product line. By spending all last Saturday in the gym with his airplane he has figured out the optimum wing adjustment and launch strategy. But he's really worried about a second problem.

The package instructions for the plane say the propeller should not be wound more than 30 times. From his flight tests last weekend my friend figures that if he only winds 30 turns he stands about a 20-percent chance of winning the bet by beating the 22.3 second time. If he winds 40 turns his chance of winning jumps to about 80 percent. But, in winding 40 turns there's a good chance the plane will break and he'll suffer a terrible loss of face. He figures it's worth $100 to him not to have *that* happen.

On the basis of his childhood experience he has identified three ways the plane can fail if 40 turns are wound on the propeller. These failure modes are:

1. The front plastic casting that holds the propeller bushing has an internal fault that cannot be seen. When turns are wound on, the casting breaks.
2. The rubber band has a weak spot that is not visible and this weak spot is either on or near the nail at the tail or on or near the hook that connects to the propeller. With 40 turns the rubber band breaks.

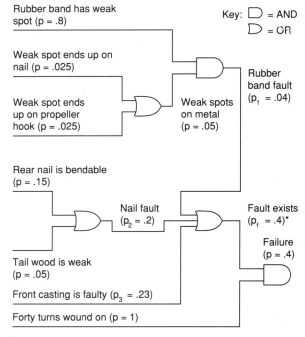

* Since all three faults can occur together, $p_f = 1 - (1 - p_1)(1 - p_2)(1 - p_3) = .4$

**Figure A. Fault tree with associated probabilities (p) for failure of a rubber-band airplane if 40 turns are wound onto the rubber band.**

3. The nail at the tail either bends or is in a weak piece of wood, which splits. After 40 turns, the rubber band slips off the tail nail. Under the rules that is an unrepairable failure.

My friend constructs a fault tree with associated probabilities [figure A].

Should he wind on 30 turns, or should he wind on 40? If he wins he gets $20, and says the "face" he'd gain would be worth about $50 to him. If he flies the plane but loses, he'd be out $20 and he says he'd pay up to $50 to avoid the loss of face. If he loses by breaking the plane he's out $20, and as noted above he values the loss of face in this case at $100. From the fault-tree analysis he has estimated the probability that the plane will fail with 40 turns to be .4. On the basis of the hundreds of flights he made as a youth he estimates the probability that the plane will fail at 30 turns to be vanishingly small and he will ignore this possibility.

One way to represent the decision he faces is with a decision tree [figure B].

The tree can be used to evaluate the expected dollar value of the two options he faces. If he flies with 30 turns there is a

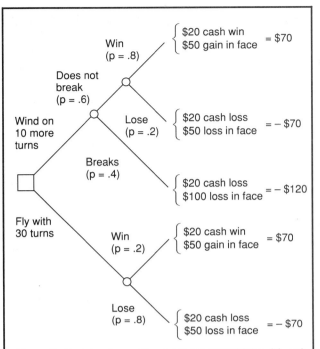

**Figure B. Decision tree with associated probabilities (p) and monetary gains or losses in rubber-band airplane bet.**

.8 chance he will lose $70 and a .2 chance he will gain $70. Thus, the "expected" dollar value of flying with 30 turns is .8(−$70) + .2($70) = −$42. In a similar way, the expected dollar value of flying with 40 turns is .4(−$120) + .8($70) = −$6. So he should wind on the 10 extra turns since this strategy has the greater expected value. The analysis can be extended to solve for the optimum number of turns.

There are several complications to consider. First, it may be that a $100 loss is more than twice as painful to my friend as a $50 loss. This phenomenon is called risk aversion and could be handled by asking him questions to learn his "utility function" for dollars. The decision tree could then be reworked with the result that the value of the options he faces would shift from their expected dollar values. In some real problems, this shift is enough to change the conclusions about which strategy represents the best solution. Though risk-aversion behavior is the most common, risk-seeking behavior is also observed in some circumstances and this can be modeled in a similar way.

There are some problems in which it is not convenient or possible to reduce everything to a single measure or attribute such as dollars.

For example, in automobiles, three attributes—safety, fuel economy, and air-pollution impact—trade off as design parameters such as vehicle weight are changed. There is a set of decision theory techniques, similar to those discussed here, for dealing with such problems with more than one measure or attribute. —*M. G. M.*

the 1930s. Most of the government regulatory agencies that today dominate our approach to risk management have been created in the last two decades.

The decision to regulate raises important questions. Most fundamental is that of government's right to regulate. Does the government have the right to impose a small involuntary cost on many of its citizens (in the form of a tax or higher prices) to make a few or even most people a little safer? Perhaps most U.S. residents would answer yes to this question if the net benefits of regulation exceeded the net costs. But there are other philosophical perspectives besides this utilitarian one. Libertarian and other philosophies that place far greater emphasis on individual rights and little on total public welfare might answer no. It is important to understand that the decision to rely on government regulation, as opposed to tort law or insurance, is an ethical decision about which not all reasonable people would agree.

Regulation, if adopted as an acceptable strategy for managing risk, can take a number of forms. Design standards can specify with great detail the way a device may be built and operated. Good examples of this are the pressure vessel code of the American Society of Mechanical Engineers and the data-encryption standard of the National Bureau of Standards. In contrast, performance standards specify the way a device must perform in a given situation.

Regulation often takes the form of a threshold, as with the exposure thresholds specified in many regulations of the Occupational Safety and Health Administration or the ambient air-pollution concentration thresholds specified by the U.S. Environmental Protection Agency under the Clean Air Act.

A rarely used approach that most economists find has great theoretical appeal is the externality tax. This requires that a person who operates a system that imposes some social cost must pay a tax roughly equal to that cost [figure 3]. This has the effect of "internalizing" the social cost, so it will be cost-effective for the operator to buy just the amount of risk-abatement technology that is socially optimal. One of the few contexts in which this approach has seen a major application is in the control of pollution in the Rhine River valley in Germany, where industries are taxed by the amount of pollution they discharge into the river.

A growing number of people across all sectors of

society appear to be reaching the conclusion that the current approaches to risk assessment and risk management are not working very well and need to be improved. The result has been a number of new and proposed institutional arrangements.

The Carter Administration made major attempts at regulatory reform. A principal vehicle was Executive Order 12044 that, among other things, stipulated that regulatory agencies had to perform an analysis for regulations that would have significant economic impact. The Reagan Administration replaced it with another (Executive Order 12291) that requires all executive agencies to prepare a regulatory impact analysis to assist them in selecting regulatory approaches in which the "potential benefits to society from the regulation outweigh the potential costs to society." Such analyses are required only for "major rules" that are likely to have an economic impact of $100 million a year or more or "significant adverse effects on competition, employment, investment productivity, innovation or on the ability of U.S.-based enterprises to compete with foreign-based enterprises in domestic or export markets; or will result in a major increase in costs or prices in the economy."

A preliminary version of the regulatory impact analysis must be submitted to the Office of Management and Budget 60 days prior to publication of a notice of proposed rule making, and a final version must be submitted 30 days before publication of the final review. Should disagreements arise between the OMB and a particular agency, they are to be resolved by the President's Task Force on Regulatory Relief, chaired by Vice President Bush.

The Reagan approach, by requiring extensive communication between the OMB and a regulatory agency, moves the country much closer to a centrally coordinated approach to government regulation. But reliance on a cost-benefit approach requires many skilled analysts. It is not clear that the Budget Office and agency budgets will allow recruitment of such staffs, nor is it clear that there are enough analysts today to do a respectable job.

In addition to these efforts in the executive branch, several congressional initiatives are in progress, including a series of bills introduced by Representative Donald Ritter (R–Pa.) and Senator Robert Dole (R–Kan.).

On the philosophical side, it is not clear that society should make all of its regulatory decisions on a cost-benefit basis. The U.S. concept of government is based in part on a system of rights, and the notion of rights is not always fully compatible with a cost-benefit approach. For example, in the case of an occupational exposure where it would take a great expense to protect a few individuals, but where many or all in society benefit from their work, the cost-benefit approach might not support the expense of protection.

## Industry's Role

Industry has often complained vociferously about the inadequacies of U.S. regulatory agencies, but it has produced relatively few constructive proposals for regulatory reform. A notable exception is a proposal developed in 1980 by the American Industrial Health Council (AIHC), an association of approximately 140 companies and 80 trade associations. The proposal is primarily concerned with carcinogenic materials, but the approach could be applied to a variety of other chronic health risks associated with industrial activity.

In the context of figure 1, the Council proposes that the assessment of effects processes be largely removed from the regulatory agencies and turned over to an independent panel of experts.

The argument is that "the panel would make a definitive assessment of the relative scientific questions, reflecting the best scientific thinking and all the evidence available at the time the assessment is made." The panel's findings would not be legally binding on an agency, but if an agency decided to reject the findings, it would carry the burden of proof. With respect to panel membership, AIHC proposes "that the very best scientifically qualified individuals be selected to serve on the panel" and that membership "not be determined or influenced by special-interest-group representation."

The proposal is not without its problems. For example, it requires that the panel come in with one of three findings: (1) That the available data indicate specified health effects; (2) That the available information indicates that specific health effects are unlikely to be experienced; or (3) That the available information is insufficient to warrant any scientifically sound conclusion on the health effects. To its credit, the AIHC argues that the fact "that no sound scientific conclusion is warranted would not necessarily mean that no regulatory action would be taken." However, restricting the panel to one of the three conclusions virtually guarantees that most answers would come back in the third category, because the available technical evidence is often fragmentary and incomplete. The available evidence may be sufficient to suggest a very rigorous norm of scientific proof.

It might be better to replace the third category with one or several that would allow the panel of experts to express their professional judgment about the probability that specified effects would occur, given the currently available evidence. However, this seemingly modest innovation in the process leads to some very fundamental problems. Except when it comes to choosing and designing the next experiment, the paradigm of science does not encourage speculation about answers. The three categories the AIHC proposed—we know there is, we know there isn't, we don't know—are the acceptable categories in the scientific paradigm. But for developing regulatory policy, these are not enough. Public policy development requires technical people to make careful, informed professional judgments about the probability of certain outcomes. The problem of how to develop institutions and procedures that will encourage our very best scientists and engineers to make such judgments, while not undermining the impartial objectivity necessary for the conduct of good science, is a serious problem that neither the regulatory nor the scientific community has adequately addressed.

A second problem with the AIHC proposal involves the separation of responsibilities between the panel and regulatory agencies. While the panel is to perform assessments of effects processes, the responsibility for assessing exposure processes is left to the regulatory agencies. However, if any agency decides to regulate, then the panel is expected to "prepare a definitive assessment of qualitative and quantitative aspects of hazards and risk," something that cannot be done without a full assessment of exposure processes—a task for which the pathologists, epidemiologists, and biostatisticians on the panel are not particularly well suited.

Despite these problems, the AIHC proposal is important because it reflects a willingness by industry to become involved in proposing regulatory institutional innovations and reforms—something it has been too reluctant to do in the past.

## Adversary Versus Consensus

One regulatory reform issue that is addressed by the proposal of the American Industrial Health Council and by several others is the fact that most of the risk-management activity in the U.S. takes place in an adversarial environment. Different parties with different interests argue things out, often in rather partisan terms, before some adjudicatory body or official. This adversarial approach is the basic model of our legal profession and the model upon which much of our government is based. It is useful and powerful but has several disadvantages.

It can consume enormous effort. It often is not a reliable way to establish matters of technical fact, failing to lead to a full, clear exploration of the system involved or to adequate treatment of uncertainty. And it often does not produce results that are consistent from problem to problem.

Rather than choosing an adversarial approach, many of the voluntary standard-setting organizations arrive at decisions through the consensus of experts. Some people have argued that we should try wider use of consensus techniques. While a number of European countries do adopt consensus-based approaches to some regulatory problems, virtually no careful, comparative international studies have been done on the strengths and limitations of alternative approaches to risk management. This is a serious oversight that government and private research agencies should move quickly to rectify.

Another way to develop a better understanding of the strengths and limitations of alternative approaches to risk management is to adopt a more pluralistic approach in this country. William Clark of Oak Ridge Associated Universities has suggested that "maximal social learning and political will could be mobilized by designing the scale of particular risk-management ventures to fit the character of the risk under consideration.

"Though we might require regional-scale regulation in . . . air-quality management," he says, "we might find that much smaller-scale regulation—and therefore more varied learning experiments and less compulsion—would be appropriate in other cases."

In thinking about improving existing risk-management systems in the U.S., several issues deserve careful attention:

1. Most existing risk-management systems do a rather poor job of characterizing and dealing with uncertainty. A few regulatory groups, however, have recognized this and are trying to develop better techniques. Most are doing little.

2. Most risk-management systems are focused too narrowly on specific classes of risks. They frequently miss the big picture. For example, flood-control measures, such as dams, reduce the risks from routine floods but often increase the risk of occasional large floods. Many efforts at ecosystem management, such as insect pest control, reflect similar difficulties. With separate air, water, and solid-waste divisions, the Environmental Protection Agency still has trouble dealing with the coupled air, water, and solid-waste effluents of large-process industries.

3. Most risk-management systems seem committed to providing immediate global solutions. We have not learned how to encourage different approaches to the same problem so we can go slowly and select the one found to work best, or so a broad approach can be adapted to local circumstances.

4. In the continuing attempt to assign blame, which the adversarial world view promotes, most risk-management systems have not become very good at adapting, at being flexible, and at being willing to admit and learn from mistakes.

5. Most risk-management systems generally do not do a very good job of promoting consensus, collaboration, and cooperation among the major actors in the area in which they are operating.

## The Risk of Worrying About Risk

Risk is a fascinating subject—so fascinating that it is easy to get completely wrapped up in it and forget that there are risks to worrying about risk. Real or imagined risks are everywhere, and ever-increasing resources are devoted to their management—resources that in some cases might otherwise go toward advancing science and technology, improving productivity, or enriching culture.

It is hard to move individuals back and forth between a risk-seeking and a risk-avoiding view of life. But that is probably not a bad thing. The danger comes from gradually moving everybody to just one end of this continuum. If people become completely preoccupied with risk reduction on a piecemeal risk-by-risk basis, they are likely to build a society that is stagnant and has very little freedom.

Yet no reasonable person would argue that society should forget about risk. Instead it must perform a continual balancing act, retaining a diversity of risk-takers and risk-avoiders, because both contribute certain strengths. Together, we must design institutions and strategies for risk management that take a wide perspective and that are flexible and adaptable. In short, people must learn that living with, managing, and adapting to risk is what life is all about.

# Questions for Thought and Discussion

1. Suppose that an economic analysis of the health risks of a baby food product has been performed and that the result is a diagram like the one in figure 3. Indicate where the desired risk level might be for (1) the producer of the baby food; (2) a public interest group that is concerned with the purity of the product; (3) a representative of the Food and Drug Administration, which regulates such products. Explain any departures of these levels from the optimal risk level. Do you agree with the presentation of the optimal risk level in figure 3?

2. How does a *revealed preference* differ from an *expressed preference*? What, in your estimation, are the relative strengths and weaknesses of each approach as a basis for determining acceptable levels of risk?

3. Table I in this paper presents four strategies for abating three examples of risk. Construct a similar table for the risk of lung disease from automobile exhaust emissions and for the risk of ecological damage from a breached oil tanker.

4. The author describes the notion of *externality tax* as having great theoretical appeal to economists, but goes on to say that it has had few major applications in this country. In general, why do you suppose this is so? In particular, why do you think that it has not been used for pollution regulation in the United States as it has in Germany?

5. Much has been said about an issue raised in this paper: the risk of worrying about risk. On one side of the argument are those who contend that there is too much concern about trivial risks; on the other side are those who maintain that unwilling citizens should not be subjected to avoidable risks, no matter how slight. Think of a small risk that has recently been in the news and explain your position on whether it is worth worrying about.

# Defining Risk

## BARUCH FISCHHOFF

*MRC Applied Psychology Unit,
Cambridge, Great Britain, and
Decision Research, A Branch of
Perceptronics, Inc., Eugene, Oregon*

## STEPHEN R. WATSON

*Emmanuel College
Cambridge University
Cambridge, Great Britain*

## CHRIS HOPE

*Cavendish Laboratory
Cambridge University
Cambridge, Great Britain*

## Defining Risk

Managing the risks of technologies has become a major topic in scientific, industrial, and public policy. It has spurred the development of some industries and prompted the demise of others. It has expanded the powers of some agencies and overwhelmed the capacity of others. It has enhanced the growth of some disciplines, distorted the paths of others. It has generated political campaigns and countercampaigns. The focal ingredient in all this has been concern over risk. Yet, the meaning of "risk" has always been fraught with confusion and controversy.

*Editors' note:* The current affiliation of Baruch Fischhoff is Carnegie Mellon University, Pittsburgh, Pennsylvania, and of Chris Hope is the University of Leeds, Leeds, Great Britain.

Some of this conflict has been overt, as when a professional body argues about the proper measure of "pollution" or "reliability" for incorporation in a health or safety standard. More often, though, the controversy is unrecognized; "risk" is used in a particular way without extensive deliberations regarding the implications of alternative uses. Typically, that particular way follows custom in the scientific discipline initially concerned with the risk.

However, the definition of "risk," like that of any other key term in policy issues, is inherently controversial. The choice of definition can affect the outcome of policy debates, the allocation of resources among safety measures, and the distribution of political power in society. The present essay begins with an analysis of the key sources of controversy in this definition. It proceeds to advance a highly flexible general approach to defining "risk." Finally, it demonstrates the approach with an analysis of the comparative risks of different energy technologies, showing that the relative "riskiness" of those technologies depends upon the definition used. No definition is advanced as the correct one, because there is no one definition that is suitable for all problems. Rather, the choice of definition is a political one, expressing someone's views regarding the importance of different adverse effects in a particular situation. Such determinations should not be the exclusive province of scientists, who have no special insight into what society should value. As a result, the present approach is designed to offer a way to generate definitions of risk suitable for many problems and value systems.

## Dimensions of Controversy

*Objectivity.* Technical experts often distinguish between "objective" and "subjective" risk. The former refers to the product of scientific research, primarily public health statistics, experimental studies, epidemiological surveys, and probabilistic risk analyses. The latter refers to non-expert perceptions of that research, embellished by whatever other considerations seize the public mind. This distinction is controversial in how it characterizes both the public and the experts.

Although it is tempting (and common) to attribute disagreements between the public and the experts to public ignorance or irrationality, closer examination often suggests a more complicated situation. Conflicts often can be traced to unrecognized disagreements about the topic, including what is meant by "risk." When the public proves misinformed, it is often for good reasons, such as receiving faulty (unclear, unbalanced) information through the news media or from the scientific community (Lichtenstein et al., 1978). In some instances, members of the lay public may even have a better understanding of specific issues (or for the definitiveness of knowledge regarding them) than do the experts (Cotgrove, 1982; Wynne, 1983).

Along with these elements of objectivity in public opinion, there are inevitably elements of subjectivity in expert estimates of risk. Within the philosophy of science, "objective" typically means something akin to "independent of observer." That is, any individual following the same procedure should reach the same conclusion. However meritorious as a goal, this sort of objectivity can rarely be achieved. Particularly in complex, novel areas, such as risk analysis, research requires the exercise of judgment. It is expert judgment, but judgment nonetheless. Even in those orderly areas for which public health statistics are available, interpretative questions must be answered before current (or even historical) risk levels can be estimated: Is there a secular trend (e.g., are we sitting on a cancer time bomb)? Is the effect of predisposing causes (e.g., poor nutrition) underestimated because deaths are typically attributed to immediate causes (e.g., pneumonia)? Are some deaths deliberately miscategorized (e.g., suicides as accidents when insurance benefits are threatened)? Total agreement on all such issues is a rarity in any active science. Thus, objectivity should always be an aspiration, but can never be an achievement of science. When public and experts disagree, it is a clash between two sets of differently informed opinions. Sciences, scientists, and definitions of risk differ greatly in how explicitly they acknowledge the role of judgment.

*Dimensionality of risk.* The risks of a technology are seldom its only consequences. No one would produce it if it did not generate some benefits for someone. No one could produce it without incurring some costs. The difference between these benefits and non-risk costs could be called its net benefit. In addition, risk itself is seldom just a single consequence. A technology may be capable of causing fatalities in several ways (e.g., by explosions and chronic toxicity), as well as inducing various forms of morbidity. It can affect plants and animals as well as humans. An analysis of "risk" needs to specify which of these dimensions will be included. In general, definitions based on a single dimension will favor technologies that do their harm in a variety of ways (as opposed to those that create a lot of one kind of problem). Although it represents particular values (and leads to decisions consonant with those values), the specification of dimensionality (like any other specification) is often the inadvertent product of convention or other forces, such as jurisdictional boundaries (Fischhoff, in press).

*Summary statistic.* For each dimension selected as relevant, some quantitative summary is needed for expressing how much of that kind of risk is created by a technology. The controversial aspects of that choice can be seen by comparing the practices of different scientists. For some, the unit of choice is the annual death toll (e.g., Zentner, 1979); for others, death per person exposed or per hour of exposure (e.g., Starr, 1969; Wilson, 1979); for others, it is the loss of life expectancy (e.g., Cohen and Lee, 1979; Reissland and Harries, 1979); for still others, lost working days (e.g., Inhaber, 1979). Crouch and Wilson (1982) have shown how the choice of unit can affect the relative riskiness of technologies; for example, today's coal mines are much less risky than those of thirty years ago in terms of accidental deaths per ton of coal, but marginally riskier in terms of accidental deaths per employee. The difference between measures is explained by increased productivity. The choice among measures is a policy question, with Crouch and Wilson suggesting that, "From a national point of view, given that a certain amount of coal has to be obtained, deaths per million tons of coal is the more appropriate measure of risk, whereas from a labor leader's point of view, deaths per thousand persons employed may be more relevant" (p. 13).

Other value questions may be seen in the units themselves. For example, loss of life expectancy places a premium on early deaths which is absent from measures that treat all deaths equally; using it means ascribing particular worth to the lives of young people. Just counting fatalities expresses indifference to whether they come immediately after mishaps or following a substantial latency period (during which it may not be clear who will die). Whatever individuals are included in a category are treated as equals; these may include beneficiaries and non-beneficiaries of the technology (reflecting an attitude toward that kind of equity), workers and members of the general public (reflecting an attitude toward that kind of voluntariness), or participants and non-participants in setting policy for the technology (reflecting an attitude toward that kind of voluntariness). Using the average of past casualties or the expectation of future fatalities means ignoring the distribution of risk over time; it treats technologies taking a steady annual toll in the same way as those that are typically benign, except for the rare catastrophic accident. When averages are inadequate, a case might be made for using one of the higher moments of the distribution of casualties over time or for incorporating a measure of the uncertainty surrounding estimates (Fischhoff, in press).

*Bounding the technology.* Willingness to count delayed fatalities means that a technology's effects

are not being bounded in time (as they are, for example, in some legal proceedings that consider the time that passes between cause, effect, discovery, and reporting). Other bounds need to be set also, either implicitly or explicitly. One is the proportion of the fuel and materials cycles to be considered: to what extent should the risks be restricted to those directly associated with the enjoyment of benefits or extended to the full range of activities necessary if those benefits are to be obtained? Crouch and Wilson (1982) offer an insightful discussion of some of these issues in the context of imported steel; the U.S. Nuclear Regulatory Commission (1983) has adopted a restrictive definition in setting safety goals for nuclear power (Fischhoff, 1983); much of the acrimony in the debates over the risks of competing energy technologies concerned treatment of the risks of back-up energy sources (Herbert et al., 1979). A second recurrent bounding problem is how far to go in considering higher-order consequences (i.e., when coping with one risk exposes people to another). A third is how to treat a technology's partial contribution to consequences, for example, when it renders people susceptible to other problems or when it accentuates other effects through synergistic processes.

*Concern.* Events that threaten people's health and safety exact a toll even if they never happen. Concern over accidents, illness, and unemployment occupy people even when they and their loved ones experience long, robust, and salaried lives. Although associated with risks, these consequences are virtual certainties. All those who know about them will respond to them in some way. In some cases, that response benefits the respondent, even if its source is an aversive event. For example, financial worries may prompt people to expand their personal skills or create socially useful innovations. Nonetheless, their resources have been diverted from other, perhaps preferred pursuits. Moreover, the accompanying stress can contribute to a variety of negative health effects, particularly when it is hard to control the threat (Elliott and Eisdorfer, 1982). Stressors not only precipitate problems of their own, but can complicate other problems and divert the psychological resources needed to cope with them. Thus, concern about a risk may hasten the end of a marriage by giving the couple one more thing to fight about and that much less energy to look for solutions.

Hazardous technologies can evoke such concern even when they are functioning perfectly. Some of the response may be focussed and purposeful, such as attempts to reduce the risk through personal and collective action. However, even that effort should be considered as a cost of the technology because that time and energy might be invested in something else (e.g., leisure, financial planning, improving professional skills) were it not for the technology. When many people are exposed to the risk (or are concerned about the exposure of their fellows), then the costs may be very extensive. Concern may have even greater impact than the actual health and safety effects. Ironically, because the signs of stress are diffuse (e.g., a few more divorces, somewhat aggravated cardiovascular problems), it is quite possible for the size of the effects to be both intolerably large (considering the benefits) and unmeasurable (by current techniques).

Including concern among the consequences of a risky technology immediately raises two additional controversial issues. One is what constitutes an appropriate level of concern. It could be argued that concern should be proportionate to physical risk. There are, however, a variety of reasons why citizens might reasonably be concerned most about hazards that they themselves acknowledge to be relatively small (e.g., they feel that an important precedent is being set, that things will get worse if not checked, or that the chances for effective action are great). The second issue is whether to hold a technology responsible for the concern evoked by people's perceptions of its risks or for the level of concern that would be evoked were they to share the best available technical knowledge. It is the former that determines actual concern; however, using it would mean penalizing some technologies for evoking unjustified concerns and rewarding others for having escaped the public eye.

## The Nature of Risky Decisions

Although a part of all risky decisions, risk is all of very few. Hazard management would be easy if risk were a substance and a technology could be characterized (and managed) effectively in terms of how much of that substance it contained (Watson, 1981). Risky decisions are, however, not about risk alone.

Rather, they are choices among options, each of which has a variety of relevant features, including a level of risk. When a technology is adopted, so is its entire package of features. Thus, it is impossible to infer from its adoption that a technology has an acceptable level of risk (Fischhoff et al., 1981; Green, 1980; Otway and von Winterfeldt, 1982). Those adopting it might prefer much less risk, but be unable to obtain it at an acceptable price. In other decisions (or even in that decision should the possibilities change), they might adopt much less risky options.

From this perspective, the most general role for a definition of risk is to provide a coherent, explicit, consistent expression of one subset of the consequences arising in risky decisions. For deliberative decisionmaking to proceed, it must be complemented by comparable conceptual analyses of the other consequences. With a clear set of concepts, it is possible to begin making the hard tradeoffs between risks and net benefits (which may include any positive value attributed to risk itself due, say, to the thrill or excitement it produces).

There are, however, some reasons for thinking about risks in isolation. One is educating the intuitions. The risks created by many technologies are so diverse that it is hard to think about them all at once. The rem and Sievert, which aggregate diverse radiation doses, attempt to serve this role. A second reason is to summarize the conclusions of policymaking that has considered other factors (Fischhoff, in press). Health and safety standards are often expressed in terms of an "acceptable level of risk," even though nonrisk costs and benefits strongly influenced how they were set (otherwise, they would be set at zero risk). That expression may enunciate a political philosophy ("we care about the public to this extent"), or it may provide an operational rule for the technical staff monitoring compliance, or it may be the only legitimate public conclusion of an agency that is mandated to manage risk (but must, in practice, consider risk-benefit tradeoffs). A third role is providing an explicit criterion for guiding and evaluating an agency's actions. A safety measure, such as a mandatory seat belt law, might have quite a different effect on "risk" if that is defined as deaths, serious injuries, or all injuries. Evaluating it fairly requires knowing what it was intended to accomplish.

## Aspects of Risk

The first step in defining "risk" is determining which consequences it should include. Because that determination depends upon the particular problem, some context must be specified in order to produce even a hypothetical example. The context adopted here is evaluating the risks of competing energy technologies, as a component of setting national energy policy. Like any other choice of context, this one renders consequences that none of the competing options create—"unimportant for present purposes"—whatever their overall importance to society. This particular choice means that the selection of consequences (like other aspects of the definition process) should reflect "society's values," rather than those of any single interest. If one wished to revise or criticize this example, that effort, too, should begin with its selection of consequences.

Figure 1(B) shows that selection. Three kinds of risky consequence are included: fatalities, concern, and morbidity. Each is meant to include consequences whose magnitude is known, even though the identity of the casualties is not. For example, fatal accidents are a risk to those exposed to motor vehicles, even though the annual death toll is quite predictable. Each is meant to exclude anything but threats to human health and safety (e.g., accompanying property and financial risks). Actually choosing among energy technologies would require consideration of the broader set of consequences appearing in figure 1(A). The general form of figure 1(B) takes a position on one of the five sources of controversy regarding risk, its dimensionality. The specific contents take a position on a second, whether to consider concern a consequence of risk, despite its not having an obvious physical or physiological measure. Positions on the others are taken below in the course of developing a procedure for expressing the risks of the energy technologies.

Figure 1(B) makes two further distinctions. One is between mortal risks to the general public and mortal risks to workers in the technology. Such a distinction is made in many industries, with safety standards for workers being (very) roughly one-tenth as stringent as those for nonworkers (Derr et al., 1983). Making the distinction here allows one to decide whether this common practice should be accepted and enshrined

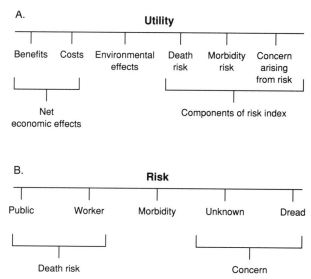

Figure 1. Possible dimensions of consequence: (A) for decision-making, (B) for risk index.

in public policy by assigning a different weight to public and worker deaths. As discussed above, distinctions might also be made on the basis of whether those who die also benefit from the technology, have consented to exposure, or lose their lives in catastrophic accidents.

The second distinction is between two kinds of concern. Studies of lay risk perceptions (Fischhoff et al., 1978; Slovic et al., in press; Vlek and Stallen, 1981) have shown that concern about technologies' risks can be predicted quite well by two "subjective dimensions of risk." These dimensions summarize a large number of individual determinants of perceived risk. They may be described as reflecting (a) the degree to which the risk is *unknown* and (b) the degree to which the risk evokes a feeling of *dread*. The former expresses aversion to uncertainty, and thus represents cognitive (or intellectual) aspects of concern, whereas the latter captures a risk's ability to evoke a visceral response. This usage takes these dimensions from the domain of prediction to the domain of prescription, from anticipating how people will respond to technologies to guiding how those technologies will be shaped.

The morbidity aspect is intended for all nonfatal injury and illness. It would also include genetic damage, whether expressed in birth defects or latent in the population. (The production of spontaneous abortions might appear here or in the previous category.) It should also include the unpleasantness of the period preceding death, which should be negligible for some accidents and considerable for some lingering illnesses.

## Constructing Risk Indices

Choosing a set of attributes to describe a risk prospect creates a vector, each element of which expresses one dimension of consequence. The definition process could be terminated at that point, leaving users to integrate the elements intuitively. Alternatively, the elements can be combined into an aggregate measure of risk in order to eliminate the costs, errors, and vagueness that come with intuitive integration. The essence of aggregation is determining the relative importance of the elements. The following is a generalized scheme that can be adapted to the needs of many problems and value systems. It is drawn from multiattribute utility theory, fuller expositions of which can be found elsewhere, for both the theory itself (Keeney and Raiffa, 1976) and its application to risk problems (Ahmed and Husseiny, 1978; Keeney, 1980; Lathrop and Watson, 1982). The present application differs from its predecessors in focussing on risk, emphasizing the full range of options, and offering no opinion as to the correct solution. The logic of such procedures is as follows:

An aggregation procedure for risks should characterize a technology by a single value such that activities having a higher value will be more risky (in the eyes of those whose values the procedure represents). After the components have been selected, the next step is their operationalization. For example, if the consequences are called $x_i$, then $x_1$ might be the number of (additional) deaths among members of the general public; $x_2$, the number of years of incapacitating illness caused; $x_3$, a measure of public concern, etc. These consequences are often called "attributes"; we will use the terms interchangeably. This operationalization creates a vector of measurements $(x_1, \ldots, x_n)$.

This vector expresses the measure, but not the worth, of those consequences. "Utility" is commonly used as a generalized unit of worth. Utility theory offers a wide variety of procedures for converting each such vector into a single number, representing its overall (un)desirability. These procedures can incorporate highly complex value systems and varying degrees of uncertainty regarding which consequences will, in fact, be experienced. In practice, though, various simplifying assumptions are adopted to render the analysis more tractable. For example, instead of explicitly modeling uncertainty (complete, say, with the elicitation of probability distributions for the values of different parameters), the analyst might treat all results as certainties. However, at the end of the analysis, each parameter will be varied through a range of plausible values to see whether such sensitivity analyses affect the previously reached conclusions.

In order to simplify the exposition, the present example makes two potentially controversial assumptions regarding the value structure. The first of these is *risk neutrality*, meaning, for example, that the certain loss of one life is just as bad as one chance in $10^5$ of $10^5$ deaths. Although it has been argued that people are particularly averse to losing many lives at once (as opposed to losing as many lives in separate accidents), this tendency appears to be due primarily to the great uncertainty that surrounds technologies capable of producing large accidents. A death is a death, whether it comes alone or with many others (Slovic et al., in press). In this definition, the uncertainty surrounding those deaths can be incorporated in the sensitivity analyses and the first "concern" attribute. The second simplifying assumption asserts that the underlying value structure is not overly complex. It is formalized as the property of *mutual preference independence*. Roughly speaking, two attributes are preference independent of all others if tradeoffs between them do not depend upon the levels of the other attributes.

If either of these assumptions seems wanting, then it is straightforward (if cumbersome) to repeat the analysis with alternative assumptions. If they seem adequate, then it is possible to express the index as

$$(1) \qquad R = \sum_{j=1}^{n} w_j y_j$$

where $y_j$ is the expected utility for attribute $j$ and $w_j$ is a weighting factor, expressing its relative importance. "Expected utility" is the product of a consequence's utility and the probability of it being incurred if a technology is pursued. For example, if $x_i$ were the number of public deaths, then $y_i$ would be the expected utility for public deaths, which would consider not only the probability for different losses, but also any changes in the significance of marginal deaths as a function of total deaths.

## Risks of Electricity Generation: Different Definitions

### Problem Description

Electricity generation is an interesting case for two reasons. One is the evidence that disagreement about the definition of key terms (including "risk") has contributed to the bitterness of many energy debates. The second is that important issues tend to generate research, producing data upon which risk estimates may be more soundly based.

In this analysis, six energy technologies are considered. Five of these, coal, hydropower, large-scale windpower, small-scale windpower, and nuclear power, can increase the supply of electricity. The sixth, energy conservation, can reduce the demand for electricity, thereby freeing existing supplies for use elsewhere.

### Attribute Definition

The five attributes of these technologies are those shown in figure 1(B). They are operationalized as follows: Both kinds of death are measured in terms of the expected number of deaths per gigawatt-year (GWyr) of electricity generated or saved. Choosing this summary statistic means taking positions on two additional dimensions of controversy regarding the definition of risk: Broad bounds are set on the technologies, so as to attribute to them all casualties incurred in conjunction with generating electricity. Deaths are just tallied, without regard for the number of years taken off each, the extent of each victim's exposure, the distribution of deaths over time, or any of the other features discussed earlier. Morbidity will be measured by expected person-days of incapacity per GWyr of electricity.

The two attributes associated with concern will be specified in terms of the technologies' ratings on the two comparable factors in psychometric studies of perceived risk (e.g., Slovic et al., in press). These studies have produced sufficiently robust results to make reliance on them conceivable; perceptions of risk have proven sufficiently good predictors of attitudes and actions for them to serve as reasonable indicators of level of concern. What is most arguable about such reliance is treating the expression of concern as evidence of adverse consequences. As discussed above, one ground for that claim is that concern itself is an adverse consequence, which should not be imposed upon people without compensating benefit; a second ground is that concern is associated with stress which is, in turn, associated with various physiological effects that are so difficult to measure that it is reasonable to use concern as a surrogate for them.

### Evaluating Consequences

Having defined the attributes, the next step is to evaluate each possible outcome on each (e.g., how bad is it to incur 10 or 100 worker deaths). In technical terms, this means defining a utility function for each attribute. A convenient way of doing so, given the assumptions made here, is to use a 100-point scale for each attribute, where 0 represents the least extreme possible consequence and 100 the most extreme possible consequence. (If both good and bad consequences were being considered, then a distinction between positive and negative scores would be necessary. Here, 100 is the worst possible outcome.) Intermediate values are defined appropriately. Although linear scaling is possible, it is not necessary. For example, for most people winning $100 will not be 10 times as satisfying as winning $10. Setting the end points of each scale requires a factual (or scientific) judgment regarding what consequences are possible. Setting the midpoints requires a value judgment regarding how those intermediate consequences are regarded.

A natural zero point for a casualty scale is zero casualties. It will be used here, recognizing that no deaths to workers, no deaths to the public, and no person-days lost are practically unachievable with any energy technology. On the basis of worst-case analyses, scores of 100 on attributes 1 and 2 are defined, respectively, as 10 public deaths and 10 occupational deaths per GWyr of electricity generated or saved. Similarly, 60,000 person-days of incapacity per GWyr would merit a score of 100 on

**Table 1. The Components of Attributes 4 and 5**

| Attribute | Score of 0 implies risk has these properties | Score of 100 implies risk has these properties |
|---|---|---|
| 4. Unknown risk | Observable<br>Known to exposed<br>Effect immediate<br>Old<br>Known to science | Not observable<br>Unknown to exposed<br>Effect delayed<br>New<br>Unknown to science |
| 5. Dread risk | Controllable<br>Not dread<br>Not global catastrophic<br>Consequences not fatal<br>Equitable<br>Individual<br>Low future risk<br>Easily reduced<br>Decreasing<br>Voluntary<br>Doesn't affect me | Uncontrollable<br>Dread<br>Global catastrophic<br>Consequences fatal<br>Not equitable<br>Catastrophic<br>High future risk<br>Not easily reduced<br>Increasing<br>Involuntary<br>Affects me |

*Source:* [Reprinted by permission of Westview Press from P. Slovic, B. Fischhoff, and S. Lichtenstein, "Characterizing Perceived Risk," in *Perilous Progress: Managing the Hazards of Technology,* edited by Robert W. Kates, Christoph Hohenemser, and Jeanne X. Kasperson. Published by Westview Press, Boulder, Colorado, 1985].

attribute 3. Intermediate scores are assigned linearly (e.g., on attributes 1 and 2, one death receives 10, two deaths receive 20, etc.), reflecting a desire to assign an equal value to each casualty (as distinct, perhaps, from the decreasing sensitivity to additional casualties that people might actually experience).

Table 1 shows the characteristics that would give a technology scores of 0 and 100 on attributes 4 (unknown risk) and 5 (dread risk). Research has shown that although no technology quite reaches either extreme, mountain climbing and handguns score close to zero on attribute 4 (at least in the U.S.A.), as do home appliances and high school football on attribute 5. At the other extreme, DNA research is rated as sufficiently unknown to receive a score in the 90s on attribute 4, while nuclear weapons do likewise on attribute 5.

## Making Tradeoffs

The final step in specifying an evaluation scheme is to assign weights reflecting the relative importance of the different attributes. As these weights reflect value judgments, disagreements are legitimate; in the present context, they are to be expected. Table 2 presents four sets of weights, each reflecting a different set of values.

Brief descriptions might help explicate the perspectives that could motivate each set's adoption. The first

rejects anything but readily measured physiological effects; treats a death as a death, whether it befalls a worker or a member of the public; views a life as equal to 6000 person-days of incapacity. Set B reflects a belief that concern is a legitimate consequence, that public deaths are twice as important as worker deaths, and that a worker death should be treated as equivalent to the loss of 6000 person-days. As Dunster (1980) argues, "it is not easy to weigh the benefits of reducing anxiety against those of saving life, but our society certainly does not require the saving of life to be given complete priority over the reduction of anxiety" (p. 127). Set C increases the importance ratio for public to occupational deaths and assigns major significance to concern. The specific weights imply a willingness to tradeoff 10 public deaths per GWyr to move from a technology causing extreme dread to one that is about average, perhaps feeling that the toll from concern-generated stress is large or that even

**Table 2. Four Possible Sets of Weights for Five Risk Attributes**

| Attributes | A | B | C | D |
|---|---|---|---|---|
| 1. Public deaths | 0.33 | 0.40 | 0.20 | 0.08 |
| 2. Occupational deaths | 0.33 | 0.20 | 0.05 | 0.04 |
| 3. Morbidity | 0.33 | 0.20 | 0.05 | 0.40 |
| 4. Unknown risk | 0 | 0.10 | 0.30 | 0.24 |
| 5. Dread risk | 0 | 0.10 | 0.40 | 0.24 |
| Sum of weights | 1 | 1 | 1 | 1 |

minor accidents in a dread technology can cause enormously costly social disruption. The D weights represent a paramount concern with the suffering of the living, whether through injury or anxiety, rather than with the number of deaths.

Whatever one's value system, the weights assigned should be very sensitive to the range of outcomes considered on each attribute. If, for example, 100 on attribute 1 meant 50 public deaths per year (rather than 10), then Set A would have to assign a larger value to attribute 1 to achieve the same effect of weighting a public and a worker death equally.

## Scoring Technologies

In order to apply this scheme to technologies, it is necessary to assess the magnitude of the consequences that each produces on each attribute. This is a scientific, not a value question. It should be informed by the best available technical knowledge. However, applying that knowledge in the present case requires the exercise of judgment, to choose, weigh, and extrapolate from existing studies. Despite having a commitment to objectivity, we cannot escape some subjectivity in attempting to derive this sort of policy-oriented advice.

Table 3 provides point estimates roughly summarizing the research reported in the following sources: Baecher et al. (1980), Birkhofer (1980), Bliss et al. (1979), Budnitz and Holdren (1976), Comar and Sagan (1976), Department of Energy (1979), Dunster (1980), Greenhalgh (1980), Hamilton (1980), Okrent (1980), Rogers and Templin (1980), and Slovic et al. (1980, in press). This literature reveals both substantial differences of opinion and substantial areas of ignorance. As two examples: The extreme values for expected occupational deaths from

coal were 0.7 and 8 deaths per GWyr of electricity generated. Very few risk data were available for either small-scale wind power or conservation; these scores were liberally adapted from knowledge of other technologies. Where available, the concern scores required the least exercise of judgment. Technologies that have been rated have proven to have rather robust scores on these dimensions, regardless of who does the rating, how the rating is carried out, and what other technologies are in the rating set (Slovic et al., in press). However, several energy technologies have yet to be evaluated in this way. Their scores were derived by conjecture. For example, the scores for conservation on attributes 4 and 5 were averages of those for home appliances and bicycles.

Given the unreliability of these estimates, any attempt to establish the risks of energy technologies would have to address the uncertainty surrounding them, with either sensitivity analyses or explicit assessment of probabilities. Given the illustrative nature of the present example, that exercise will be foregone as misplaced imprecision.

## Computing Risk

Using these values and Set A's weights, Expression (1) shows the risk from coal to be 0.33(80) + 0.33(30) + 0.33(20) = 42.9. Other scores are computed similarly and displayed in figure 2. Because the scores are standardized to range from 0 to 100 and the weights to sum to 1.0, it is possible to compare scores across technologies and across weighting schemes. That comparison shows that the riskiness of coal, small-scale windpower, and conservation vary little across these four sets of weights, whilst those for hydro, large-scale windpower, and particularly nuclear power vary greatly. Thus, if one accepts the

**Table 3. The Scores of Six Technologies of Five Risk Attributes by One Expert**

| Attribute | Coal | Hydro | Large scale wind | Small scale wind | Nuclear | Conservation |
|---|---|---|---|---|---|---|
| 1. Public deaths | 80 | 10 | 20 | 5 | 10 | 5 |
| 2. Occupational deaths | 30 | 20 | 10 | 30 | 5 | 10 |
| 3. Morbidity | 20 | 20 | 40 | 50 | 10 | 40 |
| 4. Unknown risk | 70 | 60 | 90 | 50 | 80 | 40 |
| 5. Dread risk | 50 | 50 | 40 | 20 | 90 | 10 |

consequence estimates of table 3, then the riskiness of these last three technologies depends upon the importance assigned to the different consequences.

Table 4 shows how this sensitivity expresses itself in terms of the relative riskiness of the six technologies. Coal, for example, ranks consistently low, whereas nuclear may be best or worst depending upon the definition used. These enormous variations occur despite complete agreement regarding the magnitude of the consequences. Thus arguments over relative risk may reflect only disagreements about values.

## Conclusion

An effective decisionmaking process, whether conducted by individuals or societies, requires agreement on basic terms. Without such conceptual clarity, miscommunication and confusion are likely. Definitional ambiguity regarding the term "risk," in particular, has spawned needless (and irresoluble) conflict over the relative riskiness of different technologies. At the same time, it has obscured the need for debate over the value issues involved in specifying what "risk" means.

The present analysis presents a framework for defining risk which directly faces those inherent conflicts. Indeed, it forces one to adopt an explicit position on each aspect of the controversy before a workable definition can be created. As a result, the specific indices of risk developed here are controversial by design. However, they are also expendable by design. The general framework is highly flexible, capable of fitting many problems and many value systems. Its use in a particular problem makes possible a diagnosis of the extent to which conflicts reflect disagreements about facts or disagreements about values. In the former case, one can hope that consensus about risk will evolve as scientific research progresses. In the latter case, consensus will only emerge if there is effective public debate about what society should value. That debate can be informed (and spurred) by ethical and policy analyses, but it cannot be resolved by them.

```
             AAAAAAAAAAAAAAAAAAAAAA
             BBBBBBBBBBBBBBBBBBBBBBBBBBB
Coal         CCCCCCCCCCCCCCCCCCCCCCCCCCCCCC
             DDDDDDDDDDDDDDDDDDDDDD

             AAAAAAAA
             BBBBBBBBBBB
Hydro        CCCCCCCCCCCCCCCCCCCCC
             DDDDDDDDDDDDDDDDDD

             AAAAAAAAAAAA
             BBBBBBBBBBBBBBBB
Large wind   CCCCCCCCCCCCCCCCCCCCCCCCC
             DDDDDDDDDDDDDDDDDDDDDDDDD

             AAAAAAAAAAAAAA
             BBBBBBBBBBBBB
Small wind   CCCCCCCCCCCCCC
             DDDDDDDDDDDDDDDDDDD

             AAAA
             BBBBBBBBBBBB
Nuclear      CCCCCCCCCCCCCCCCCCCCCCCCCCCCCCC
             DDDDDDDDDDDDDDDDDDDDDDD

             AAAAAAAAA
             BBBBBBBBB
Conservation CCCCCCCCCC
             DDDDDDDDDDDDDD
             .........:.........:.........:.........:
             0        20        40        60        80
                                                RISK →
```

**Figure 2. The risk indices of six technologies on four sets of weights.**

Applying this potentially rich procedure required a series of simplifying assumptions. These included taking only five risky consequences from the vector

**Table 4. The Risk-To-Human-Health Rankings of Six Technologies Given Four Sets of Weights on One Set of Five Attributes and the Scores of One Expert**

| Rank | | Set of Weights: A | B | C | D |
|---|---|---|---|---|---|
| Best | 1. | *Nuclear* | Conservation | Conservation | Conservation |
| | 2. | Hydro | Hydro | Small wind | Hydro |
| | 3. | Conservation | *Nuclear* | Hydro | Small wind |
| | 4. | Large wind | Small wind | Large wind | Coal |
| | 5. | Small wind | Large wind | Coal | *Nuclear* |
| Worst | 6. | Coal | Coal | *Nuclear* | Large wind |

of possibilities, asserting risk neutrality and mutual preference independence, and representing effect magnitude by point estimates. Despite these restrictions, this illustrative analysis showed that the relative riskiness of different energy technologies is quite sensitive to how risk is defined.

The emphasis here has been on the logic of the analysis, rather than on its content. Making a definitive statement regarding the risks of competing energy technologies would require definitive estimates of both the magnitude and the importance of those consequences for a particular society. Neither was attempted here. An additional caution is that even a definitive statement about risk would have no necessary implications for most policymaking. People do not accept risks, but technologies, one of whose significant features may be their risks. Developing an index of risk allows systematic treatment of one aspect of those decisions, but only one aspect. Analogous treatments of other consequence domains would be needed to complete the picture.

Developing a definition of risk requires a variety of explicit value judgments. Choosing to express risk in a numerical index may itself make a statement of values. The present exposition emphasized the possibilities that an index offers for including different people's values in policymaking. However, it may also be used to exclude the people themselves from the policymaking process, with policy experts serving as self-appointed spokespeople for what the public wants. Even if careful research is conducted to identify the public values that are to be incorporated in society's risk index, such technical recognition need not substitute for active, personal participation. If it is used that way, then the index may be blamed for the faults of a political process that can tolerate public opinion, but not the public.

## Acknowledgments

This research was supported by the National Science Foundation under Grant PRA-8116925 to Perceptronics, Inc. Any opinions, findings and conclusions expressed in this article are those of the authors and do not necessarily reflect the views of the National Science Foundation. We give our thanks to Vincent Covello, Jack Dowie, Ken Hammond, Paul Slovic, and Ola Svenson for perceptive comments on previous drafts.

## References

Ahmed, S. and Husseiny, A. A. (1978). "A multivariate utility approach for selection of energy sources," *Energy* 3: 669–700.

Baecher, G. B., Pate, M. E. and de Neufville, R. (1980). "Risk of dam failure in benefit-cost analysis," *Water Resources Research* 16: 449–456.

Birkhofer, A. (1980). "The German risk study for nuclear power plants," *IAEA Bulletin* 22 (5/6): 23–33.

Bliss, C., Clifford, P., Goldgraben, G., Graf-Webster, E., Krickenberger, K., Maher, H. and Zimmerman, N. (1979). *Accidents and Unscheduled Events Associated with Non-Nuclear Energy Resources and Technology.* Washington, D.C.: MITRE Corporation for Environmental Protection Agency.

Budnitz, R. J. and Holdren, J. P. (1976). "Social and environmental costs of energy systems," *Annual Review of Energy* 1: 553–580.

Cohen, B. and Lee, I. (1979). "A catalog of risks," *Health Physics* 36: 707–722.

Comar, C. L. and Sagan, L. A. (1976). "Health effects of energy production and conversion," *Annual Review of Energy* 1: 581–660.

Cotgrove, A. (1982). *Catastrophe or Cornucopia? The Environment, Politics and the Future.* New York: Wiley.

Crouch, E. A. C. and Wilson, R. (1982). *Risk/Benefit Analysis.* Cambridge, Mass.: Ballinger.

Department of Energy (1979). *Energy Technologies for the United Kingdom, Vol. I and II.* London: HMSO.

Derr, P., Goble, R., Kasperson, R. E. and Kates, R. W. (1983). "Responding to the double standard of worker/public protection," *Environment* 25 (6): 6–11, 35–36.

Dunster, J. H. (1980). "The approach of a regulatory authority to the concept of risk," *IAEA Bulletin* 22 (5/6): 123–128.

Elliott, G. R. and Eisdorfer, C., (Eds.) (1982). *Stress and Human Health.* New York: Springer Verlag.

Fischhoff, B. (1983). "Acceptable risk: The case of nuclear power," *Journal of Policy Analysis and Management* 2: 559–575.

Fischhoff, B. (1984). "Standard setting standards," *Management Science* in press [30: 823–843].

Fischhoff, B., Lichtenstein, S., Slovic, P., Derby, S. L. and Keeney, R. L. (1981). *Acceptable Risk.* New York: Cambridge University Press.

Fischhoff, B., Slovic, P. and Lichtenstein, S. (1983). "The public vs. The experts: Perceived vs. actual disagreements about the risks of nuclear power," in V. Covello, G. Flamm, J. Rodericks, and R. Tardiff (Eds.), *Analysis of Actual vs. Perceived Risks.* New York: Plenum.

Fischhoff, B., Slovic, P., Lichtenstein, S., Read, S. and Combs, B. (1978). "How safe is safe enough? A psychometric study of attitudes towards technological risks and benefits," *Policy Sciences* 8: 127–152.

Green, C. H. (1980). "Risk: Attitudes and beliefs," in D. V. Canter (Ed.), *Behaviour in Fires*. Chichester: Wiley.

Greenhalgh, G. (1980). *The Necessity for Nuclear Power*. London: Graham & Trotman.

Hamilton, L. D. (1980). "Comparative risks from different energy systems: Evolutions of the methods of studies," *IAEA Bulletin* 22 (5/6): 35–71.

Herbert, J. H., Swanson, L. and Reddy, P. (1979). "A risky business," *Environment* 21 (6): 28–33.

Inhaber, H. (1979). "Risk with energy from conventional and nonconventional sources," *Science* 203: 718–723.

Keeney, R. L. (1980). "Evaluating alternatives involving potential fatalities," *Operations Research* 28: 188–205.

Keeney, R. L. and Raiffa, H. (1976). *Decisions with Multiple Objectives. Preferences and Value Trade-offs*. New York: Wiley.

Lathrop, J. W. and Watson, S. R. (1982). "Decision analysis for the evaluation of risk in nuclear waste management," *Journal of the Operational Research Society* 3: 407–418.

Lichtenstein, S., Slovic, P., Fischhoff, B., Layman, M. and Combs, B. (1978). "Judged frequency of lethal events," *Journal of Experimental Psychology: Human Learning and Memory* 4: 551–578.

Okrent, D. (1980). "Comment on social risk," *Science* 208: 372–375.

Otway, H. and von Winterfeldt, D. (1982). "Beyond acceptable risk: On the social acceptability of technologies," *Policy Sciences* 14: 247–256.

Reissland, J. and Harries, V. (1979). "A scale for measuring risks," *New Scientist* 83: 809–811.

Rogers, D. W. O. and Templin, R. J. (1980). "Errors in a risk assessment of renewable resources," *Energy* 5: 101–103.

Slovic, P., Fischhoff, B. and Lichtenstein, S. (1980). "Facts and fears: Understanding perceived risk," in R. Schwing and W. A. Albers Jr. (Eds.), *Societal Risk Assessment: How Safe is Safe Enough?* New York: Plenum.

Slovic, P., Fischhoff, B. and Lichtenstein, S. (1984). "Characterizing perceived risk," in R. W. Kates and C. Hohenemser, (Eds.), *Technological Hazard Management . . .* in press. [*Editors' note:* Published in 1985 by Westview Press, Boulder, Colorado, under the title: *Perilous Progress: Managing the Hazards of Technology,* edited by R. W. Kates, C. Hohenemser, and J. X. Kasperson.]

Slovic, P., Lichtenstein, L. and Fischhoff, B. (1984). "Modeling the societal impact of fatal accidents," *Management Science* in press [30: 464–474].

Starr, C. (1969). "Social benefit versus technological risk," *Science* 165: 1232–1238.

U.S. Nuclear Regulatory Commission (1982). *Safety Goals for Nuclear Power Plants: A Discussion Paper*. NUREG-0880. Washington, D.C.: The Commission.

Vlek, C. A. J. and Stallen, P. J. M. (1981). "Risk perception in the small and in the large," *Organizational Behavior and Human Performance* 28: 235–271.

Watson, S. R. (1981). "On risks and acceptability," *Journal of the Radiological Protection Society* 1: 21–25.

Wilson, R. (1979). "Analyzing the daily risks of life," *Technology Review* 81 (4): 40–46.

Wynne, B. (1983). "Institutional mythologies and dual societies in the management of risk," in H. C. Kunreuther and E. V. Ley (Eds.), *The Risk Analysis Controversy*. New York: Springer Verlag.

Zentner, R. D. (1979). "Hazards in the chemical industry," *Chemical and Engineering News* 57 (45): 25–27, 30–34.

# Questions for Thought and Discussion

1. Fischhoff, Watson, and Hope state that defining risk is a political act, but then make little mention of politics in this paper. Interpret and illustrate this statement.

2. Identify a prominent health risk and a prominent safety risk. In each case describe the benefits and nonrisk costs, and identify the units of a suitable summary statistic for expressing the risk.

3. Given the information in table 1 in this paper, list four activities other than those mentioned in the discussion of the table (mountain climbing, using handguns, using home appliances, playing high school football, DNA research, and stockpiling nuclear weapons), one for each of the following attributes: low unknown risk, high unknown risk, low dread risk, high dread risk.

4. Exposures to environmental carcinogens are regulated in the United States so that the associated lifetime risk of cancer from any one of them is on the order of $10^{-6}$ to $10^{-5}$ (1 in 1 million to 1 in 100,000), which corresponds to about 3 to 30 cases per year nationwide (assuming all U.S. residents are exposed). In contrast, 45,000 to 50,000 persons are killed in motor vehicle accidents in the United States each year and about 8,000 are killed by handguns. What does this say about the set of weights (as in table 2) that is used in setting public policy with regard to these risks?

5. What are the principal net economic effects (benefits and costs), environmental effects, and risk components (death risk, morbidity risk, and concern arising from risk) associated with each of the six energy technologies discussed in this paper?

# Risk Analysis: Understanding "How Safe Is Safe Enough?"

STEPHEN L. DERBY
*Stanford University*
*Stanford, California*

RALPH L. KEENEY
*Woodward–Clyde Consultants*
*San Francisco, California*

## 1. Introduction

"How safe is safe enough?" is a question that is increasingly asked about the hazards from using technology in our society.[1] "Are Nuclear Power Plants Really Safe?," "11,000 Industrial Waste Disposal Sites a Hazard to U.S. Citizens," and "Saccharin May Cause Cancer" are common headlines. Such examples illustrate various types of technology that evoke the question "How safe is safe enough?" Additional examples are easy to suggest. However, clearly defining what is meant by "How safe is safe enough?," which is our purpose here, is not at all easy.

This paper presents a characterization of the problem. We recognize, however, that there is likely no universal agreement on this characterization. Our purpose is to explain the problem as clearly as possible. This is done in several ways. We define the essential ingredients of "How safe is safe enough?" problems, the generic features often complicating these problems, and the types of "solutions" that are appropriate. Several types of answers that, in general, are not appropriate, but are often suggested as simple panaceas for the general problem, are identified. Complicating technical characteristics of the problem that render it so unwieldy for analysis; the effects of complicating social, political, and ethical features that limit the usefulness of analysis; and the role of regulatory agencies in determining acceptable risk are also considered.

## 2. What Is Acceptable Risk?

Perhaps the biggest difficulty with determining "How safe is safe enough?" is that it, per se, is not a problem. Therefore, it does not have *an* answer. Rather "How safe is safe enough?" is a catchy phrase to identify a component in many complex sociotechnical decision problems. The answer in any particular problem depends on many things.

Before addressing acceptable risk, let us define risk. Risk as used in this paper will refer to the possibility of consequences involving mortality, morbidity, or injury to members of the public. Collectively, these will be referred to as safety risks.

The question of "How safe is safe enough?" addresses the acceptability of such risks from using technology that may endanger the safety of the public. The answer to this question requires that both

*Editors' note:* The current affiliation of Stephen L. Derby is the Strategic Decision Group, Menlo Park, California, and of Ralph L. Keeney is the University of Southern California, Los Angeles.

technical and social aspects be carefully considered.

The key aspect of acceptable risk problems is that the solution is found by a decision among alternatives. The generic problem involves choosing the best combination of advantages and disadvantages from among several alternatives. The particular decision problems that concern us in this paper have one disadvantage relating to safety risks of the public. In selecting the alternative, the various pros and cons should be weighed in some responsible manner. The level of risk associated with the chosen alternative is then, by definition, acceptable. Stated differently, the risk associated with the best alternative is safe enough.

Before proceeding, let us try to clear up one possible source of confusion. Acceptable risk is not necessarily the level of risk with which we are happy. We all would prefer less risk to more risk if all other consequences were held fixed. However, this is never the case. In a situation with no alternatives, then the level of safety associated with the only course of action is by definition acceptable, no matter how disagreeable the situation. Said another way, acceptable risk is the risk associated with the best of the available alternatives, not with the best of the alternatives which we would hope to have available.

## 3.  The Kinds of Choices That Determine Acceptable Risk

We want to point out that one of the alternatives in the acceptable risk problem may be not to utilize a new technological device or technology, that is, the general problem area includes the no-go option. For instance, the options of utilizing nuclear power or not utilizing nuclear power are two choices that properly constitute enough alternatives for determining acceptable risk for that specific context. Public safety may be threatened either through direct or indirect effects of the alternatives which are chosen. For instance, if all nuclear power plants and all coal power plants were shut down, then the direct safety risks to the public from these facilities would obviously be zero. However, in this case the chosen no-go option would have risks to the public as a result of no energy. Whether or not it is appropriate and necessary to include indirect threats (such as from no energy) to the safety of the public is dependent on the particular problem context and the alternatives being examined.

This problem domain includes a wide array of governmental and institutional activities. It would certainly include most of those large-scale projects in the energy area, such as the building of energy facilities and the acquisition, fabrication, and transportation of fuels. Many of the problems addressed by the Food and Drug Administration also have a "How safe is safe enough?" aspect. Problems concerning the nation's transportation systems, medical facilities, and work places (factories and office buildings) also are to be included. There are very few large-scale problems in a modern society that do not involve both technological and social concerns, a choice of alternatives, and the possibility of risk to the safety of members of the public.[2]

One special case of the general problem is particularly interesting, the case where the advantages of each of the alternatives are considered equal. These kinds of alternatives then differ only by their disadvantages. Often the important disadvantages are the economic costs and the impacts on safety to the public. In such problems, one could decide to pay more money for safety devices in order to improve safety. The choice of a best alternative then involves finding the best combination of costs and safety. An example occurs after one has decided to build a nuclear plant of a particular size. The major questions then are where to build it and how to design its safety features.

## 4.  Addressing the Acceptable-Risk Decision Problem

The nature of determining acceptable risk is such that one must address it with a prescriptive analysis. That is, the analysis helps identify how safe the chosen alternative *should* be. This could be based on many things, including behavioral descriptions of how people respond to various types of risks.

Given this prescriptive orientation, addressing "How safe is safe enough?" is not the same as determining how safe various alternatives are. The relationship between the two is seen more clearly by considering the complete analysis process. This process has five interdependent steps that should be followed in evaluating the alternatives in any decision problem:

1. Define the alternatives.

2. Specify the objectives and measures of effectiveness to indicate the degree to which they are achieved.

3. Identify the possible consequences of each alternative.

4. Quantify the values for the various consequences.

5. Analyze the alternatives to select the best choice.

Step 3 of the process identifies the safety of the various alternatives. This risk assessment is not all of the activity carried out under Step 3. Consequences other than those concerning safety (i.e., benefits and other costs) also must be specified.

## 5. The Answer to the Problem

The answer to "How safe is safe enough?" depends on the five steps of analysis presented above. Acceptable risk is determined by what alternatives are available, what objectives must be achieved, the possible consequences of the alternatives, and the values to be used.

For a simple example of how alternatives affect acceptable risk, consider figure 1a. We assume here for illustration that the benefits of all the alternatives are identical. The differences among alternatives are only in their financial cost and level of risk and 0 is the best level for each of these dimensions. If just alternatives K and L are available, then the choice is between high cost with low risk and low cost with high risk. The acceptable risk would be the level of risk associated with the particular alternative chosen, either K or L.

If another alternative M were introduced into the problem, then clearly M with lower cost and lower risk would be preferred to either K or L. Consequently, acceptable risk is now the safety level of alternative M. This risk is different from the level associated with the other alternatives. Clearly the appropriate level of risk depends on the alternatives available.

Figure 1b illustrates how acceptable risk changes with what objectives are achieved. For this illustration, only alternatives K and L are assumed available. If the objective is solely to minimize the risk then alternative K would be chosen. The acceptable risk

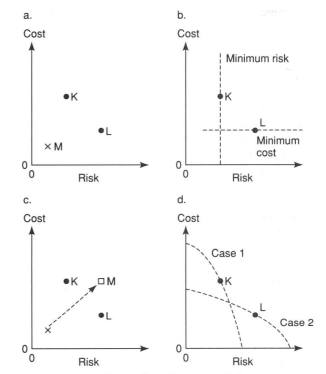

Figure 1. Acceptable risk depends on many factors.

would then be the risk level associated with K. On the other hand, if the objective is solely to minimize the cost, then alternative L would be chosen. Acceptable risk under this objective would be the risk level for L. Each objective leads to choosing different alternatives. In each case the acceptable risk changes with the objective used to make the choice.

Figure 1c illustrates the effect on determining acceptable risk from new information on the possible consequences. We assume that alternative M determines the acceptable risk, as in figure 1a. However, additional information provided by experience, research, development, or analysis reveals that the initial assessment of alternative M must be revised. Instead of confirming that M had lower cost and lower risk than both alternatives K and L, the new information shows that M has both the high cost of K and the high risk of L. The acceptable risk is now determined by the choice between K and L, and is no longer determined by the alternative M.

Figure 1d illustrates the effect of values and preferences on the choice between K and L. In this figure different preferences for trading-off increased cost for lower risk are represented by the two curves. In case 1, the trade-off curve reflects the willingness to incur large costs to reduce risk by small amounts. Alternative *K* then is the most attractive choice with this preference. In case 2, the trade-off curve reflects less of a willingness to increase costs in exchange for specific reductions in risk. This preference selects alternative *L* as the best choice. Since acceptable risk is determined by the choice among the two alternatives, these different preferences change the answer to "How safe is safe enough?"

Comparing figure 1b and figure 1d gives an interesting insight into the sensitivity of acceptable risk to different values and preferences. Both the objective of only minimizing risk with no trade-off for cost in figure 1b and the cost-risk trade-off curve of case 1 in figure 1d choose the same alternative *K* as the best choice. Both preferences lead to determining the same acceptable risk since no other alternatives are available that further reduce the risk. Likewise, the objective for only minimizing cost in figure 1b and the preference for the cost-risk tradeoff of case 2 in figure 1d lead to the same acceptable risk.

These illustrations clearly show that determining acceptable risk depends on many factors that can change over time due to new technology and experience and can differ from one person or group to another.

## 6. What the Answer Is Not

In a sense, it should not be necessary to discuss what is not the answer to "How safe is safe enough?" However, because of the general complexity of the question and because of the confusion about the fact that acceptable risk is itself not a problem, many oversimplified "solutions" have been suggested. Our purpose is in this section to dispel some of these misconceptions.

One common misconception is extremely clear. An inappropriate answer to "How safe is safe enough?" is that no risk to safety should be tolerated.[3] Alternatives are simply not available that have no risks. At first thought, an alternative that completely bans a particular substance or a technological development would seem to cause no risk to safety. If banned, of course, this substance or development would cause no risk. However, in these

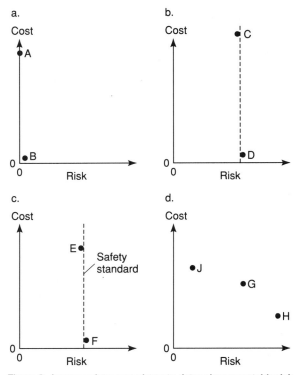

**Figure 2. Inappropriate procedures to determine acceptable risk.**

circumstances there can be risks with the lack of that particular substance or device. A recent example illustrates the point. Laboratory testing gave regulators justification to ban saccharin because it may cause cancer. Unfortunately, the lack of saccharin could also cause increased risks to diabetes patients and others who must avoid other sweeteners like sugar and honey.

Graphically this point may be illustrated as shown in figure 2a. As before, let us suppose that we have measures for both economic costs and risk and that 0 is the best for each of these dimensions. In this example assume that benefits are fixed at some high level for each of the alternatives. Suppose now that we have alternatives *A* and *B* as indicated in the figure, where *A* has a very high cost but no risk problems, and *B* has an extremely low cost and just a minute but actual risk problem. Many individuals may say that *B* should be preferred to *A* even though *A* has no risk. The minute increase in risk may be worth the large decrease in economic costs.

Another example of an inappropriate answer to

"How safe is safe enough?" is that the alternative should be as safe as possible. This can be illustrated with a slight alteration of figure 2a to give us figure 2b. Here again we have assumed the benefits for alternatives C and D are identical and that the costs associated with alternative C are extremely high and there is a risk problem. With alternative D, the costs are very low but there is a risk problem slightly larger than that with alternative C. In this circumstance, it is clear that D could be adjusted such that its risk is so close to that of C that D would be preferred and yet C would have a smaller risk.

In some simplified "solutions," the answer to "How safe is safe enough?" is stated by a small number (like $10^{-7}$), which refers to the probability that an individual will be a fatality due to the cause under consideration in a particular year. Figure 2c illustrates that a solution such as this would, in general, be inappropriate. Suppose the risk standard is stated as a fixed small number, and that alternatives E and F, respectively, lie slightly to the more safe side and to the less safe side of that standard. Assume the benefits with alternatives E and F are identical and that alternative E has a cost substantially larger than alternative F. Then as in previous examples, alternative F would be preferred to alternative E by many individuals, and yet the choice of the alternative which met the risk standard would lead to a choice of alternative E.

Another argument indicating why a number like $10^{-7}$/person/year risk would be inappropriate is that it does not address all the aspects of risk which may be important.[4] For instance, aside from just individual risk, it may be important for society to minimize the probability of accidents involving large numbers of people. For instance, if there is a 1 in 10,000 chance of 10,000 people dying in one situation and a 1 in 10 chance of 10 people dying in another, the average risk per person is the same for each. However, it may be that society would not like to evaluate these two situations to be the same. Even if the answer to the problem posed above were expanded to include standards on the risk per person and standards on the maximum number of people that might be involved in an accident, for the same reason as indicated in figure 2c, although requiring an explanation in more than one dimension, those safety standards collectively would not represent an appro-priate answer to "How safe is safe enough?"

Another incomplete answer would be to specify the value trade-offs between the measures in question. For instance, in figure 2d it may be that alternative G is preferred to H even though G would necessarily double the cost of H to reduce the risk by about one-fourth. At the same time one might prefer G to J even though the change from G to J "buys" more safety for less cost than the change from H to G. This choice could be consistent. The ranges over which the impacts fall may be large enough such that different value trade-offs between the costs and risk may be appropriate in different parts of the ranges. In figure 2d, as in the other figures, we have assumed that the benefits accruing from each of alternatives G, H, and J are the same.

## 7. Complicating Technical Features of the Problem

Technical features render it both more appropriate and more difficult to conduct a thorough analysis to help reach an answer to "How safe is safe enough?" We explain the complexity of these technical features in categories that correspond to the steps of analysis outlined earlier.

First, in many situations it is simply not that obvious what all the alternative courses of action are. Some obvious choices will be available, but the question remains whether a more creative alternative can be identified that is much better. The identification or creation of such an alternative could result in significant improvements on costs and safety, as well as benefits. Whether or not to seek such an alternative is itself a possible course of action.

Second, the objectives and their appropriate measures are not clear in most problems. Stating simply that we want to maximize benefits and minimize costs (including detrimental safety effects) is too general to be useful. However, when the question becomes more specific, the problem becomes much more difficult. For example, if the choice is whether or not to ban nuclear power, a great number of objectives can be appropriate. These objectives may consider many concerns that vary from the defense of the country, increased jobs and employment, obtaining energy independence, and the implications of foreign control of government policy to consideration of concerns for environmental, social, safety, and economic issues.

Identifying the appropriate measures for all of these objectives is not an easy task.

Third, in any complex problem there are a great number of uncertainties. The consequences of any particular course of action are by no means well known. Furthermore, in many circumstances there are no tests that eliminate all of the major uncertainties present in the problem. With the large array of possible consequences, the number of disciplines that have information relevant to those consequences is large. In major choices, such as those concerning the energy sources of the nation, essentially all of the scientific disciplines can have knowledge relevant to the problem. Collecting and meaningfully using all the data that is available is a formidable task.

Fourth, assessing the structure of values and preferences appropriate for evaluating consequences is very involved. The first question to be resolved is whose values should be used? Should they come from the elected officials representing the public, from the regulatory authorities charged with various responsibilities to control safety hazards, from the public themselves, or from a combination of the above? How should these values be determined? Perhaps more importantly, regardless of how well this is done, there will be major conflicts in values among various people. How should these conflicting values be reconciled? To further complicate this assessment is the possibility that values and preferences for difficult and unpleasant consequences are not clear or well formed for many people. Having a person or group understand their own values concerning unfamiliar consequences is as much of a challenge as identifying whose values should be used.

There is certainly no prescribed course of action for addressing these complexities. To address them requires a good deal of ingenuity and creativity. Furthermore, one hopes that these complications are addressed systematically and openly for those problems so critical to our society.

## 8. Complicating Social, Political, and Ethical Features of Determining Acceptable Risk

These complicating features are issues that surround the use of technical analysis in evaluating acceptable risk. The heart of the social, political, and ethical complications is the fact that collective action must be taken on risk management alternatives. Critical issues address what process will be used for making the decision and who or what organization should make the decision. In our society, the answers to these questions are determined in the political process, not by scientific analysis. There are a number of factors that make the political process somewhat cumbersome in determining acceptable risk.

Regardless of how the decision is made, it is meant to represent how millions of members of the public would individually make the acceptable risk decision. The collective decision is meant to reflect both the judgments and perceptions of each person and their values. However, different individuals may have widely varying judgments and perceptions and very diverse values. There are no simple solutions to the collective decision in such a case. Furthermore, such decisions cannot please everyone. In any specific case, regardless of the level of acceptable risk resulting from the decision, many individuals can be quite disappointed and disagree with the alternative chosen.

A factor that worsens these problems is the high level of technical details that are involved in most acceptable-risk decisions. In most cases, these details are either not known or understood by members of the general public. In many of the cases, there is no way that the public can become completely informed. The fact is that one must be a specialist to understand many of the technicalities. Furthermore, there are enough technical details that no one individual can be a specialist in all aspects of an acceptable-risk decision.

That the solution to an acceptable risk decision is not based solely on technical considerations complicates how such a decision should be made. We have referred to this fact several times in this paper when we mentioned that the social aspects of the problem were critical. As such, the general guidelines for how the decision should be made must address not only the technological details but the social realities also.

There are numerous ethical constraints on the entire decision process. Some of these come directly from the charters of various governmental organizations and others are historical in nature. Ethical constraints mean that there are certain alternatives and certain decision processes that just cannot be followed. For example, a decision process that excluded the participation of the people who would bear the risk from technological hazards is unethical. Since the ethics for our form of representative democracy are based on the consideration of the rights of individuals, such a decision process conflicts with our basic approach to government. Ethics are also involved with the political question of choosing the person or group that is responsible for making acceptable risk decisions. This responsibility carries with it the understanding that the collective decision process will be representative and consistent with our political ethics and social values and thus acceptable to the public. In order to determine acceptable risk with collective decisions, the decision process itself must be acceptable.

## 9. The Role of a Regulatory Agency in Determining Acceptable Risk

In many situations in our society, the general responsibility for making acceptable risk decisions rests with the regulatory agency. However, the legislative charters for these regulatory agencies often state general, vague objectives for what the agency should do. However, these charters never clearly state what the specific objectives of the agency should be or how to measure or achieve the regulatory objectives. These critical questions are left open for the agency to decide for itself, often outside the effects of public participation.

In principle the regulatory agency provides a mechanism for the collective decisions that must be made on acceptable risk. The typical mechanism is that the regulators identify specific technical alternatives for managing the risk of the potential technological hazards. Then information on these risks for each of the alternatives is gathered and a recommendation or ruling is made. This ruling has the effect of either choosing the alternative or specifying guidelines by which it should be chosen by others. Rarely are the technical complications discussed earlier explicitly addressed in the detail which can be useful. In particular, most of the focus is on trying to classify the possible consequences of the alternatives that are identified. Formal analysis to help identify different objectives to clearly explain the uncertain consequences, and to clearly indicate the value structure are often only implicitly considered.

The primary role of a regulatory agency is to bridge the gap between its general charter objectives and the specific regulations and rules that the agency uses. Systematic, scientific analysis has many appealing features for aiding (not for replacing) the regulatory agency in its decision process. To exploit this potential value from analysis requires that the technical features and the social, political, and ethical aspects complicating the problem be explicitly recognized and addressed in those analyses.

## 10. Conclusion

This paper is meant to provide an understanding of the main features present in acceptable risk problems. Based on this, one conclusion is that there is no single "How safe is safe enough?" problem and thus, the search for its answer is fruitless. Rather, the problem appears in many different contexts and solutions for each case are context and problem specific. Because of the complexity surrounding such problems, we suggest that formal analysis has a significant potential to aid our society in selecting responsible courses of action concerning risks. Such analyses must, however, explicitly address the problem complexity if this potential is to be exploited. An appraisal of alternative approaches to provide this assistance is found in Fischhoff et al.[5] which presents the larger investigation upon which this paper is based.

### Acknowledgments

This work was performed for the U.S. Nuclear Regulatory Commission under NRC Interagency Agreement 40-550-75. We thank B. Fischhoff, S. Lichtenstein and P. Slovic for many comments on earlier drafts of this paper.

### References

1. See, for example, C. L. Comar, *Science* **203**, 319 (1979); A. Wildavsky, *American Scientist* **67**, 32 (1979); T. Alexander, *Fortune* **94** (February 26, 1979) and *Environmental Science and Technology* **13**, 146 (1979).

2. P. Slovic, S. Lichtenstein, and B. Fischhoff, in *Proceedings of the Second International Scientific Forum on an Acceptable World Energy Future*, A. Perlmutter, O. K. Kadiroglu, and L. Scott, eds. (Ballinger, Cambridge, Mass., 1979).

3. An amendment to the Food, Drug, and Cosmetic Act referred to as the Delaney Clause forbids the use of any additive shown to cause cancer in humans or animals. This is de facto a "zero-risk" attitude.

4. See, for example, W. Barnaby, *Nature* **276**, 554 (1978).

5. B. Fischhoff, S. Lichtenstein, P. Slovic, S. L. Derby, R. L. Keeney, *Acceptable Risk* (Cambridge University Press, New York, 1981).

# Questions for Thought and Discussion

1. Derby and Keeney maintain that regardless of the nature of the hazard, acceptable risk is defined as the risk associated with the best of the available alternatives. Others believe that in some cases it is better to set a standard for acceptable risk, that is, a level of risk that cannot be exceeded. What is your view?

2. To what degree do you think federal regulatory agencies such as EPA should allow for the cost of regulatory compliance—e.g., by raising the acceptable-risk level if the cost is very high—when determining the acceptable level of risk from a pesticide in food? What other considerations should be taken into account?

3. An example in this paper says that if there is a 1-in-10,000 chance of 10,000 people dying in one situation and a 1-in-10 chance of 10 people dying in another, the average risk per person is the same. Yet some research has shown that there may be more public concern about the first case because the consequence is catastrophic. Why do you suppose this is so?

4. The authors point out that the charters for regulatory agencies "often state general, vague objectives for what the agency should do . . . [and] critical questions are left open for the agency to decide for itself. . . ." What are the main advantages and disadvantages of legislative charters being so unspecific when it comes to deciding how safe is safe enough?

# PART 2

# RISK COMPARISONS

# Introduction ——————————————

Putting any risk in perspective requires that it be compared with other risks. The process of comparing a risk to other familiar risks makes evaluation easier, especially when decisions regarding the acceptability of the other risks have already been made.

Practical difficulties arise, of course, when comparing risks that are infinitesimally small or estimates that are not very precise, which is often the case. Furthermore, when making risk comparisons it is essential to address all of the relevant quantitative and qualitative considerations, not just the relative numerical values of the risk estimates. For instance, other things being equal, a risk that is taken voluntarily is easier to accept than one that is not. Research also shows that when the perception of a risk differs from its measurement, it is oftentimes because of associated factors such as the nature of the circumstances that contribute to the risk (e.g., whether it is controllable or not) and the nature of the potential consequence (e.g., whether it is familiar or not).

The papers in part 2, while very different from one another, all use risk comparisons to make their separate points about the acceptability and perception of risks. Comparisons also have a place in helping us decide individually and collectively which risks are negligible, which are not negligible but are nevertheless acceptable, and which of the unacceptable ones deserve the most attention.

Written in a humorous vein, Richard Wilson's "Analyzing the Daily Risks of Life" focuses on the risks of some of life's most trivial activities, serving to illustrate the point that in spite of increased life expectancies, our society has become obsessed with small risks, perhaps because there are fewer large risks to worry about. Wilson has drawn up a novel list of twenty-three activities, ranging from eating 40 tablespoons of peanut butter to traveling 6 minutes by canoe, that have comparable risks: each is estimated to increase the chance of death by 1 in 1 million.

The paper entitled "Rating the Risks" reports on the findings of a risk comparison experiment conducted by three of the leading researchers in risk perception. Authors Paul Slovic, Baruch Fischhoff, and Sarah Lichtenstein found that a group of experts ranked a given list of risks according to the associated annual fatality rates, whereas lay people considered other attributes of the risks—for example, the potential loss of a large number of lives in a single disaster—when formulating their rankings. This result may be one of the reasons why many laypersons who are unfamiliar with the technology of nuclear power fear it more than the technical experts do.

Bruce N. Ames, Renae Magaw, and Lois Swirsky Gold compare quantitative estimates of the risks of cancer from natural and man-made chemicals in "Ranking Possible Carcinogenic Hazards." They acknowledge that we must rely on animal tests to identify carcinogens but argue that there is no proven justification for extrapolating from animal results to make quantitative predictions of human risk. Instead, they propose that carcinogens be ranked on the basis of the ratio of (a) human exposure to each substance to (b) the dose that causes cancer in animals. They conclude that the risks from environmental pollution, pesticides, and other residues of man-made chemicals are lower than the risks from naturally occurring chemicals in the food supply.

Accompanying this last paper is a technical comment by Samuel S. Epstein and Joel B. Swartz and their cosigners and a rejoinder by authors Ames and Gold. This exchange demonstrates the sharp disagreement among experts as to whether man-made chemicals and pollutants are major causes of cancer.

# Analyzing the Daily Risks of Life

RICHARD WILSON
*Harvard University*
*Cambridge, Massachusetts*

The world seems a very hazardous place. Every day the newspapers announce that some chemical has been found to be carcinogenic, or some catastrophic accident has occurred in some far-off place. This leads some of us to hanker after a simpler world where there are fewer risks to life. But does such a world really exist?

If we look back at the world of a century ago, we find that expectation of life was 50 years; now it is 70 years. Therefore the sum of all the risks to which we are now exposed must be less than it was. We find that many of the large risks of the last century have been eliminated, leaving us conscious of a myriad of small risks, most of which have always existed.

The moment I climb out of bed I start taking risks. As I drowsily turn on the light I feel a slight tingle; my house is old with old wiring and there is a small risk of electrocution. Every year 500 people are electrocuted in the United States. I take a shower, and as I reach for the soap, I wonder about the many chemicals it contains. Are they all good for the skin, as the advertisements claim? My clothes have been cleaned with the best bleaching detergent. Most bleaching agents contain a chemical that fluoresces

slightly in the sunlight to enhance the whiteness. Does this make bleaches carcinogenic?

I ponder this risk as I walk down to breakfast, taking care not to fall upon the stairs. Falls kill 16,000 people per year—mostly in domestic accidents. Shall I drink coffee or tea with my breakfast? Both contain caffeine, a well-known stimulant which may be carcinogenic. I have a sweet tooth; do I use sugar which makes me fat and gives me heart disease, or saccharin which we now know causes cancer? It is better to abstain.

After breakfast I make a sandwich for lunch. My son likes peanut butter. But improperly stored peanuts can develop a mold which produces a potent carcinogen—aflatoxin. In Africa and Southeast Asia, where aflatoxin appears more frequently, it has been blamed for numerous cases of liver cancer. In our (less natural) society storage facilities are better, so the risk is less—but it is not zero.

I prefer meat. But Americans, like other prosperous people, eat too much meat. It is not certain, but a meat-heavy diet probably contributes to cancer of the colon.

I live seven miles from work and can commute by car, by bicycle, or by bus. Which has the lowest risk? To travel by bicycle would keep my weight down, and bicycle riding does not cause pollution—but statistics show that it is more likely to involve me in an accident. And since a bicyclist is unprotected, fatal

"Analyzing the Daily Risks of Life" by Richard Wilson is reprinted, with permission, from *Technology Review*, copyright 1979, vol. 81, no. 4 (February 1979), pp. 41–46.

accidents are also frequent on a bicycle. A car would be safer, but a bus is safest. I am happy that I no longer have to choose between a horse and a canoe; both are more dangerous (per mile) than a bicycle.

As I approach Boston, I see the urban haze caused by air pollution. There are toxic parts of air pollution which are not visible, as well. The risk to life caused by air pollution is high. Asthma victims have known this for a long time and fled the industrialized eastern United States for the purer air of the West. A press release from a government laboratory states that air pollution kills 20,000 people a year in the eastern United States. Air pollution, though still bad, has been reduced in most cities.

I remember the pea-soup London fogs of my youth caused by burning soft coal, where I could not see ahead ten feet; and the infamous week in December of 1952 where 3,000 people died from air pollution in four days.

I go to a committee meeting in a small, unventilated room. Although I don't smoke tobacco, half of the committee does, and I am exposed to the poison which causes 40 per cent of all cancers and kills 15 per cent of all Americans. Even though I breathe less tobacco smoke than my smoking colleagues, I often get a headache. One of my friends, who is more allergic than I, wears goggles at work.

At mid-morning I take a drink of water. The water tastes of chlorine, showing that the city's sanitation engineers use chlorine to kill microbes in the water. By such methods the country has nearly wiped out cholera and typhus. But the chlorine reacts with organic matter in the water to produce many known carcinogens. One of them, chloroform, is produced in a concentration of 100 parts per billion; enough to present a health hazard.

My office walls are brick and cinder block. Both contain radioactive materials, and radiation can increase my risk of cancer. One of these radioactive materials, radon, is a gas which is not chemically active. It is released by the brick and I can breathe it, which accentuates the hazard. I could prevent the release of this radioactive gas by painting the walls with thick epoxy paint to seal them, but that would introduce another risk. As the epoxy paint cures, it emits gaseous chemicals which are themselves carcinogenic. Which is worse?

Radiation enters all of my life. State law requires that I have a regular chest x-ray to see whether I have tuberculosis and may convey that dread disease to my students. But this adds to my risk of cancer from radiation. Is it correct for society to demand that I accept this risk, even to protect the rest of society from a greater one?

I frequently travel to meetings. Should I go by car, bus, train, or airplane? Thirty years ago the statistics were clear; the airplane was far more dangerous than all the others, since many airplanes crashed. Now, for journeys of 1,000 miles or more, air travel is the safest. But airplane travel causes an often-ignored radiation hazard, exposure to cosmic radiation from outer space. Airplanes fly at 30,000 feet, and at that altitude cosmic radiation exposure is 40 times what it is at sea level. Even a vacation trip to the high altitudes of Colorado and Wyoming can increase cosmic ray exposure. Sunlight at these altitudes, and excessive exposure even at sea level, showers us with ultraviolet light, which causes skin cancer.

These are personal concerns, and it might be argued that they are of no concern to anyone else, since I can avoid some of them. But in doing so I may well cause problems for others in society.

In the bad old days of my childhood we burnt coal in the house. If I heat my house by electricity I will not personally pollute the air with the products of fossil fuel burning; but these may still be produced at the power plant. One hopes the electric company is more careful about these pollutants than my parents used to be.

Whether I burn the coal myself or let the electric utility company do so, coal miners must still go underground. Anyone who has read *How Green Is My Valley* knows that 100 years ago coal mining was one of the most dangerous occupations. Even though mine safety has improved, it still has hazards: 156 out of every 100,000 miners were killed in accidents in 1972 in the United States. Yet accidents are not the worst hazard of coal mining: 800 miners yearly contract the dread black lung disease—coal workers' pneumoconiosis—from inhaling coal dust. One quarter of all American miners working in 1977 will probably contract this disease during their lifetimes. As an environmentalist I hate to see the beautiful western states laid bare by strip mining; but do I have a right to allow miners to die by refusing to let them work above ground?

Our society has a quirk which is fostered by our news media. We are far more concerned with infrequent large accidents than with numerous small accidents which, in total, cause many more deaths. Congress was prompted into insisting on better mine ventilation to prevent black lung disease only after a much smaller number were killed in a single accident. A single accident of a school bus receives more newspaper coverage than the thousands of children killed yearly in automobile accidents.

This obsession with large accidents is getting worse. We are apprehensive at the *thought* of a large accident in a nuclear power plant, although none has happened so far, and experts are optimistic that none will ever happen. Nor is the fear unique to nuclear power. We bring to the United States considerable quantities of liquefied natural gas (LNG) and worry about the possibility of the ship leaking and blowing up. LNG *has* caused problems in the past; 30 years ago, an LNG tank, one-tenth the size of modern ones, collapsed and killed 133 people. We now know why this tank collapsed, and new tanks will not collapse in the same way since the metal from which the tanks are made has been changed.

## Comparing the Risks We Face

There are those who would try to eliminate all known risks and would try to force this by law. This sounds plausible, but it creates an incentive for ignorance, not an incentive for safety. Under this procedure if we do not know whether something is risky and close our eyes to the possibility of risk, no one will bother us. On the other hand, if we look carefully and find there is a risk—even though it is small—some regulatory agency may stop us.

It would be a better policy to try to measure our risks quantitatively, and to give an *upper limit* on a risk when there is uncertainty. Then we could compare risks and decide which to accept or reject. I suspect most of us would decide to reduce the largest risks first.

To compare risks we must calculate them. As I prepared the table (on this page), I realized that an increased risk of death of one in a million is often seen as acceptable, but people instinctively think about large risks. I list here several actions which increase the chance of death in any year by one in a million.

**Risks Which Increase Chance of Death by 0.000001***

| | |
|---|---|
| Smoking 1.4 cigarettes | Cancer, heart disease |
| Drinking 1/2 liter of wine | Cirrhosis of the liver |
| Spending 1 hour in a coal mine | Black lung disease |
| Spending 3 hours in a coal mine | Accident |
| Living 2 days in New York or Boston | Air pollution |
| Travelling 6 minutes by canoe | Accident |
| Travelling 10 miles by bicycle | Accident |
| Travelling 300 miles by car | Accident |
| Flying 1000 miles by jet | Accident |
| Flying 6000 miles by jet | Cancer caused by cosmic radiation |
| Living 2 months in Denver on vacation from N.Y. | Cancer caused by cosmic radiation |
| Living 2 months in average stone or brick building | Cancer caused by natural radioactivity |
| One chest x-ray taken in a good hospital | Cancer caused by radiation |
| Living 2 months with a cigarette smoker | Cancer, heart disease |
| Eating 40 tablespoons of peanut butter | Liver cancer caused by aflatoxin B |
| Drinking Miami drinking water for 1 year | Cancer caused by chloroform |
| Drinking 30 12 oz. cans of diet soda | Cancer caused by saccharin |
| Living 5 years at site boundary of a typical nuclear power plant in the open | Cancer caused by radiation |
| Drinking 1000 24 oz. soft drinks from recently banned plastic bottles | Cancer from acrylonitrile monomer |
| Living 20 years near PVC plant | Cancer caused by vinyl chloride (1976 standard) |
| Living 150 years within 20 miles of a nuclear power plant | Cancer caused by radiation |
| Eating 100 charcoal broiled steaks | Cancer from benzopyrene |
| Risk of accident by living within 5 miles of a nuclear reactor for 50 years | Cancer caused by radiation |

*(1 part in 1 million)

Of course, if the risk of death in one year is increased, the risk of dying from another cause in a later year is decreased. The average expectation of life is shortened. Accidents often occur early in life, and life may be shortened 30 years by a typical accident. Cancer, black lung, and bronchitis kill later in life, and life is shortened only about 15 years.

Therefore, a risk of 0.000001 (or $10^{-6}$) shortens life on the average by $30 \times 10^{-6}$ years, or 15 minutes if it is an accident risk, 8 minutes if it is a risk of fatal illness.

I illustrate what this table means by calculating examples. In the United States 627 billion cigarettes were made in 1975. This is enough for 3,000 per person (including children), or a little less than half a pack a day. It is estimated that 15 per cent of all Americans ( 30 per cent of all smokers) die from lung or other cancers or heart disease due to smoking. We described this as an average lifetime risk of 0.15. Dividing by the 70-year lifetime gives a yearly risk of 0.002 or $2 \times 10^{-3}$; dividing again by 3,000 gives a risk per cigarette of $0.7 \times 10^{-6}$. It is amusing to note that smoking a cigarette takes ten minutes and reduces the expectation of life by five minutes.

Human affairs are much more random than we like to think. One boy playing on a street can be killed by a passing car while his playmates are unharmed. All were equally at risk before the accident, but only one died. Similarly, one out of three lifetime smokers dies of cancer or heart disease because of the habit; the rest are unaffected and die of other causes. Moreover, those that die of cancer and heart disease do so at different ages. We have no way of telling which particular smokers will die of cancer, so we say that all are equally at risk.

It has been shown that those who smoke 40 cigarettes a day are ten times more likely to develop cancer as those who smoke four cigarettes a day. Perhaps there is a level of consumption where the risk becomes zero, but we cannot measure that low. It is easier to assume that every cigarette contributes the same amount to the total risk.

Brookhaven National Laboratory recently estimated that 20,000 Americans die every year from air pollution east of the Mississippi. This is partly due to sulphur emitted from burning coal and oil, and measurements suggest that the sulphate particulates spread themselves roughly uniformly over town and country. About 100 million Americans are exposed to this dirty air, so the average risk is 20,000/100,000,000 every year or $2 \times 10^{-4}$ or 0.0002. Two days in New York City give a risk smaller by 2/365 or about $10^{-6}$ (one in a million).

Recent aircraft accident statistics tell us that aircraft in the United States carry passengers 100 billion passenger-miles every year and only about 100 people a year are killed in airplane crashes. This gives a risk of one in a million for one thousand miles of flight.

Professor Norman G. Rasmussen of M.I.T. made a study of nuclear reactor accident probabilities for the Nuclear Regulatory Commission. He concluded that a reactor accident involving loss of life is very unlikely. The chance of an accident with more than 1,000 deaths is less than one in 100 million per year of operation for each reactor. Most of these would be among the 20,000 or so people living within five miles of the reactor. So the probability of an individual living near a reactor being killed in a large accident is 1/2000 million. But those close by might also suffer in smaller accidents which, even though still unlikely, are more probable, leading to a risk of 1/50 million for persons living close to reactors.

Other more dangerous radiation hazards, such as natural radioactivity in brick, cosmic radiation, and diagnostic x-rays, are calculated by measuring the radiation dose and dividing it by the measured effect of large doses. The risks of these commonly accepted radiation hazards are far greater than those estimated for nuclear power.

I find these comparisons help me evaluate risks, and I imagine that they may help others do so, as well. But the most important use of these comparisons must be to help the decisions we make, as a nation, to improve our health and reduce our accident rate.

## Taxing a Risk

Economists are fond of using taxation to control human affairs. Indeed, the invention of money by Croesus made a great simplification in the relationships in society. One suggestion, then, is to tax anyone who introduces a risk into society. This tax could pay for medical care, for compensating society for the loss of services, etc. The question arises: How much should the tax be? I suggest, as a basis for discussion, that this tax be at the rate of $1 million for every life that is lost by this extra risk, or one dollar for a risk of one in a million. Conversely, anyone that can save a life by an expenditure of $1 million must be encouraged to do so.

For example, the manufacturer who panders to

the bad habit of cigarette smoking would pay an increased tax of 70 cents *per cigarette*. This is more than enough to pay the societal cost of cigarette smoking (hospital costs, fire hazards, reduced working time), which is variously estimated at from $1 to $2 per pack. Other taxes—five cents per diet soda—are less dramatic and might have to be accompanied by a tax of five cents on other sodas as well to prevent a switch to sugar.

These taxes might be earmarked to pay for risk reductions such as converting an existing sanitation system to using ozone instead of chlorine for sanitation, to avoid the production of chloroform.

Whether we quantify these risks or not, we must and do constantly make decisions about them. We do this as individuals, and our politicians make these decisions for us on a larger scale. What we are not doing, and need to do, is comparing the risks of various activities and then reducing the largest risks—which may not be the obvious ones.

After calculating these risks all day, I go home. I am still faced with decisions about risks. If I cook a meal in the microwave oven and the door doesn't fit tightly, I will be exposed to microwaves. It has recently been claimed that microwaves, even at low concentrations, give people nervous problems. Or I can use the gas stove, but the burning gas can fill my kitchen with both noxious carbon monoxide and nitrogen oxides.

Just as I go to bed I take a glass of beer. Alcohol causes cirrhosis of the liver and has been associated with oral and other cancers. However, the relaxing effect of the beer will reduce my stresses and permit a good night's sleep. This will prolong my life and is worth the risk.

The beer is in a green glass bottle which contains chromium, a small amount of which enters the beer. Chromium is a known carcinogen when ingested in moderate quantities, but it must not be avoided altogether because it is essential to life in small concentrations. How much chromium should I take to minimize the risk? Is the amount in the beer too much? Should I drink the beer from a plastic bottle? A plastic bottle suitable for beer has just been banned because a trace of the chemical from which the plastic was made could dissolve in the contents, and there is a suspicion that the chemical is carcinogenic.

I ponder this decision as I put on my pajamas. Are the pajamas inflammable? There is always a small risk of fire starting while I am in bed. Is the risk of being burnt in a fire greater or smaller than the risk of cancer caused by a flame retardant such as TRIS?

I remember the truism "more people die in bed than anywhere else," so at least I'm in the right place.

# Questions for Thought and Discussion

1. What specific data would you seek out, where would you look for it, and what calculations would you perform to verify or disprove the author's statement that horse and canoe are "more dangerous (per mile) than a bicycle"?

2. Of the risks listed in the table in this paper, what are the first, second, and third most worrisome to you personally? Why did you select these? Of the other risks in the list that also affect you personally, select the one that is *least* worrisome and explain your relative lack of concern.

3. Are any of the following risks discussed by Wilson surprising to you: aflatoxin in peanut butter, chloroform in drinking water, radioactive materials in brick, radiation in airplanes? Do you think you should change your behavior regarding any of these four risks? Why or why not? What do you think the government should do about these risks, if anything?

4. Governmental intervention to reduce risks takes one of two general forms: *regulation,* which requires that certain uses or exposures be restricted or eliminated, or *warning,* which requires that the public be informed of the risk via messages on packages, in advertisements, or in public service announcements. For each risk in the table, indicate whether the government uses regulation or warning and describe the specific nature of the interventions of which you are aware. In each case, offer an explanation as to why these forms of intervention are used or why there has been no intervention.

# Rating the Risks

**PAUL SLOVIC, BARUCH FISCHHOFF, and SARAH LICHTENSTEIN**
*Decision Research, A Branch of Perceptronics, Inc., Eugene, Oregon*

People respond to the hazards they perceive. If their perceptions are faulty, efforts at public and environmental protection are likely to be misdirected. In order to improve hazard management, a risk assessment industry has developed over the last decade which combines the efforts of physical, biological, and social scientists in an attempt to identify hazards and measure the frequency and magnitude of their consequences.[1]

For some hazards extensive statistical data is readily available; for example, the frequency and severity of motor vehicle accidents are well documented. For other familiar activities, such as the use of alcohol and tobacco, the hazardous effects are less readily discernible and their assessment requires complex epidemiological and experimental studies. But in either case, the hard facts go only so far and then human judgment is needed to interpret the findings and determine their relevance for the future.

Other hazards, such as those associated with recombinant DNA research or nuclear power, are so

*Editors' note:* The current affiliation of Baruch Fischhoff is Carnegie Mellon University, Pittsburgh, Pennsylvania.

new that risk assessment must be based on theoretical analyses such as fault trees (see figure 1), rather than on direct experience. While sophisticated, these analyses, too, include a large component of human judgment. Someone, relying on educated intuition, must determine the structure of the problem, the consequences to be considered, and the importance of the various branches of the fault tree.

Once the analyses have been performed, they must be communicated to the various people who are actually responsible for dealing with the hazards, including industrialists, environmentalists, regulators, legislators, and voters. If these people do not see, understand, or believe these risk statistics, then distrust, conflict, and ineffective hazard management can result.

## Judgmental Biases

When lay people are asked to evaluate risks, they seldom have statistical evidence on hand. In most cases they must rely on inferences based on what they remember hearing or observing about the risk in question. Recent psychological research has identified a number of general inferential rules that people seem to use in such situations.[2] These judgmental rules, known technically as *heuristics*, are employed to reduce difficult mental tasks to simpler ones. Although valid in some circumstances, in others they can lead to large and persistent biases with serious implications for risk assessment.

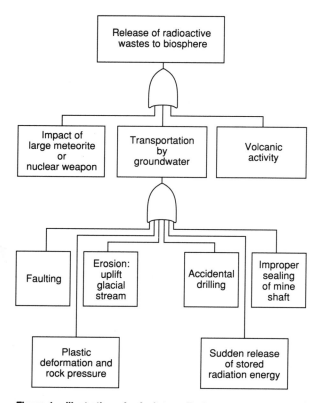

**Figure 1.   Illustration of a fault tree. Fault trees are used most often to characterize hazards for which direct experience is not available. The tree shown here indicates the various ways in which radioactive material might accidentally be released from nuclear wastes buried within a salt deposit. To read this tree, start with the bottom row of possible initiating events, each of which can lead to the transportation of radioactivity by groundwater. This transport can in turn release radioactivity to the biosphere. As indicated by the second level of boxes, release of radioactivity can also be produced directly (without the help of groundwater) through the impact of a large meteorite, a nuclear weapon, or a volcanic eruption. Fault trees may be used to map all relevant possibilities and to determine the probability of the final outcome. To accomplish this latter goal, the probabilities of all component stages, as well as their logical connections, must be completely specified.**
*Source:* P. E. McGrath, "Radioactive Waste Management," Report EURFNR 1204, Karlsruhe, Germany, 1974. [Redrawn, with permission.]

## Availability

One heuristic that has special relevance for risk perception is known as "availability."[3] People who use this heuristic judge an event as likely or frequent if instances of it are easy to imagine or recall. Frequently occurring events are generally easier to imagine and recall than rare events. Thus, availability is often an appropriate cue. However, availability is also affected by numerous factors unrelated to fre-

**Table 1.  Bias in Judged Frequency of Death**

| Most overestimated | Most underestimated |
|---|---|
| All accidents | Smallpox vaccination |
| Motor vehicle accidents | Diabetes |
| Pregnancy, child-birth, and abortion | Stomach cancer |
| Tornadoes | Lightning |
| Flood | Stroke |
| Botulism | Tuberculosis |
| All cancer | Asthma |
| Fire and flames | Emphysema |
| Venomous bite or sting | |
| Homicide | |

[*Editors' note:* To control for the general systematic biases observed in figure 2, these comparisons are made relative to the curved line.]

quency of occurrence. For example, a recent disaster or a vivid film such as "Jaws" can seriously distort risk judgments.

Availability-induced errors are illustrated by several recent studies in which we asked college students and members of the League of Women Voters to judge the frequency of various causes of death, such as smallpox, tornadoes, and heart disease.[4] In one study, these people were told the annual death toll for motor vehicle accidents in the United States (50,000); they were then asked to estimate the frequency of forty other causes of death. In another study, participants were given two causes of death and asked to judge which of the two is more frequent. Both studies showed people's judgments to be moderately accurate in a global sense; that is, people usually knew which were the most and least frequent lethal events. However, within this global picture, there was evidence that people made serious misjudgments, many of which seemed to reflect availability bias.

Figure 2 compares the judged number of deaths per year with the actual number according to public health statistics. If the frequency judgments were accurate, they would equal the actual death rates, and all data points would fall on the straight line making a 45 degree angle with the axes of the graph. In fact, the points are scattered about a curved line that sometimes lies above and sometimes below the line of accurate judgment. In general, rare causes of death were overestimated and common causes of death were underestimated. As a result, while the actual death toll varied over a range of one million,

Estimated number of
deaths per year

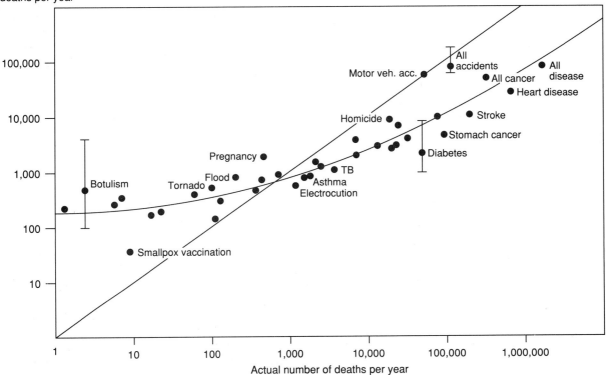

Actual number of deaths per year

**Figure 2.** **Relationship between judged frequency and the actual number of deaths per year for 41 causes of death. If judged and actual frequencies were equal, the data would fall on the straight line. The points, and the curved line fitted to them, represent the averaged responses of a large number of lay people. While people were approximately accurate, their judgments were systematically distorted. As described in the text, both the compression of the scale and the scatter of the results indicate this. To give an idea of the degree of agreement among subjects, vertical bars are drawn to depict the 25th and 75th percentile of individual judgment for botulism, diabetes, and all accidents. Fifty percent of all judgments fall between these limits. The range of responses for the other 37 causes of death was similar.**

*Source:* This figure is taken from Lichtenstein et al. (note 4). [Redrawn with permission. © 1978 American Psychological Association.]

average frequency judgments varied over a range of only a thousand.

In addition to this general bias, many important specific biases were evident. For example, accidents were judged to cause as many deaths as diseases, whereas diseases actually take about fifteen times as many lives. Homicides were incorrectly judged to be more frequent than diabetes and stomach cancer. Homicides were also judged to be about as frequent as stroke, although the latter actually claims about eleven times as many lives. Frequencies of death from botulism, tornadoes, and pregnancy (including childbirth and abortion) were also greatly overestimated.

Table 1 lists the lethal events whose frequencies were most poorly judged in our studies. In keeping

with availability considerations, overestimated items were dramatic and sensational whereas underestimated items tended to be unspectacular events which claim one victim at a time and are common in nonfatal form.

In the public arena the availability heuristic may have many effects. For example, the biasing effects of memorability and imaginability may pose a barrier to open, objective discussions of risk. Consider an engineer demonstrating the safety of subterranean nuclear waste disposal by pointing out the improbability of each branch of the fault tree in figure 1. Rather than reassuring the audience, the presentation might lead individuals to feel that "I didn't realize there were so many things that could go wrong." The very

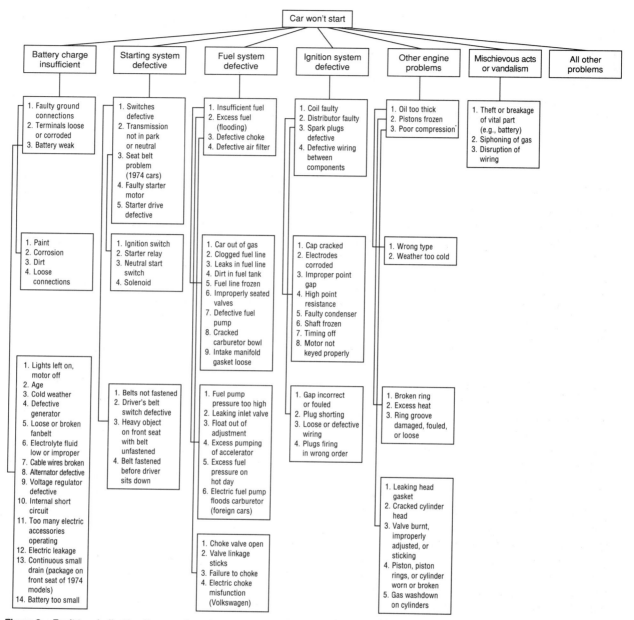

**Figure 3.    Fault tree indicating the ways in which a car might fail to start. It was used by the authors to study whether people are sensitive to the completeness of this type of presentation. Omission of large sections of the diagram was found to have little influence on the judged degree of completeness. In effect, what was out of sight was out of mind. Professional automobile mechanics did not do appreciably better on the test than did lay people.**
*Source:* This diagram is from Fischhoff, Slovic, and Lichtenstein (note 5). [Redrawn with permission. © 1978 American Psychological Association.]

discussion of any low-probability hazard may increase the judged probability of that hazard regardless of what the evidence indicates.

In other situations, availability may lull people

into complacency. In a recent study,[5] we presented people with various versions of a fault tree showing the "risks" of starting a car. Participants were asked to judge the completeness of the representation (reproduced in figure 3). Their estimate of the proportion of no-starts falling in the category labeled "all other problems" was about the same when looking at the full tree of figure 3 or at versions in which half of the branches were deleted. Such pruning should have dramatically increased the judged likelihood of "all other problems." However, it did not. In keeping with the availability heuristic, what was out of sight was effectively out of mind.

## Overconfidence

A particularly pernicious aspect of heuristics is that people are typically very confident about judgments based on them. For example, in a follow-up to the study on causes of death, participants were asked to indicate the odds that they were correct in their judgment about which of two lethal events was more frequent.[6] Odds of 100:1 or greater were given often (25 percent of the time). However, about one out of every eight answers associated with such extreme confidence was wrong (fewer than 1 in 100 would have been wrong if the odds had been appropriate). About 30 percent of the judges gave odds greater than 50:1 to the incorrect assertion that homicides are more frequent than suicides. The psychological basis for this unwarranted certainty seems to be people's insensitivity to the tenuousness of the assumptions upon which their judgments are based (in this case, the validity of the availability heuristic). Such overconfidence is dangerous. It indicates that we often do not realize how little we know and how much additional information we need about the various problems and risks we face.

Overconfidence manifests itself in other ways as well. A typical task in estimating failure rates or other uncertain quantities is to set upper and lower bounds so that there is a 98 percent chance that the true value lies between them. Experiments with diverse groups of people making many different kinds of judgments have shown that, rather than 2 percent of true values falling outside the 98 percent confidence bounds, 20 percent to 50 percent do so.[7] People think that they can estimate such values with much greater precision than is actually the case.

Unfortunately, experts seem as prone to overconfidence as lay people. When the fault tree study described above was repeated with a group of pro-fessional automobile mechanics, they, too, were insensitive to how much had been deleted from the tree. Hynes and Vanmarcke[8] asked seven "internationally known" geotechnical engineers to predict the height of an embankment that would cause a clay foundation to fail and to specify confidence bounds around this estimate that were wide enough to have a 50 percent chance of enclosing the true failure height. None of the bounds specified by these experts actually did enclose the true failure height. The multi-million dollar Reactor Safety Study (the "Rasmussen Report"),[9] in assessing the probability of a core melt in a nuclear reactor, used a procedure for setting confidence bounds that has been found in experiments to produce a high degree of overconfidence. Related problems led the recent review committee, chaired by H. W. Lewis of the University of California, Santa Barbara, to conclude that the Reactor Safety Study greatly overestimated the precision with which the probability of a core melt could be assessed.[10]

Another case in point is the 1976 collapse of the Teton Dam. The Committee on Government Operations has attributed this disaster to the unwarranted confidence of engineers who were absolutely certain they had solved the many serious problems that arose during construction.[11] Indeed, in routine practice, failure probabilities are not even calculated for new dams even though about 1 in 300 fails when the reservoir is first filled. Further anecdotal evidence of overconfidence may be found in many other technical risk assessments. Some common ways in which experts may overlook or misjudge pathways to disaster include:

- Failure to consider the ways in which human errors can affect technological systems. Example: The disastrous fire at the Brown's Ferry Nuclear Plant was caused by a technician checking for an air leak with a candle, in violation of standard operating procedures.
- Overconfidence in current scientific knowledge. Example: The failure to recognize the harmful effects of X-rays until societal use had become widespread and largely uncontrolled.
- Insensitivity to how a technological system functions as a whole. Example: Though the respiratory risk of fossil-fueled power plants has been recognized for some time, the related effects of acid

rains on ecosystems were largely missed until very recently.

• Failure to anticipate human response to safety measures. Example: The partial protection offered by dams and levees gives people a false sense of security and promotes development of the flood plain. When a rare flood does exceed the capacity of the dam, the damage may be considerably greater than if the flood plain had been unprotected. Similarly, "better" highways, while decreasing the death toll per vehicle mile, may increase the total number of deaths because they increase the number of miles driven.

### Desire for Certainty

Every technology is a gamble of sorts and, like other gambles, its attractiveness depends on the probability and size of its possible gains and losses. Both scientific experiments and casual observation show that people have difficulty thinking about and resolving the risk/benefit conflicts even in simple gambles. One way to reduce the anxiety generated by confronting uncertainty is to deny that uncertainty. The denial resulting from this anxiety-reducing search for certainty thus represents an additional source of overconfidence. This type of denial is illustrated by the case of people faced with natural hazards, who often view their world as either perfectly safe or as predictable enough to preclude worry. Thus, some flood victims interviewed by Kates[12] flatly denied that floods could ever recur in their areas. Some thought (incorrectly) that new dams and reservoirs in the area would contain all potential floods, while others attributed previous floods to freak combinations of circumstances, unlikely to recur. Denial, of course, has its limits. Many people feel that they cannot ignore the risks of nuclear power. For these people, the search for certainty is best satisfied by outlawing the risk.

Scientists and policy makers who point out the gambles involved in societal decisions are often resented for the anxiety they provoke. Borch[13] noted how annoyed corporate managers get with consultants who give them the probabilities of possible events instead of telling them exactly what will happen. Just before a blue-ribbon panel of scientists reported that they were 95 percent certain that cyclamates do not cause cancer, Food and Drug Administration Commissioner Alexander Schmidt said, "I'm looking for a clear bill of health, not a wishy-washy, iffy answer on cyclamates."[14] Senator Edmund Muskie has called for "one-armed" scientists who do not respond "on the one hand, the evidence is so, but on the other hand . . ." when asked about the health effects of pollutants.[15]

The search for certainty is legitimate if it is done consciously, if the remaining uncertainties are acknowledged rather than ignored, and if people realize the costs. If a very high level of certainty is sought, those costs are likely to be high. Eliminating the uncertainty may mean eliminating the technology and foregoing its benefits. Often some risk is inevitable. Efforts to eliminate it may only alter its form. We must choose, for example, between the vicissitudes of nature on an unprotected flood plain and the less probable, but potentially more catastrophic, hazards associated with dams and levees.

### Analyzing Judgments of Risk

In order to be of assistance in the hazard management process, a theory of perceived risk must explain people's extreme aversion to some hazards, their indifference to others, and the discrepancies between these reactions and experts' recommendations. Why, for example, do some communities react vigorously against locating a liquid natural gas terminal in their vicinity despite the assurances of experts that it is safe? Why do other communities situated on flood plains and earthquake faults or below great dams show little concern for the experts' warnings? Such behavior is doubtless related to how people assess the *quantitative* characteristics of the hazards they face. The preceding discussion of judgmental processes was designed to illuminate this aspect of perceived risk. The studies reported below broaden the discussion to include more *qualitative* components of perceived risk. They ask, when people judge the risk inherent in a technology, are they referring only to the (possibly misjudged) number of people it could kill or also to other, more qualitative features of the risk it entails?

### Quantifying Perceived Risk

In our first studies, we asked four different groups of people to rate thirty different activities and technolo-

Table 2. Ordering of Perceived Risk for 30 Activities and Technologies[a]

| | Group 1: LOWV | Group 2: college students | Group 3: Active Club members | Group 4: experts |
|---|---|---|---|---|
| Nuclear power | 1 | 1 | 8 | 20 |
| Motor vehicles | 2 | 5 | 3 | 1 |
| Handguns | 3 | 2 | 1 | 4 |
| Smoking | 4 | 3 | 4 | 2 |
| Motorcycles | 5 | 6 | 2 | 6 |
| Alcoholic beverages | 6 | 7 | 5 | 3 |
| General (private) aviation | 7 | 15 | 11 | 12 |
| Police work | 8 | 8 | 7 | 17 |
| Pesticides | 9 | 4 | 15 | 8 |
| Surgery | 10 | 11 | 9 | 5 |
| Fire fighting | 11 | 10 | 6 | 18 |
| Large construction | 12 | 14 | 13 | 13 |
| Hunting | 13 | 18 | 10 | 23 |
| Spray cans | 14 | 13 | 23 | 26 |
| Mountain climbing | 15 | 22 | 12 | 29 |
| Bicycles | 16 | 24 | 14 | 15 |
| Commercial aviation | 17 | 16 | 18 | 16 |
| Electric power | 18 | 19 | 19 | 9 |
| Swimming | 19 | 30 | 17 | 10 |
| Contraceptives | 20 | 9 | 22 | 11 |
| Skiing | 21 | 25 | 16 | 30 |
| X-rays | 22 | 17 | 24 | 7 |
| High school & college football | 23 | 26 | 21 | 27 |
| Railroads | 24 | 23 | 20 | 19 |
| Food preservatives | 25 | 12 | 28 | 14 |
| Food coloring | 26 | 20 | 30 | 21 |
| Power mowers | 27 | 28 | 25 | 28 |
| Prescription antibiotics | 28 | 21 | 26 | 24 |
| Home appliances | 29 | 27 | 27 | 22 |
| Vaccinations | 30 | 29 | 29 | 25 |

[a]The ordering is based on the geometric mean risk ratings within each group. Rank 1 represents the most risky activity or technology.

gies according to the present risk of death from each.[16] Three of these groups were from Eugene, Oregon; they included 30 college students, 40 members of the League of Women Voters (LOWV), and 25 business and professional members of the "Active Club." The fourth group was composed of 15 persons selected nationwide for their professional involvement in risk assessment. This "expert" group included a geographer, an environmental policy analyst, an economist, a lawyer, a biologist, a biochemist, and a government regulator of hazardous materials.

All these people were asked, for each of the thirty items, "to consider the risk of dying (across all U.S. society as a whole) as a consequence of this activity or technology." In order to make the evaluation task easier, each activity appeared on a 3″ × 5″ card. Respondents were told first to study the items individually, thinking of all the possible ways someone might die from each (e.g., fatalities from non-nuclear electricity were to include deaths resulting from the mining of coal and other energy production activities as well as electrocution; motor-vehicle fatalities were to include collisions with bicycles and pedestrians). Next, they were to order the items from least to most risky and then assign numerical risk values by giving a rating of 10 to the least risky item and making the other ratings accordingly. They were also given additional suggestions, clarifications and encouragement to do as accurate a job as possible. For example, they were told "A rating of 12 indicates that that

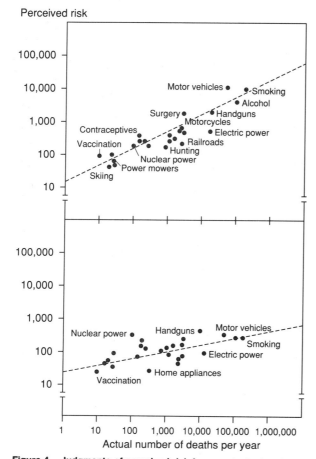

**Figure 4.  Judgments of perceived risk for experts (top) and lay people (bottom) plotted against the best technical estimates of annual fatalities for 25 technologies and activities. Each point represents the average responses of the participants. The dashed lines are the straight lines that best fit the points. The experts' risk judgments are seen to be more closely associated with annual fatality rates than are the lay judgments.**

ences as well. Active Club members viewed pesticides and spray cans as relatively much safer than did the other groups. Nuclear power was rated as highest in risk by the LOWV and student groups, but only eighth by the Active Club. The students viewed contraceptives and food preservatives as riskier and swimming and mountain climbing as safer than did the other lay groups. Experts' judgments of risk differed markedly from the judgments of lay persons. The experts viewed electric power, surgery, swimming, and X-rays as more risky than the other groups, and they judged nuclear power, police work, and mountain climbing to be much less risky.

## What Determines Risk Perception?

What do people mean when they say that a particular technology is quite risky? A series of additional studies was conducted to answer this question.

*Perceived risk compared to frequency of death.* When people judge risk, as in the previous study, are they simply estimating frequency of death? To answer this question, we collected the best available technical estimates of the annual number of deaths from each of the thirty activities included in our study. For some cases, such as commercial aviation and handguns, there is good statistical evidence based on counts of known victims. For other cases, such as the lethal potential of nuclear or fossil-fuel power plants, available estimates are based on uncertain inferences about incompletely understood processes. For still others, such as food coloring, we could find no estimates of annual fatalities.

For the 25 cases for which we found technical estimates for annual frequency of death, we compared these estimates with perceived risk. Results for experts and the LOWV sample are shown in figure 4 (the results for the other lay groups were quite similar to those from the LOWV sample). The experts' mean judgments were so closely related to the statistical or calculated frequencies that it seems reasonable to conclude that they viewed the risk of an activity or technology as synonymous with its annual fatalities. The risk judgments of lay people, however, showed only a moderate relationship to the annual frequencies of death,[17] raising the possibility that, for them, risk may not be synonymous with fatalities. In particular, the perceived risk from nuclear power was

item is 1.2 times as risky as the least risky item (i.e., 20 percent more risky). A rating of 200 means that the item is 20 times as risky as the least risky item, to which you assigned a 10. . . ." They were urged to cross-check and adjust their numbers until they believed they were right.

Table 2 shows how the various groups ranked the relative riskiness of these 30 activities and technologies. There were many similarities between the three groups of lay persons. For example, each group believed that motorcycles, other motor vehicles, and handguns were highly risky, and that vaccinations, home appliances, power mowers, and football were relatively safe. However, there were strong differ-

**Table 3. Fatality Estimates and Disaster Multipliers for 30 Activities and Technologies**

| Activity or technology | Technical fatality estimates | Geometric mean fatality estimates, average year | | Geometric mean multiplier, disastrous year | |
|---|---|---|---|---|---|
| | | LOWV | Students | LOWV | Students |
| Smoking | 150,000 | 6,900 | 2,400 | 1.9 | 2.0 |
| Alcoholic beverages | 100,000 | 12,000 | 2,600 | 1.9 | 1.4 |
| Motor vehicles | 50,000 | 28,000 | 10,500 | 1.6 | 1.8 |
| Handguns | 17,000 | 3,000 | 1,900 | 2.6 | 2.0 |
| Electric power | 14,000 | 660 | 500 | 1.9 | 2.4 |
| Motorcycles | 3,000 | 1,600 | 1,600 | 1.8 | 1.6 |
| Swimming | 3,000 | 930 | 370 | 1.6 | 1.7 |
| Surgery | 2,800 | 2,500 | 900 | 1.5 | 1.6 |
| X-rays | 2,300 | 90 | 40 | 2.7 | 1.6 |
| Railroads | 1,950 | 190 | 210 | 3.2 | 1.6 |
| General (private) aviation | 1,300 | 550 | 650 | 2.8 | 2.0 |
| Large construction | 1,000 | 400 | 370 | 2.1 | 1.4 |
| Bicycles | 1,000 | 910 | 420 | 1.8 | 1.4 |
| Hunting | 800 | 380 | 410 | 1.8 | 1.7 |
| Home appliances | 200 | 200 | 240 | 1.6 | 1.3 |
| Fire fighting | 195 | 220 | 390 | 2.3 | 2.2 |
| Police work | 160 | 460 | 390 | 2.1 | 1.9 |
| Contraceptives | 150 | 180 | 120 | 2.1 | 1.4 |
| Commercial aviation | 130 | 280 | 650 | 3.0 | 1.8 |
| Nuclear power | 100[a] | 20 | 27 | 107.1 | 87.6 |
| Mountain climbing | 30 | 50 | 70 | 1.9 | 1.4 |
| Power mowers | 24 | 40 | 33 | 1.6 | 1.3 |
| High school & college football | 23 | 39 | 40 | 1.9 | 1.4 |
| Skiing | 18 | 55 | 72 | 1.9 | 1.6 |
| Vaccinations | 10 | 65 | 52 | 2.1 | 1.6 |
| Food coloring | —[b] | 38 | 33 | 3.5 | 1.4 |
| Food preservatives | —[b] | 61 | 63 | 3.9 | 1.7 |
| Pesticides | —[b] | 140 | 84 | 9.3 | 2.4 |
| Prescription antibiotics | —[b] | 160 | 290 | 2.3 | 1.6 |
| Spray cans | —[b] | 56 | 38 | 3.7 | 2.4 |

[a]Technical estimates for nuclear power were found to range between 16 and 600 annual fatalities. The geometric mean of these estimates was used here.
[b]Estimates were unavailable.

disproportionately high compared to its estimated number of fatalities.

*Lay fatality estimates.* Perhaps lay people based their risk judgments on annual fatalities, but estimated their numbers inaccurately. To test this hypothesis, we asked additional groups of students and LOWV members "to estimate how many people are likely to die in the U.S. in the next year (if the next year is an average year) as a consequence of these thirty activities and technologies." We asked our student and LOWV samples to consider all sources of death associated with these activities.

The mean fatality estimates of LOWV members and students are shown in columns 2 and 3 of table 3. If lay people really equate risk with annual fatalities, one would expect that their own estimates of annual fatalities, no matter how inaccurate, would be very similar to their judgments of risk. But this was not so. There was a moderate agreement between their annual fatality estimates and their risk judgments, but there were important exceptions. Most notably, nuclear power had the *lowest* fatality estimate and the *highest* perceived risk for both LOWV members and students. Overall, lay people's risk perceptions were no more closely related to their own fatality estimates than they were to the technical estimates (figure 4).

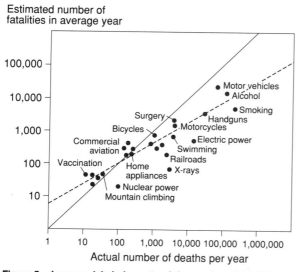

Estimated number of fatalities in average year

**Figure 5.** Lay people's judgments of the number of fatalities in an average year plotted against the best estimates of annual fatalities for 25 activities and technologies. The solid line indicates accurate judgment, while the dashed line best fits the data points. These results have much the same character as those shown in figure 2 for a different collection of hazards. Low frequencies were overestimated and high ones were underestimated. The overall relationship is marred by specific biases (e.g., the underestimation of fatalities associated with railroads, X-rays, electric power, and smoking).

These results lead us to reject the idea that lay people wanted to equate risk with annual fatality estimates but were inaccurate in doing so. Instead, we are led to believe that lay people incorporate other considerations besides annual fatalities into their concept of risk.

Some other aspects of lay people's fatality estimates are of interest. One is that they were moderately accurate. The relationship between the LOWV members' fatality estimates and the best technical estimates is plotted in figure 5. The lay estimates showed the same overestimation of those items that cause few fatalities and underestimation of those resulting in the most fatalities that was apparent in figure 2 for a different collection of hazards. Also as in figure 2, the moderate overall relationship between lay and technical estimates was marred by specific biases (e.g., the underestimation of fatalities associated with railroads, X-rays, electric power, and smoking).

*Disaster potential.* The fact that the LOWV members and students assigned very high risk values to nuclear power along with very low estimates of its annual fatality rates is an apparent contradiction. One possible explanation is that LOWV members expected nuclear power to have a low death rate in an average year but considered it to be a high risk technology because of its potential for disaster.

In order to understand the role played by expectations of disaster in determining lay people's risk judgments, we asked these same respondents to give a number for each activity and technology indicating how many times more deaths would occur if next year were "particularly disastrous" rather than average. The averages of these multipliers are shown in table 3. For most activities, people saw little potential for disaster. For the LOWV sample all but five of the multipliers were less than 3, and for the student sample all but six were less than 2. The striking exception in both cases is nuclear power, with a geometric mean disaster multiplier in the neighborhood of 100.

For any individual an estimate of the expected number of fatalities in a disastrous year could be obtained by applying the disaster multiplier to the estimated fatalities for an average year. When this was done for nuclear power, almost 40 percent of the respondents expected more than 10,000 fatalities if next year were a disastrous year. More than 25 percent expected 100,000 or more fatalities. These extreme estimates can be contrasted with the Reactor Safety Study's conclusion that the maximum credible nuclear accident, coincident with the most unfavorable combination of weather and population density, would cause only 3,300 prompt fatalities.[18] Furthermore, that study estimated the odds against an accident of this magnitude occurring during the next year (assuming 100 operating reactors) to about 2,000,000:1.

Apparently, disaster potential explains much or all of the discrepancy between the perceived risk and frequency of death values for nuclear power. Yet, because disaster plays only a small role in most of the thirty activities and technologies we have studied, it provides only a partial explanation of the perceived risk data.

*Qualitative characteristics.* Are there other determinants of risk perceptions besides frequency estimates? We asked experts, students, LOWV members, and Active Club members to rate the thirty technologies and activities on nine qualitative characteristics

## Dimensions of Risk

The ratings of the nine risk characteristics listed in table 4 tended to be related to each other across the thirty activities and technologies. For example, risks faced voluntarily were typically judged well known and controllable. These interrelations were sufficiently high to suggest that all the ratings could be explained in terms of a few basic dimensions of risk. Application of a statistical technique known as factor analysis confirmed this hypothesis.

The nine characteristics of perceived risk could be collapsed into two underlying dimensions, each reflecting a specific combination of the original nine characteristics. The vertical dimension appears to discriminate between high and low levels of technological sophistication. The horizontal dimension primarily reflects the likelihood of a mishap being fatal.

Activities and technologies appearing close to one another in the two-dimensional space shown above have similar profiles over the original nine risk characteristics, while activities and technologies distant from one another have markedly different profiles. These dimensions are strongly related to people's attitudes regarding the acceptability of the risks posed by a technology.

Although the data presented in the above figure are from LOWV members, analyses for each of the other three groups produced remarkably similar figures, suggesting that there may be substantial agreement in our culture regarding the qualitative dimensions of risk.

**Table 4. Risk Characteristics Rated by LOWV Members and Students**

**Voluntariness of risk**
Do people face this risk voluntarily? If some of the risks are voluntarily undertaken and some are not, mark an appropriate spot towards the center of the scale.

| risk assumed voluntarily | 1 | 2 | 3 | 4 | 5 | 6 | 7 | risk assumed involuntarily |

**Immediacy of effect**
To what extent is the risk of death immediate—or is death likely to occur at some later time?

| effect immediate | 1 | 2 | 3 | 4 | 5 | 6 | 7 | effect delayed |

**Knowledge about risk**
To what extent are the risks known precisely by the persons who are exposed to those risks?

| risk level known precisely | 1 | 2 | 3 | 4 | 5 | 6 | 7 | risk level not known |

To what extent are the risks known to science?

| risk level known precisely | 1 | 2 | 3 | 4 | 5 | 6 | 7 | risk level not known |

**Control over risk**
If you are exposed to the risk, to what extent can you, by personal skill or diligence, avoid death?

| personal risk can't be controlled | 1 | 2 | 3 | 4 | 5 | 6 | 7 | personal risk can be controlled |

**Newness**
Is this risk new and novel or old and familiar?

| new | 1 | 2 | 3 | 4 | 5 | 6 | 7 | old |

**Chronic/Catastrophic**
Is this a risk that kills people one at a time (chronic risk) or a risk that kills large numbers of people at once (catastrophic risk)?

| chronic | 1 | 2 | 3 | 4 | 5 | 6 | 7 | catastrophic |

**Common/Dread**
Is this a risk that people have learned to live with and can think about reasonably calmly, or is it one that people have great dread for—on the level of a gut reaction?

| common | 1 | 2 | 3 | 4 | 5 | 6 | 7 | dread |

**Severity of consequences**
When the risk from the activity is realized in the form of a mishap or illness, how likely is it that the consequence will be fatal?

| certain not to be fatal | 1 | 2 | 3 | 4 | 5 | 6 | 7 | certain to be fatal |

that have been hypothesized to be important.[19] These ratings scales are described in table 4.

Examination of "risk profiles" based on mean ratings for the nine characteristics proved helpful in understanding the risk judgments of lay people. Nuclear power, for example, had the dubious distinction of scoring at or near the extreme on all of the characteristics associated with high risk. Its risks were seen as involuntary, delayed, unknown, uncontrollable, unfamiliar, potentially catastrophic, dreaded, and severe (certainly fatal). Its spectacular and unique risk profile is contrasted in figure 6 with non-nuclear electric power and with another radiation technology, X-rays, both of whose risks were judged to be much lower. Both electric power and X-rays were judged much more voluntary, less catastrophic, less dreaded, and more familiar than nuclear power.

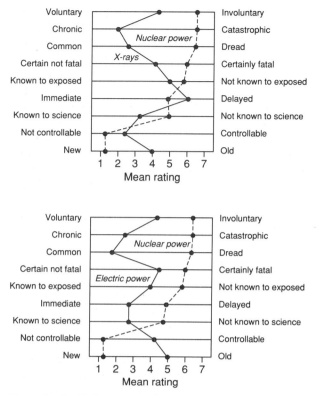

**Figure 6. Qualitative characteristics of perceived risk for nuclear power and related technologies. In the right-hand diagram, risk profiles for nuclear power and non-nuclear electric power are compared. In the [top] diagram, nuclear power and X-rays are compared. Each profile consists of nine dimensions rated on a seven point scale. The instructions that elicited these responses are reproduced in table 4. The perceived qualities of nuclear power are dramatically different from the comparison technologies.**
The source of this data is the LOWV sample studied by Fischhoff et al. (note 16).

Across all thirty items, ratings of dread and of the severity of consequences were found to be closely related to lay persons' perceptions of risk. In fact, ratings of dread and severity along with the subjective fatality estimates and the disaster multipliers in table 3 enabled the risk judgments of the LOWV and student groups to be predicted almost perfectly.[20] Experts' judgments of risk were not related to any of the nine qualitative risk characteristics.[21]

*Judged seriousness of death.* In a further attempt to improve our understanding of perceived risk, we examined the hypothesis that some hazards are feared more than others because the deaths they produce are much "worse" than deaths from other activities. We thought, for example, that deaths from

risks imposed involuntarily, from risks not under one's control, or from hazards that are particularly dreaded might be given greater weight in determining people's perceptions of risk.

However, when we asked students and LOWV members to judge the relative "seriousness" of a death from each of the thirty activities and technologies, the differences were slight. The most serious forms of death (from nuclear power and handguns) were judged to be only about two to four times worse than the least serious forms of death (from alcoholic beverages and smoking). Furthermore, across all thirty activities, judged seriousness of death was not closely related to perceived risk of death.

### Reconciling Divergent Opinions

Our data show that experts and lay people have quite different perceptions about how risky certain technologies are. It would be comforting to believe that these divergent risk judgments would be responsive to new evidence so that, as information accumulates, perceptions would converge towards one "appropriate" view. Unfortunately, this is not likely to be the case. As noted earlier in our discussion of availability, risk perception is derived in part from fundamental modes of thought that lead people to rely on fallible indicators such as memorability and imaginability.

Furthermore, a great deal of research indicated that people's beliefs change slowly and are extraordinarily persistent in the face of contrary evidence.[22] Once formed, initial impressions tend to structure the way that subsequent evidence is interpreted. New evidence appears reliable and informative if it is consistent with one's initial belief; contrary evidence is dismissed as unreliable, erroneous, or unrepresentative. Thus, depending on one's predispositions, intense effort to reduce a hazard may be interpreted to mean either that the risks are great or that the technologists are responsive to the public's concerns. Likewise, opponents of a technology may view minor mishaps as near catastrophes and dismiss the contrary opinions of experts as biased by vested interests.

From a statistical standpoint, convincing people that the catastrophe they fear is extremely unlikely is difficult under the best conditions. Any mishap could be seen as proof of high risk, whereas demonstrating safety would require a massive amount of evidence.[23] Nelkin's case history of a nuclear siting controversy[24] provides a good example of the inability of technical arguments to change opinions. In that debate each side capitalized on technical ambiguities in ways that reinforced its own position.

## The Fallibility of Judgment

Our examination of risk perception leads us to the following conclusions:

- Cognitive limitations, coupled with the anxieties generated by facing life as a gamble, cause uncertainty to be denied, risks to be distorted, and statements of fact to be believed with unwarranted confidence.

- Perceived risk is influenced (and sometimes biased) by the imaginability and memorability of the hazard. People may, therefore, not have valid perceptions even for familiar risks.

- Our expert's risk perceptions correspond closely to statistical frequencies of death. Lay people's risk perceptions were based in part upon frequencies of death, but there were some striking discrepancies. It appears that for lay people, the concept of risk includes qualitative aspects such as dread and the likelihood of a mishap being fatal. Lay people's risk perceptions were also affected by catastrophic potential.

- Disagreements about risk should not be expected to evaporate in the presence of "evidence." Definitive evidence, particularly about rare hazards, is difficult to obtain. Weaker information is likely to be interpreted in a way that reinforces existing beliefs.

The significance of these results hinges upon one's acceptance of our assumption that subjective judgments are central to the hazard management process. Our conclusions mean little if one can assume that there are analytical tools which can be used to assess most risks in a mechanical fashion and that all decision makers have perfect information and the know-how to use it properly. These results gain in importance to the extent that one believes, as we do, that expertise involves a large component of judgment, that the facts are not all in (or obtainable) regarding many important hazards, that people are often poorly informed or misinformed, and that they respond not just to numbers but also to qualitative aspects of hazards.

Whatever role judgment plays, its products should be treated with caution. Research not only demonstrates that judgment is fallible, but it shows that the degree of fallibility is often surprisingly great and that faulty beliefs may be held with great confidence.

When it can be shown that even well-informed lay people have difficulty judging risks accurately, it is tempting to conclude that the public should be removed from the hazard-management process. The political ramifications of such a transfer of power to a technical elite are obvious. Indeed, it seems doubtful that such a massive disenfranchisement is feasible in any democratic society.

Furthermore, this transfer of decision making would seem to be misguided. For one thing, we have no assurance that experts' judgments are immune to biases once they are forced to go beyond their precise knowledge and rely upon their judgment. Although judgmental biases have most often been demonstrated with lay people, there is evidence that the cognitive functioning of experts is basically like that of everyone else.

In addition, in many if not most cases effective hazard management requires the cooperation of a large body of lay people. These people must agree to do without some things and accept substitutes for others; they must vote sensibly on ballot measures and for legislators who will serve them as surrogate hazard managers; they must obey safety rules and use the legal system responsibly. Even if the experts were much better judges of risk than lay people, giving experts an exclusive franchise on hazard management would involve substituting short-term efficiency for the long-term effort needed to create an informed citizenry.

For those of us who are not experts, these findings pose an important series of challenges: to be better informed, to rely less on unexamined or unsupported judgments, to be aware of the qualitative aspects that strongly condition risk judgments, and to be open to new evidence that may alter our current risk perceptions.

For the experts, our findings pose what may be a more difficult challenge: to recognize their own cognitive limitations, to temper their assessments of risk with the important qualitative aspects of risk that influence the responses of lay people, and somehow to create ways in which these considerations can find expression in hazard management without, in the process, creating more heat than light.

## Acknowledgments

The authors wish to express their appreciation to Christoph Hohenemser, Roger Kasperson, and Robert Kates for their many helpful comments and suggestions. This work was supported by the National Science Foundation under Grant ENV77-15332 to Perceptronics, Inc. Any opinions, findings and conclusions, or recommendations expressed herein are those of the authors and do not necessarily reflect the views of the National Science Foundation.

## Notes

1. We have not attempted here to review all the important research in this area. Interested readers should see C. H. Green, "Risk: Attitudes and Beliefs," in D. V. Canter (ed.), **Behavior in Fires,** forthcoming; R. W. Kates, **Risk Assessment of Environmental Hazard,** Wiley, New York, 1978; and H. J. Otway, D. Maurer, and K. Thomas, "Nuclear Power: The Question of Public Acceptance," **Futures,** April (1978).

2. A. Tversky and D. Kahneman, "Judgment under Uncertainty: Heuristics and Biases," **Science,** 185 (1974): 1124–1131.

3. A. Tversky and D. Kahneman, "Availability: A Heuristic for Judging Frequency and Probability," **Cognitive Psychology,** 4 (1973): 207–232.

4. S. Lichtenstein, P. Slovic, B. Fischhoff, M. Layman, and B. Combs, "Judged Frequency of Lethal Events," **Journal of Experimental Psychology: Human Learning and Memory,** 4 (1978): 551–578.

5. B. Fischhoff, P. Slovic, and S. Lichtenstein, "Fault Trees: Sensitivity and Estimated Failure Probabilities to Problem Representation," **Journal of Experimental Psychology: Human Perception and Performance,** 4 (1978): 342–355.

6. B. Fischhoff, P. Slovic, and S. Lichtenstein, "Knowing with Certainty: The Appropriateness of Extreme Confidence," **Journal of Experimental Psychology: Human Perception and Performance,** 3 (1977): 552–564.

7. S. Lichtenstein, B. Fischhoff, and L. D. Phillips, "Calibration of Probabilities: The State of the Art," in H. Jungermann and G. de Zeeuw (eds.), **Decision Making and Change in Human Affairs,** D. Reidel, Dordrecht, The Netherlands, 1977.

8. M. Hynes and E. Vanmarcke, "Reliability of Embankment Performance Predictions," **Proceedings of the ASCE Engineering Mechanics Division Specialty Conference,** University of Waterloo Press, Waterloo, Ontario, Canada, 1976.

9. U.S. Nuclear Regulatory Commission, **Reactor Safety Study: An Assessment of Accident Risks in U.S. Commercial Nuclear Power Plants,** WASH 1400 (NUREG-75/014), Washington, D.C., October 1975.

10. U.S. Nuclear Regulatory Commission, "Risk Assessment Review Group Report to the U.S. Nuclear Regulatory Commission," **NUREG/CR-0400,** September 1978.

11. U.S. Government, **Teton Dam Disaster,** Committee on Government Operations, Washington, D.C., 1976.

12. R. W. Kates, "Hazard and Choice Perception in Flood Plain Management," **Research Paper 78,** Department of Geography, University of Chicago, Chicago, 1962.

13. K. Borch, **The Economics of Uncertainty,** Princeton University Press, Princeton, N.J., 1968.

14. **Eugene Register-Guard,** "Doubts Linger on Cyclamate Risks," January 14, 1976.

15. E. E. David, "One-Armed Scientists?" **Science,** 189 (1975): 891.

16. B. Fischhoff, P. Slovic, S. Lichtenstein, S. Read, and B. Combs, "How Safe is Safe Enough? A Psychometric Study of Attitudes towards Technological Risks and Benefits," **Policy Sciences,** 8 (1978): 127–152; P. Slovic, B. Fischhoff, and S. Lichtenstein, "Expressed Preferences," unpublished manuscript, Decision Research, Eugene, Oregon, 1978.

17. The correlations between perceived risk and the annual frequencies of death were .92 for the experts and .62, .50, and .56 for the League of Women Voters, students, and Active Club samples, respectively.

18. U.S. Nuclear Regulatory Commission, note 9 above.

19. W. Lowrance, **Of Acceptable Risk,** Wm. Kaufman, Los Altos, California, 1976.

20. The multiple correlation between the risk judgments of the LOWV members and students and a linear combination of their fatality estimates, disaster multipliers, dread ratings, and severity ratings was .95.

21. A secondary finding was that both experts and lay persons believed that the risks from most of the activities were better known to science than to the individuals at risk. The experts believed that the discrepancy in knowledge was particularly great for vaccinations, X-rays, antibiotics, alcohol, and home appliances. The only activities whose risks were judged better known to those exposed were mountain climbing, fire fighting, hunting, skiing, and police work.

22. L. Ross, "The Intuitive Psychologist and His Shortcomings," in L. Berkowitz (ed.), **Advances in Social Psychology,** Academic Press, New York, 1977.

23. A. E. Green and A. J. Bourne, **Reliability Technology,** Wiley Interscience, New York, 1972.

24. D. Nelkin, "The Role of Experts on a Nuclear Siting Controversy," **Bulletin of the Atomic Scientists,** 30 (1974): 29–36.

# Questions for Thought and Discussion

1. Based on the concepts described in this paper—which are grouped under the headings judgmental biases, analyzing judgments of risk, and the fallibility of judgment—offer an explanation of why, despite the risk, many Californians choose to live close to known earthquake faults.

2. How do the news media influence people's perceptions of the risks of catastrophes? What suggestions would you make to a television network executive about reducing distortions that might contribute to these risk perceptions?

3. This paper was published before the risk of contracting AIDS began to attract widespread public attention. Rate the qualitative characteristics of this risk using the seven-point scale in table 4, and plot your ratings on a diagram like the one in figure 6. Compare the resulting profile with that of nuclear power and X-rays shown in figure 6. What important characteristics of the risk of AIDS, if any, are not included in this analysis?

4. Construct an argument in favor of a political system in which the determination of the social acceptability of new technologies is left entirely up to a body of the foremost risk experts. Then construct an argument opposing such a system.

# Ranking Possible Carcinogenic Hazards

BRUCE N. AMES
*University of California at Berkeley*

RENAE MAGAW and LOIS SWIRSKY GOLD
*Lawrence Berkeley Laboratory, Berkeley, California*

Epidemiologists estimate that at least 70% of human cancer would, in principle, be preventable if the main risk and antirisk factors could be identified (*1*). This is because the incidence of specific types of cancer differs markedly in different parts of the world where people have different life-styles. For example, colon and breast cancer, which are among the major types of cancer in the United States, are quite rare among Japanese in Japan, but not among Japanese-Americans. Epidemiologists are providing important clues about the specific causes of human cancer, despite inherent methodological difficulties. They have identified tobacco as an avoidable cause of about 30% of all U.S. cancer deaths and of an even larger number of deaths from other causes (*1, 2*). Less specifically, dietary factors, or their absence, have been suggested in many studies to contribute to a substantial proportion of cancer deaths, though the intertwined risk and antirisk factors are being identified only slowly

(*1, 3, 4*). High fat intake may be a major contributor to colon cancer, though the evidence is not as definitive as that for the role of saturated fat in heart disease or of tobacco in lung cancer. Alcoholic beverage consumption, particularly by smokers, has been estimated to contribute to about 3% of U.S. cancer deaths (*1*) and to an even larger number of deaths from other causes. Progress in prevention has been made for some occupational factors, such as asbestos, to which workers used to be heavily exposed, with delayed effects that still contribute to about 2% of U.S. cancer deaths (*1, 5*). Prevention may also become possible for hormone-related cancers such as breast cancer (*1, 6*), or virus-related cancers such as liver cancer (hepatitis B) and cancer of the cervix (papilloma virus HPV16) (*1, 7*).

Animal bioassays and in vitro studies are also providing clues as to which carcinogens and mutagens might be contributing to human cancer. However, the evaluation of carcinogenicity in rodents is expensive and the extrapolation to humans is difficult. (*8–11*). We will use the term "possible hazard" for estimates based on rodent cancer tests and "risk" for those based on human cancer data (*10*).

Extrapolation from the results of rodent cancer tests done at high doses to effects on humans exposed to low doses is routinely attempted by regulatory agencies when formulating policies attempting to prevent future cancer. There is little sound scientific basis for this type of extrapolation, in part due to our lack of knowledge about mechanisms of cancer induction, and it is viewed with great unease by many epidemiologists and toxicologists (5, 9–11). Nevertheless, to be prudent in regulatory policy, and in the absence of good human data (almost always the case), some reliance on animal cancer tests is unavoidable. The best use of them should be made even though few, if any, of the main avoidable causes of human cancer have typically been the types of man-made chemicals that are being tested in animals (10). Human cancer may, in part, involve agents such as hepatitis B virus, which causes chronic inflammation; changes in hormonal status; deficiencies in normal protective factors (such as selenium or β-carotene) against endogenous carcinogens (12); lack of other anticarcinogens (such as dietary fiber or calcium) (4); or dietary imbalances such as excess consumption of fat (3, 4, 12) or salt (13).

There is a need for more balance in animal cancer testing to emphasize the foregoing factors and natural chemicals as well as synthetic chemicals (12). There is increasing evidence that our normal diet contains many rodent carcinogens, all perfectly natural or traditional (for example, from the cooking of food) (12), and that no human diet can be entirely free of mutagens or agents that can be carcinogenic in rodent systems. We need to identify the important causes of human cancer among the vast number of minimal risks. This requires knowledge of both the amounts of a substance to which humans are exposed and its carcinogenic potency.

Animal cancer tests can be analyzed quantitatively to give an estimate of the relative carcinogenic potencies of the chemicals tested. We have previously published our Carcinogenic Potency Database, which showed that rodent carcinogens vary in potency by more than 10 millionfold (14).

This article attempts to achieve some perspective on the plethora of possible hazards to humans from exposure to known rodent carcinogens by establishing a scale of the possible hazards for the amounts of various common carcinogens to which humans might be chronically exposed. We view the value of our calculations not as providing a basis for absolute human risk assessment, but as a guide to priority setting. One problem with this type of analysis is that few of the many natural chemicals we are exposed to in very large amounts (relative to synthetic chemicals) have been tested in animals for carcinogenicity. Thus, our knowledge of the background levels of human exposure to animal carcinogens is fragmentary, biased in favor of synthetic chemicals, and limited by our lack of knowledge of human exposures.

## Ranking of Possible Carcinogenic Hazards

Since carcinogens differ enormously in potency, a comparison of possible hazards from various carcinogens ingested by humans must take this into account. The measure of potency that we have developed, the $TD_{50}$, is the daily dose rate (in milligrams per kilogram) to halve the percent of tumor-free animals by the end of a standard lifetime (14). Since the $TD_{50}$ (analogous to the $LD_{50}$*) is a dose rate, the lower the $TD_{50}$ value the more potent the carcinogen. To calculate our index of possible hazard we express each human exposure (daily lifetime dose in milligrams per kilogram) as a percentage of the rodent $TD_{50}$ dose (in milligrams per kilogram) for each carcinogen. We call this percentage HERP [Human Exposure dose/Rodent Potency dose]. The $TD_{50}$ values are taken from our ongoing Carcinogenic Potency Database (currently 3500 experiments on 975 chemicals), which reports the $TD_{50}$ values estimated from experiments in animals (14). Human exposures have been estimated from the literature as indicated. As rodent data are all calculated on the basis of lifetime exposure at the indicated daily dose rate (14), the human exposure data are similarly expressed as lifelong daily dose rates even though the human exposure is likely to be less than daily for a lifetime.

It would be a mistake to use our HERP index as a direct estimate of human hazard. First, at low dose rates human susceptibility may differ systematically from rodent susceptibility. Second, the general shape of the dose-response relationship is not known. A

*Editors' note: $LD_{50}$ (lethal dose–fifty) is defined as the dose of a substance that causes death in 50 percent of the exposed test animals.

linear dose response has been the dominant assumption in regulating carcinogens for many years, but this may not be correct. If the dose responses are not linear but are actually quadratic or hockey-stick shaped or show a threshold, then the actual hazard at low dose rates might be much less than the HERP values would suggest. An additional difficulty is that it may be necessary to deal with carcinogens that differ in their mechanisms of action and thus in their dose-response relationship. We have therefore put an asterisk next to HERP values for carcinogens that do not appear to be active through a genotoxic (DNA damaging or mutagenic) mechanism ([15]) so that comparisons can be made within the genotoxic or nongenotoxic classes.

Table 1 presents our HERP calculations of possible cancer hazards in order to compare them within several categories so that, for example, pollutants of possible concern can be compared to natural carcinogens in the diet. A convenient reference point is the possible hazard from the carcinogen chloroform in a liter of average (U.S.) chlorinated tap water, which is close to a HERP of 0.001%. Chloroform is a by-product of water chlorination, which protects us from pathogenic viruses and bacteria.

*Contaminated water.* The possible hazards from carcinogens in contaminated well water [for example, Santa Clara ("Silicon") Valley, California, or Woburn, Massachusetts] should be compared to the possible hazard of ordinary tap water (table 1). Of 35 wells shut down in Santa Clara Valley because of their supposed carcinogenic hazard, only two have HERP values greater than ordinary tap water. Well water is not usually chlorinated and typically lacks the chloroform present in chlorinated tap water. Water from the most polluted well (HERP = 0.004% per liter for trichloroethylene), as indicated in table 1, has a HERP value orders of magnitude less than for the carcinogens in an equal volume of cola, beer, or wine. Its HERP value is also much lower than that of many of the common natural foods that are listed in table 1, such as the average peanut butter sandwich. Caveats for any comparisons are given below. Since the consumption of tap water is only about 1 or 2 liters per day, the animal evidence provides no good reason to expect that chlorination of water or current levels of man-made pollution of water pose a significant carcinogenic hazard.

*Pesticide residues.* Intake of man-made pesticide residues from food in the United States, including residues of industrial chemicals such as polychlorinated biphenyls (PCBs), averages about 150 µg/day. Most (105 µg) of this intake is composed of three chemicals (ethylhexyl diphenyl phosphate, malathion, and chlorpropham) shown to be noncarcinogenic in tests in rodents ([16]). A carcinogenic pesticide residue in food of possible concern is DDE, the principal metabolite (>90%) of DDT ([16]). The average U.S. daily intake of DDE from DDT (HERP = 0.0003%) is equivalent to the HERP of the chloroform in one glass of tap water and thus appears to be insignificant compared to the background of natural carcinogens in our diet (table 1). Even daily consumption of 100 times the average intake of DDE/DDT or PCBs would produce a possible hazard that is small compared to other common exposures shown in table 1.

*Nature's pesticides.* We are ingesting in our diet at least 10,000 times more by weight of natural pesticides than of man-made pesticide residues ([12]). These are natural "toxic chemicals" that have an enormous variety of chemical structures, appear to be present in all plants, and serve to protect plants against fungi, insects, and animal predators ([12]). Though only a few are present in each plant species, they commonly make up 5 to 10% of the plant's dry weight ([12]). There has been relatively little interest in the toxicology or carcinogenicity of these compounds until quite recently, although they are by far the main source of "toxic chemicals" ingested by humans. Only a few dozen of the thousands present in the human diet have been tested in animal bioassays, and only some of these tests are adequate for estimating potency in rodents ([14]). A sizable proportion of those that have been tested are carcinogens, and many others have been shown to be mutagens ([12]), so it is probable that many more will be found to be carcinogens if tested. Those shown in table 1 are: estragole (HERP = 0.1% for a daily 1 g of dried basil), safrole (HERP = 0.2% for a daily natural root beer), symphytine (a pyrrolizidine alkaloid, 0.03% for a daily cup of comfrey tea), comfrey tablets sold in health food stores (6.2% for a daily dose), hydrazines in mushrooms (0.1% for one daily raw mushroom), and allyl isothiocyanate (0.07% for a daily 5 g of brown mustard).

Plants commonly produce very much larger

**Table 1. Ranking Possible Carcinogenic Hazards.** *Potency of carcinogens:* A number in parentheses indicates a $TD_{50}$ value not used in HERP calculation because it is the less sensitive species; (−) = negative in cancer test. (+) = positive for carcinogenicity in test(s) not suitable for calculating a $TD_{50}$; (?) = is not adequately tested for carcinogenicity. $TD_{50}$ values shown are averages calculated by taking the harmonic mean of the $TD_{50}$'s of the positive tests in that species from the Carcinogenic Potency Database. Results are similar if the lowest $TD_{50}$ value (most potent) is used instead. For each test the target site with the lowest $TD_{50}$ value has been used. The average $TD_{50}$ has been calculated separately for rats and mice, and the more sensitive species is used for calculating the possible hazard. The database, with references to the source of the cancer tests, is complete for tests published through 1984 and for the National Toxicology Program bioassays through June 1986 (*14*). We have not indicated the route of exposure or target sites or other particulars of each test, although these are reported in the database. *Daily human exposure:* We have tried to use average or reasonable daily intakes to facilitate comparisons. In several cases, such as contaminated well water or factory exposure to EDB, this is difficult to determine, and we give the value for the worst found and indicate pertinent information in the References and Notes. The calculations assume a daily dose for a lifetime; where drugs are normally taken for only a short period we have bracketed the HERP value. For inhalation exposures we assume an inhalation of 9,600 liters per 8 hours for the workplace and 10,800 liters per 14 hours for indoor air at home. *Possible hazard:* The amount of rodent carcinogen indicated under carcinogen dose is divided by 70 kg to give a milligram per kilogram of human exposure, and this human dose is given as the percentage of $TD_{50}$ dose in the rodent (in milligrams per kilogram) to calculate the *H*uman *E*xposure/*R*odent *P*otency index (HERP).

| Possible hazard: HERP (%) | Daily human exposure | Carcinogen dose per 70-kg person | Potency of carcinogen: $TD_{50}$ (mg/kg) Rats | Mice | References |
|---|---|---|---|---|---|
| | ENVIRONMENTAL POLLUTION | | | | |
| 0.001* | Tap water, 1 liter | Chloroform, 83 µg (U.S. average) | (119) | 90 | *96* |
| 0.004* | Well water, 1 liter contaminated (worst well in Silicon Valley) | Trichloroethylene, 2800 µg | (−) | 941 | *97* |
| 0.0004* | Well water, 1 liter contaminated, Woburn | Trichloroethylene, 267 µg | (−) | 941 | *98* |
| 0.0002* | | Chloroform, 12 µg | (119) | 90 | |
| 0.0003* | | Tetrachloroethylene, 21 µg | 101 | (126) | |
| 0.008* | Swimming pool, 1 hour (for child) | Chloroform, 250 µg (average pool) | (119) | 90 | *99* |
| 0.6 | Conventional home air (14 hour/day) | Formaldehyde, 598 µg | 1.5 | (44) | *100* |
| 0.004 | | Benzene, 155 µg | (157) | 53 | |
| 2.1 | Mobile home air (14 hour/day) | Formaldehyde, 2.2 mg | 1.5 | (44) | *28* |
| | PESTICIDE AND OTHER RESIDUES | | | | |
| 0.0002* | PCBs: daily dietary intake | PCBs, 0.2 µg (U.S. average) | 1.7 | (9.6) | *101* |
| 0.0003* | DDE/DDT: daily dietary intake | DDE, 2.2 µg (U.S. average) | (−) | 13 | *16* |
| 0.0004 | EDB: daily dietary intake (from grains and grain products) | Ethylene dibromide, 0.42 µg (U.S. average) | 1.5 | (5.1) | *102* |
| | NATURAL PESTICIDES AND DIETARY TOXINS | | | | |
| 0.003 | Bacon, cooked (100 g) | Dimethylnitrosamine, 0.3 µg | (0.2) | 0.2 | *40* |
| 0.006 | | Diethylnitrosamine, 0.1 µg | 0.02 | (+) | |
| 0.003 | Sake (250 ml) | Urethane, 43 µg | (41) | 22 | *24* |
| 0.03 | Comfrey herb tea, 1 cup | Symphytine, 38 µg (750 µg of pyrrolizidine alkaloids) | 1.9 | (?) | *103* |
| 0.03 | Peanut butter (32 g; one sandwich) | Aflatoxin, 64 ng (U.S. average, 2 ppb) | 0.003 | (+) | *18* |
| 0.06 | Dried squid, broiled in gas oven (54 g) | Dimethylnitrosamine, 7.9 µg | (0.2) | 0.2 | *37* |
| 0.07 | Brown mustard (5 g) | Allyl isothiocyanate, 4.6 mg | 96 | (−) | *47* |
| 0.1 | Basil (1 g of dried leaf) | Estragole, 3.8 mg | (?) | 52 | *48* |
| 0.1 | Mushroom, one raw (15 g) (*Agaricus bisporus*) | Mixture of hydrazines, and so forth | (?) | 20,300 | *104* |
| 0.2 | Natural root beer (12 ounces; 354 ml) (now banned) | Safrole, 6.6 mg | (436) | 56 | *105* |
| 0.008 | Beer, before 1979 (12 ounces; 354 ml) | Dimethylnitrosamine, 1 µg | (0.2) | 0.2 | *38* |
| 2.8* | Beer (12 ounces; 354 ml) | Ethyl alcohol, 18 ml | 9110 | (?) | *23* |
| 4.7* | Wine (250 ml) | Ethyl alcohol, 30 ml | 9110 | (?) | *23* |
| 6.2 | Comfrey-pepsin tablets (nine daily) | Comfrey root, 2700 mg | 626 | (?) | *103* |
| 1.3 | Comfrey-pepsin tablets (nine daily) | Symphytine, 1.8 mg | 1.9 | (?) | |
| | FOOD ADDITIVES | | | | |
| 0.0002 | AF-2: daily dietary intake before banning | AF-2 (furylfuramide), 4.8 µg | 29 | (131) | *44* |
| 0.06* | Diet Cola (12 ounces; 354 ml) | Saccharin, 95 mg | 2143 | (−) | *106* |
| | DRUGS | | | | |
| [0.3] | Phenacetin pill (average dose) | Phenacetin, 300 mg | 1246 | (2137) | *51* |
| [5.6] | Metronidazole (therapeutic dose) | Metronidazole, 2000 mg | (542) | 506 | *107* |
| [14] | Isoniazid pill (prophylactic dose) | Isoniazid, 300 mg | (150) | 30 | *108* |
| 16* | Phenobarbital, one sleeping pill | Phenobarbital, 60 mg | (+) | 5.5 | *50* |
| 17* | Clofibrate (average daily dose) | Clofibrate, 2000 mg | 169 | (?) | *52* |
| | OCCUPATIONAL EXPOSURE | | | | |
| 5.8 | Formaldehyde: Workers' average daily intake | Formaldehyde, 6.1 mg | 1.5 | (44) | *109* |
| 140 | EDB: Workers' daily intake (high exposure) | Ethylene dibromide, 150 mg | 1.5 | (5.1) | *55* |

*Asterisks indicate HERP from carcinogens thought to be nongenotoxic.

amounts of their natural toxins when damaged by insects or fungi (12). For example, psoralens, light-activated carcinogens in celery, increase 100-fold when the plants are damaged by mold and, in fact, can cause an occupational disease in celery-pickers and in produce-checkers at supermarkets (12, 17).

Molds synthesize a wide variety of toxins, apparently as antibiotics in the microbiological struggle for survival: over 300 mycotoxins have been described (18). They are common pollutants of human food, particularly in the tropics. A considerable percentage of those tested have been shown to be mutagens and carcinogens: some, such as aflatoxin and sterigmatocystin, are among the most potent known rodent carcinogens. The potency of aflatoxin in different species varies widely; thus, a bias may exist as the HERP uses the most sensitive species. The aflatoxin content of U.S. peanut butter averages 2 ppb, which corresponds to a HERP of 0.03% for the peanut butter in an average sandwich (table 1). The Food and Drug Administration (FDA) allows ten times this level (HERP = 0.3%), and certain foods can often exceed the allowable limit (18). Aflatoxin contaminates wheat, corn (perhaps the main source of dietary aflatoxin in the United States), and nuts, as well as a wide variety of stored carbohydrate foodstuffs. A carcinogenic, though less potent, metabolite of aflatoxin is found in milk from cows that eat moldy grain.

There is epidemiologic evidence that aflatoxin is a human carcinogen. High intake in the tropics is associated with a high rate of liver cancer, at least among those chronically infected with the hepatitis B virus (19, 20). Considering the potency of those mold toxins that have been tested and the widespread contamination of food with molds, they may represent the most significant carcinogenic pollution of the food supply in developing countries. Such pollution is much less severe in industrialized countries, due to refrigeration and modern techniques of agriculture and storage, including use of synthetic pesticides and fumigants.

*Preparation of foods and beverages* can also produce carcinogens. Alcohol has been shown to be a human carcinogen in numerous epidemiologic studies (1, 21). Both alcohol and acetaldehyde, its major metabolite, are carcinogens in rats (22,23). The carcinogenic potency of ethyl alcohol in rats is remark-ably low (23), and it is among the weakest carcinogens in our database. However, human intake of alcohol is very high (about 18 g per beer), so that the possible hazards shown in table 1 for beer and wine are large (HERP = 2.8% for a daily beer). The possible hazard of alcohol is enormous relative to that from the intake of synthetic chemical residues. If alcohol (20), trichloroethylene, DDT, and other presumptive nongenotoxic carcinogens are active at high doses because they are tumor promoters, the risk from low doses may be minimal.

Other carcinogens are present in beverages and prepared foods. Urethane (ethyl carbamate), a particularly well-studied rodent carcinogen, is formed from ethyl alcohol and carbamyl phosphate during a variety of fermentations and is present in Japanese sake (HERP = 0.003%), many types of wine and beer, and in smaller amounts in yogurt and bread (24). Another fermentation product, the dicarbonyl aldehyde methylglyoxal, is a potent mutagen and was isolated as the main mutagen in coffee (about 250 µg in one cup). It was recently shown to be a carcinogen, though not in a test suitable for calculating a $TD_{50}$ (25). Methylglyoxal is also present in a variety of other foods, such as tomato puree (25, 26). Diacetyl (2,3-butanedione), a closely related dicarbonyl compound, is a fermentation product in wine and a number of other foods and is responsible for the aroma of butter. Diacetyl is a mutagen (27) but has not been tested for carcinogenicity.

Formaldehyde, another natural carcinogenic and mutagenic aldehyde, is also present in many common foods (22, 26–28). Formaldehyde gas caused cancer only in the nasal turbinates of the nose-breathing rodents and even though formaldehyde is genotoxic, the dose response was nonlinear (28, 29). Hexamethylenetetramine, which decomposes to formaldehyde in the stomach, was negative in feeding studies (30). The effects of oral versus inhalation exposure for formaldehyde remain to be evaluated more thoroughly.

As formaldehyde is almost ubiquitous in foods, one can visualize various formaldehyde-rich scenarios. Daily consumption of shrimp (HERP = 0.09% per 100 g) (31), a sandwich (HERP of two slices of bread = 0.4%) (22), a cola (HERP = 2.7%) (32), and a beer (HERP = 0.2%) (32) in various combinations could provide as much formaldehyde as living in some mobile homes (HERP = 2.1%; table 1). Formaldehyde is also generated in animals metabolically, for example, for methoxy compounds that humans ingest in considerable amounts from plants. The level of formaldehyde reported in normal human blood is strikingly high (about 100 $\mu M$ or 3000 ppb) (33) suggesting that detoxification mechanisms are important.

The cooking of food generates a variety of mutagens and carcinogens. Nine heterocyclic amines, isolated on the basis of their mutagenicity from proteins or amino acids that were heated in ways that occur in cooking, have now been tested; all have been shown to be potent carcinogens in rodents (34). Many others are still being isolated and characterized (34). An approximate HERP of 0.02% has been calcuated by Sugimura et al. for the daily intake of these nine carcinogens (34). Three mutagenic nitropyrenes present in diesel exhaust have now been shown to be carcinogens (35), but the intake of these carcinogenic nitropyrenes has been estimated to be much higher from grilled chicken than from air pollution (34, 36). The total amount of browned and burnt material eaten in a typical day is at least several hundred times more than that inhaled from severe air pollution (12).

Gas flames generate $NO_2$, which can form both the carcinogenic nitropyrenes (35, 36) and the potently carcinogenic nitrosamines in food cooked in gas ovens, such as fish or squid (HERP = 0.06%; table 1) (37). We suspect that food cooked in gas ovens may be a major source of dietary nitrosamines and nitropyrenes, though it is not clear how significant a risk these pose. Nitrosamines were ubiquitous in beer and ale (HERP = 0.008%) and were formed from $NO_2$ in the gas flame-heated air used to dry the malt. However, the industry has switched to indirect heating, which resulted in markedly lower levels (<1 ppb) of dimethylnitrosamine (38). The dimethylnitrosamine found in human urine is thought to be formed in part from $NO_2$ inhaled from kitchen air (39). Cooked bacon contains several nitrosamines (HERP = 0.009%) (40).

*Oxidation of fats and vegetable oils* occurs during cooking and also spontaneously if antioxidant levels are low. The result is the formation of peroxides, epoxides, and aldehydes, all of which appear to be rodent carcinogens (8, 12, 27). Fatty acid hydroperoxides (present in oxidized oils) and cholesterol

epoxide have been shown to be rodent carcinogens (though not in tests suitable for calculating a $TD_{50}$). Dried eggs contain about 25 ppm of cholesterol epoxide (a sizable amount), a result of the oxidation of cholesterol by the $NO_2$ in the drying air that is warmed by gas flames (12).

Normal oxidation reactions in fruit (such as browning in a cut apple) also involve production of peroxides. Hydrogen peroxide is a mutagenic rodent carcinogen that is generated by oxidation of natural phenolic compounds that are quite widespread in edible plants. A cup of coffee contains about 750 $\mu g$ of hydrogen peroxide (25); however, since hydrogen peroxide is a very weak carcinogen (similar in potency to alcohol), the HERP for drinking a daily cup of coffee would be very low [comparable to DDE/DDT, PCBs, or ethylene dibromide (EDB) dietary intakes]. Hydrogen peroxide is also generated in our normal metabolism; human blood contains about 5 $\mu M$ hydrogen peroxide and 0.3 $\mu M$ of the cholesterol ester of fatty acid hydroperoxide (41). Endogenous oxidants such as hydrogen peroxide may make a major contribution to cancer and aging (42).

*Caloric intake,* which could be considered the most striking rodent carcinogen ever discovered, is discussed remarkably little in relation to human cancer. It has been known for about 40 years that increasing the food intake in rats and mice by about 20% above optimal causes a remarkable decrease in longevity and a striking increase in endocrine and mammary tumors (43). In humans, obesity (associated with high caloric intake) leads to increased levels of circulating estrogens, a significant cause of endometrial and gall bladder cancer. The effects of moderate obesity on other types of human cancer are less clear (1).

*Food additives* are currently screened for carcinogenicity before use if they are synthetic compounds. AF-2 (HERP = 0.0002%), a food preservative, was banned in Japan (44). Saccharin (HERP = 0.06%) is currently used in the United States (the dose-response in rats, however, is clearly sublinear) (45). The possible hazard of diethylstilbestrol residues in meat from treated farm animals seems miniscule relative to endogenous estrogenic hormones and plant estrogens (46). Some natural carcinogens are also widely used as additives, such as allyl isothiocyanate (47), estragole (48), and alcohol (23).

*Air pollution.* A person inhales about 20,000 liters of air in a day; thus, even modest contamination of the atmosphere can result in inhalation of appreciable doses of a pollutant. This can be seen in the possible hazard in mobile homes from formaldehyde (HERP = 2.1%) or in conventional homes from formaldehyde (HERP = 0.6%) or benzene (HERP = 0.004%; table 1). Indoor air pollution is, in general, worse than outdoor air pollution, partly because of cigarette smoke. The most important indoor air pollutant may be radon gas. Radon is a natural radioactive gas that is present in the soil, gets trapped in houses, and gives rise to radioactive decay products that are known to be carcinogenic for humans (*49*). It has been estimated that in 1 million homes in the United States that level of exposure to products of radon decay may be higher than that received by today's uranium miners. Two particularly contaminated houses were found that had a risk estimated to be equivalent to receiving about 1200 chest x-rays a day (*49*). Approximately 10% of the lung cancer in the United States has been tentatively attributed to radon pollution in houses (*49*). Many of these cancers might be preventable since the most hazardous houses can be identified and modified to minimize radon contamination.

General outdoor air pollution appears to be a small risk relative to the pollution inhaled by a smoker: one must breathe Los Angeles smog for a year to inhale the same amount of burnt material that a smoker (two packs) inhales in a day (*12*), though air pollution is inhaled starting from birth. It is difficult to determine cancer risk from outdoor air pollution since epidemiologists must accurately control for smoking and radon.

*Some common drugs* shown in table 1 give fairly high HERP percentages, primarily because the dose ingested is high. However, since most medicinal drugs are used for only short periods while the HERP index is a daily dose rate for a lifetime, the possible hazard would usually be markedly less. We emphasize this in table 1 by bracketing the numbers for these shorter exposures. Phenobarbital (HERP = 16%) was investigated thoroughly in humans who had taken it for decades, and there was no convincing evidence that it caused cancer (*50*). There is evidence of increased renal cancer in long-term human ingestion of phenacetin, an analgesic (*51*). Acetaminophen, a metabolite of phenacetin, is one of the most widely used over-the-counter pain killers. Clofibrate (HERP = 17%) is used as a hypolipidemic agent and is thought to be carcinogenic in rodents because it induces hydrogen peroxide production through peroxisome proliferation (*52*).

*Occupational exposures* can be remarkably high, particularly for volatile carcinogens, because about 10,000 liters of air are inhaled in a working day. For formaldehyde, the exposure to an average worker (HERP = 5.8%) is higher than most dietary intakes. For a number of volatile industrial carcinogens, the ratio of the permitted exposure limit [U.S. Occupational Safety and Health Administration (OSHA)] in milligrams per kilogram to the $TD_{50}$ has been calculated; several are close to the $TD_{50}$ in rodents and about two-thirds have permitted HERP values >1% (*53*). The possible hazard estimated for the actual exposure levels of the most heavily exposed EDB workers is remarkably high, HERP = 140% (table 1). Though the dose may have been somewhat overestimated (*54*), it was still comparable to the dose causing cancer in half the rodents. An epidemiologic study of these heavily exposed EDB workers who inhaled EDB for over a decade did not show any increase in cancer, though because of the limited duration of exposure and the relatively small numbers of people monitored the study would not have detected a small effect (*54, 55*). OSHA still permits exposures above the $TD_{50}$ level. California, however, lowered the permitted level over 100-fold in 1981. In contrast with these heavy workplace exposures, the Environmental Protection Agency (EPA) has banned the use of EDB for fumigation because of the residue levels found in grain (HERP = 0.0004%).

## Uncertainties in Relying on Animal Cancer Tests for Human Prediction

*Species variation.* Though we list a possible hazard if a chemical is a carcinogen in a rat but not in a mouse (or vice versa), this lack of agreement raises the possibility that the risk to humans is nonexistent. Of 392 chemicals in our database tested in both rats and mice, 226 were carcinogens in at least one test, but 96 of these were positive in the mouse and negative in the rat or vice versa (*56*). This discordance occurs despite the fact that rats and mice are very closely related and have short life-spans. Qualitative extrapolation of cancer risks from rats or mice to humans, a very dissimilar long-lived species, is unlikely to be as reliable. Conversely, important human carcinogens may not be detected in standard tests in rodents; this was true for a long time for both tobacco smoke and alcohol, the two largest identified causes of neoplastic death in the United States.

For many of the chemicals considered rodent carcinogens, there may be negative as well as positive tests. It is difficult to deal with negative results satisfactorily for several reasons, including the fact that some chemicals are tested only once or twice, while others are tested many times. The HERP index ignores negative tests. Where there is species variation in potency, use of the more sensitive species, as is generally done and as is done here, could introduce a tendency to overestimate possible hazards; however, for most chemicals that are positive in both species, the potency is similar in rats and mice (57). The HERP may provide a rough correlate of human hazard from chemical exposure; however, for a given chemical, to the extent that the potency in humans differs from the potency in rodents, the relative hazard would be different.

*Quantitative uncertainties.* Quantitative extrapolation from rodents to humans, particularly at low doses, is guesswork that we have no way of validating (1, 5, 10, 11, 58). It is guesswork because of lack of knowledge in at least six major areas: (i) the basic mechanisms of carcinogenicity; (ii) the relation of cancer, aging, and life-span (1, 10, 42, 59); (iii) the timing and order of the steps in the carcinogenic process that are being accelerated; (iv) species differences in metabolism and pharmacokinetics; (v) species differences in anticarcinogens and other defenses (1, 60); and (vi) human heterogeneity—for example, pigmentation affects susceptibility to skin cancer from ultraviolet light. These sources of uncertainty are so numerous, and so substantial, that only empirical data will resolve them, and little of this is available.

*Uncertainties due to mechanism in multistage carcinogenesis.* Several steps (stages) are involved in chemical carcinogenesis, and the dose-response curve for a carcinogen might depend on the particular stage(s) it accelerates (58), with multiplicative effects if several stages are affected. This multiplicative effect is consistent with the observation in human cancer that synergistic effects are common. The three steps of carcinogenesis that have been analyzed in most detail are initiation (mutation), promotion, and progression, and we discuss these as an aid to understanding aspects of the dose-response relation.

Mutation (or DNA damage) as one stage of the carcinogenic process is supported by various lines of evidence: association of active forms of carcinogens with mutagens (61), the changes in DNA sequence of oncogenes (62), genetic predisposition to cancer in human diseases such as retinoblastoma (63) or DNA-repair deficiency diseases such as xeroderma pigmentosum (64). The idea that genotoxic carcinogens might show a linear dose-response might be plausible if only the mutation step of carcinogenesis was accelerated and if the induction of repair and defense enzymes were not significant factors (65).

Promotion, another step in carcinogenesis, appears to involve cell proliferation, or perhaps particular types of cell proliferation (66), and dose-response relations with apparent thresholds, as indicated by various lines of evidence: (i) The work of Trosko *et al.* (67) on promotion of carcinogenesis due to interference with cell-cell communication, causing cell proliferation. (ii) Rajewsky's and other work indicating initiation by some carcinogenic agents appears to require proliferating target cells (68). (iii) The work of Farber *et al.* (69) on liver carcinogenesis supports the idea that cell proliferation (caused by partial hepatectomy or cell killing) can be an important aspect of hepatocarcinogenesis. They have also shown for several chemicals that hepatic cell killing shows a toxic threshold with dose. (iv) Work on carcinogenesis in the pancreas, bladder and stomach (70), and other tissues (58) is also consistent with results on the liver (71, 72) though the effect of cell proliferation might be different in tissues that normally proliferate. (v) The work of Mirsalis *et al.* (71) suggests that a variety of nongenotoxic agents are hepatocarcinogens in the B6C3F1 mouse (commonly used in cancer tests) because of their toxicity. Other studies on chloroform and trichloroethylene also support this interpretation (72, 73). Cell proliferation resulting from the cell killing in the mouse liver shows a threshold with dose (71). Also relevant is the extraordinarily high spontaneous rates of liver tumors (21% carcinomas, 10% adenomas) in the male B6C3F1 mouse (74). These spontaneous tumors have a mutant *ras* oncogene, and thus the livers in these mice appear to be highly initiated (mutated) to start with (75). (vi) Oncogenes: As Weinberg (62) has pointed out, "Oncogene-bearing cells surrounded by normal neighbors do not grow into a large mass if they carry only a single oncogene. But if the normal neighbors are removed . . . by killing them with a cytotoxic drug . . . then a single oncogene often suffices." (vii) Cell killing, as well as mutation, appears to be an important aspect of radiation carcinogenesis (76).

Promotion has also been linked to the production of oxygen radicals, such as from phagocytic cells (77). Since chronic cell killing would usually involve imflammatory reactions caused by neutrophils, one would commonly expect chemicals tested at the maximally tolerated dose (MTD) to be promoters because of the chronic inflammation.

Progression, another step in carcinogenesis, leading to selection for invasiveness and metastases, is not well understood but can be accelerated by oxygen radicals (78).

Chronic cell toxicity caused by dosing at the MTD in rodent cancer bioassays thus not only could cause inflammation and cell proliferation, but also should be somewhat mutagenic and clastogenic to neighboring cells because of the release of oxygen radicals from phagocytosis (12, 79, 80). The respiratory burst from phagocytic neutrophils releases the same oxidative mutagens produced by radiation (77, 79). Thus, animal cancer tests done at the MTD of a chemical might commonly stimulate all three steps in carcinogenesis and be positive because the chemical caused chronic cell killing and inflammation with some mutagenesis. Some of the considerable human evidence for chronic inflammation contributing to carcinogenesis and also some evidence for and against a general effect of inflammation and cytotoxicity in rodent carcinogenesis have been discussed (81).

Another set of observations may also bear on the question of toxicity and extrapolation. Wilson, Crouch, and Zeise (82) have pointed out that among carcinogens one can predict the potency in high-dose animal cancer experiments from the toxicity (the $LD_{50}$) of the chemical, though one cannot predict whether the substance is a carcinogen. We have shown that carcinogenic potency values are bounded by the MTD (57). The evidence from our database suggests that the relationship between $TD_{50}$ and

MTD has a biological as well as a statistical basis (57). We postulate that a just sublethal level of a carcinogen causes cell death, which allows neighboring cells to proliferate, and also causes oxygen radical production from phagocytosis and thus chronic inflammation, both important aspects of the carcinogenic process (57). The generality of this relationship and its basis needs further study.

If most animal cancer tests done at the MTD are partially measuring cell killing and consequent cell proliferation and phagocytic oxygen radical damage as steps in the carcinogenic process, one might predict that the dose-response curves would generally be nonlinear. For those experiments in our database for which life table data (14) were available, a detailed analysis (83) shows that the dose-response relationships are more often consistent with a quadratic (or cubic) model than with a linear model.

Experimentally, it is very difficult to discriminate between the various extrapolation models at low doses (11, 58). However, evidence to support the idea that a nonlinear dose-response relationship is the norm is accumulating for many nongenotoxic and some genotoxic carcinogens. Dose-response curves for saccharin (45), butylated hydroxyanisole [BHA (84)], and a variety of other nongenotoxic carcinogens appear to be nonlinear (85). Formaldehyde, a genotoxic carcinogen, also has a nonlinear dose response (28, 29). The data for both bladder and liver tumors in the large-scale study on acetyl-aminofluorene, a genotoxic chemical, could fit a hockey stick–shaped curve, though a linear model, with a decreased effect at lower dose rates when the total dose is kept constant (86), has not been ruled out.

Carcinogens effective at both mutating and killing cells (which includes most mutagens) could be "complete" carcinogens and therefore possibly more worrisome at doses far below the MTD than carcinogens acting mainly by causing cell killing or proliferation (15). Thus, all carcinogens are not likely to be directly comparable, and a dose of 1/100 the $TD_{50}$ (HERP = 1%) might be much more of a carcinogenic hazard for the genotoxic carcinogens dimethylnitrosamine or aflatoxin than for the apparently nongenotoxic carcinogens trichloroethylene, PCBs, or alcohol (HERP values marked with asterisks in table 1). Short-term tests for mutagenicity (61, 87) can have a role to play, not only in understanding mechanisms, but also in getting a more realistic view of the background levels of potential genotoxic carcinogens in the world. Knowledge of mechanism of action and comparative metabolism in rodents and humans might help when estimating the relative importance of various low-dose exposures.

Human cancer, except in some occupational or medicinal drug exposures, is not from high (just subtoxic) exposures to a single chemical but is rather from several risk factors often combined with a lack of antirisk factors (60); for example, aflatoxin (a potent mutagen) combined with an agent causing cell proliferation, such as hepatitis B virus (19). High salt [a possible risk factor in stomach cancer (13)] and high fat [a possible risk factor in colon cancer (4)] both appear to be effective in causing cell killing and cell proliferation.

Risk from carcinogenesis is not linear with time. For example, among regular cigarette smokers the excess annual lung cancer incidence is approximately proportional to the fourth power of the duration of smoking (88). Thus, if human exposures in table 1 are much shorter than the lifetime exposure, the possible hazard may be markedly less than linearly proportional.

A key question about animal cancer tests and regulatory policy is the percentage of tested chemicals that will prove to be carcinogens (89). Among the 392 chemicals in our database that were tested in both rats and mice, 58% are positive in at least one species (14). For the 64 "natural" substances in the group, the proportion of positive results is similar (45%) to the proportion of positive results in the synthetic group (60%). One explanation offered for the high proportion of positive results is that more suspicious chemicals are being tested (for example, relatives of known carcinogens), but we do not know if the percentage of positives would be low among less suspicious chemicals. If toxicity is important in carcinogenicity, as we have argued, then at the MTD a high percentage of all chemicals might be classified as "carcinogens."

## The Background of Natural Carcinogens

The object of this article is not to do risk assessment on naturally occurring carcinogens or to worry people unduly about an occasional raw mushroom or beer, but to put the possible hazard of man-made carcinogens in proper perspective and to point out that we lack the knowledge to do low-dose "risk assessment." We also are almost completely ignorant of the carcinogenic potential of the enormous background of natural chemicals in the world. For example, cholinesterase inhibitors are a common class of pesticides, both man-made and natural. Solanine and chaconine (the main alkaloids in potatoes) are cholinesterase inhibitors and were introduced generally into the human diet about 400 years ago with the dissemination of the potato from the Andes. They can be detected in the blood of almost all people (12, 90). Total alkaloids are present at a level of 15,000 $\mu$g per 200-g potato with not a large safety factor (about sixfold) from the toxic level for humans (91). Neither alkaloid has been tested for carcinogenicity. By contrast, malathion, the main synthetic organophosphate cholinesterase inhibitor in our diet (17 $\mu$g/day) (16), is not a carcinogen in rodents.

The idea that nature is benign and that evolution has allowed us to cope perfectly with the toxic chemicals in the natural world is not compelling for several reasons: (i) there is no reason to think that natural selection should eliminate the hazard of carcinogenicity of a plant toxin that causes cancer in old age past the reproductive age, though there could be selection for resistance to the acute effects of particular carcinogens. For example, aflatoxin, a mold toxin that presumably arose early in evolution, causes cancer in trout, rats, mice, and monkeys, and probably people, though the species are not equally sensitive. Many of the common metal salts are carcinogens (such as lead, cadmium, beryllium, nickel, chromium, selenium, and arsenic) despite their presence during all of evolution. (ii) Given the enormous variety of plant toxins, most of our defenses may be general defenses against acute effects, such as shedding the surface lining of cells of our digestive and respiratory systems every day; protecting these surfaces with a mucin layer; having detoxifying enzymes that are often inducible, such as cytochrome P-450, conjugating enzymes, and glutathione transferases; and having DNA repair enzymes, which would be useful against a wide variety of ingested toxic chemicals, both natural and synthetic. Some human cancer may be caused by interfering with these normal protective systems. (iii) The human diet has changed drastically in the last few thousand years, and most of us are eating plants (such as coffee, potatoes, tomatoes, and kiwi fruit) that our ancestors did not. (iv) Normal metabolism produces radiomimetic mutagens and carcinogens, such as hydrogen peroxide and other reactive forms of oxygen. Though we have defenses against these agents, they still may be major contributors to aging and cancer. A wide variety of external agents may disturb this balance between damage and defense (12, 42).

## Implications for Decision-Making

For all of these considerations, our scale is not a scale of risks to humans but is only a way of setting priorities for concern, which should also take into account the numbers of people exposed. It should be emphasized that it is a linear scale and thus may overestimate low potential hazards if, as we argue above, linearity is not the normal case, or if nongenotoxic carcinogens are not of very much concern at doses much below the toxic dose.

Thus, it is not scientifically credible to use the results from rodent tests done at the MTD to directly estimate human risks at low doses. For example, an EPA "risk assessment" (92) based on a succession of worst case assumptions (several of which are unique to EDB) concluded that EDB residues in grain (HERP = 0.0004%) could cause 3 cases of cancer in 1000 people (about 1% of all U.S. cancer). A consequence was the banning of the main fumigant in the country. It would be more reasonable to compare the possible hazard of EDB residues to that of other common possible hazards. For example, the aflatoxin in the average peanut butter sandwich, or a raw mushroom, are 75 and 200 times, respectively, the possible hazard of EDB. Before banning EDB, a useful substance with rather low residue levels, it might be reasonable to consider whether the hazards of the alternatives, such as food irradiation, or the consequences of banning, such as increased mold contamination of grain, pose less risk to society. Also, there is a disparity between OSHA not regulating worker exposures at a HERP of 140%, while the EPA bans the substance at a HERP of 0.0004%. In addition, the FDA allows a possible hazard up to a HERP of 0.3% for peanut butter (20 ppb), and there is no warning about buying comfrey pills.

Because of the large background of low-level carcinogenic and other (93) hazards, and the high costs of regulation, priority setting is a critical first step. It is important not to divert society's attention away from the few really serious hazards, such as tobacco or saturated fat (for heart disease), by the pursuit of hundreds of minor or nonexistent hazards. Our knowledge is also more certain about the enormous toll of tobacco—about 350,000 deaths per year (1, 2).

There are many trade-offs to be made in all technologies. Trichloroethylene and tetrachloroethylene (perchloroethylene) replaced hazardous flammable solvents. Modern synthetic pesticides displaced lead arsenate, which was a major pesticide before the modern chemical era. Lead and arsenic are both natural carcinogens. There is also a choice to be made between using synthetic pesticides and raising the level of plants' natural toxins by breeding. It is not clear that the latter approach, even where feasible, is preferable. For example, plant breeders produced an insect-resistant potato, which has to be withdrawn from the market because of its acute toxicity to humans due to a high level of the natural plant toxins solanine and chaconine (12).

This analysis on the levels of synthetic pollutants in drinking water and of synthetic pesticide residues in foods suggests that this pollution is likely to be a minimal carcinogenic hazard relative to the background of natural carcinogens. This result is consistent with the epidemiologic evidence (1). Obviously prudence is desirable with regard to pollution, but we do need to work out some balance between chemophobia with its high costs to the national wealth, and sensible management of industrial chemicals (94).

Human life expectancy continues to lengthen in industrial countries, and the longest life expectancy in the world is in Japan, an extremely crowded and industrialized country. U.S. cancer death rates, except for lung cancer due to tobacco and melanoma due to ultraviolet light, are not on the whole increasing and have mostly been steady for 50 years. New progress in cancer research, molecular biology, epidemiology, and biochemical epidemiology (95) will probably continue to increase the understanding necessary for lengthening life-span and decreasing cancer death rates.

## References and Notes

1. R. Doll and R. Peto, *The Causes of Cancer* (Oxford Univ. Press, Oxford, England, 1981).

2. *Smoking and Health: A Report of the Surgeon General.* Department of Health, Education and Welfare

Publication No. (PHS) 79-50066 (Office of the Assistant Secretary for Health, Washington, DC, 1979).

3. G. J. Hopkins and K. K. Carroll, *J. Environ. Pathol. Toxicol. Oncol.* **5**, 279 (1985); J. V. Joossens, M. J. Hill, J. Geboers, Eds., *Diet and Human Carcinogenesis* (Elsevier, Amsterdam, 1985); I. Knudsen, Ed., *Genetic Toxicology of the Diet* (Liss, New York, 1986); Committee on Diet, Nutrition and Cancer, Assembly of Life Sciences, National Research Council, *Diet, Nutrition and Cancer* (National Academy Press, Washington, DC, 1982).

4. R. P. Bird, R. Schneider, D. Stamp, W. R. Bruce, *Carcinogenesis* **7**, 1657 (1986); H. L. Newmark *et al.*, in *Large Bowel Cancer*, vol. 3 in *Cancer Research Monographs*, A. J. Mastromarino and M. G. Brattain, Eds. (Praeger, New York, 1985), pp. 102–130; E. A. Jacobson, H. L. Newmark, E. Bright-See, G. McKeown-Eyssen, W. R. Bruce, *Nutr. Rep. Int.* **30**, 1049 (1984); M. Buset, M. Lipkin, S. Winawer, S. Swaroop, E. Friedman, *Cancer Res.* **46**, 5426 (1986).

5. D. G. Hoel, R. A. Merrill, F. P. Perera, Eds., *Banbury Report 19. Risk Quantitation and Regulatory Policy* (Cold Spring Laboratory, Cold Spring Harbor, NY, 1985).

6. B. E. Henderson *et al.*, *Cancer Res.* **42**, 3232 (1982).

7. R. Peto and H. zur Hausen, Eds., *Banbury Report 21. Viral Etiology of Cervical Cancer* (Cold Spring Harbor Laboratory, Cold Spring Harbor, NY, 1986); F.-S. Yeh *et al.*, *Cancer Res.* **45**, 872 (1985).

8. International Agency for Research on Cancer, *IARC Monographs on the Evaluation of the Carcinogenic Risk of Chemicals to Humans* (International Agency for Research on Cancer, Lyon, France, 1985), vol. 39.

9. D. A. Freedman and H. Zeisel, *From Mouse to Man: The Quantitative Assessment of Cancer Risks* (Tech. Rep. No. 79, Department of Statistics, University of California, Berkeley, 1987).

10. R. Peto, in *Assessment of Risk from Low-Level Exposure to Radiation and Chemicals*, A. D. Woodhead, C. J. Shellabarger, V. Pond, A. Hollaender, Eds. (Plenum, New York and London, 1985), pp. 3–16.

11. S. W. Samuels and R. H. Adamson, *J. Natl. Cancer Inst.* **74**, 945 (1985); E. J. Calabrese, *Drug Metab. Rev.* **15**, 505 (1984).

12. B. N. Ames, *Science* **221**, 1256 (1983); *ibid.* **224**, 668, 757 (1984).

13. H. Ohgaki *et al.*, *Gann* **75**, 1053 (1984); S. S. Mirvish, *J. Natl. Cancer Inst.* **71**, 630 (1983); J. V. Joossens and J. Geboers, in *Frontiers in Gastrointestinal Cancer*, B. Levin and R. H. Riddell, Eds. (Elsevier, Amsterdam, 1984), pp. 167–183; T. Hirayama, *Jpn. J. Clin. Oncol.* **14**, 159 (1984); C. Furihata *et al.*, *Biochem. Biophys. Res. Commun.* **121**, 1027 (1984).

14. R. Peto, M. C. Pike, L. Bernstein, L. S. Gold, B. N. Ames, *Environ. Health Perspect.* **58**, 1 (1984); L. S. Gold *et al.*, *ibid.*, p. 9; L. S. Gold *et al.*, *ibid.* **67**, 161 (1986); L. S. Gold *et al.*, *ibid.*, in press.

15. G. M. Williams and J. H. Weisburger, in *Casarett and Doull's Toxicology. The Basic Science of Poisons*, C. D. Klaassen, M. O. Amdur, J. Doull, Eds. (Macmillan, New York, ed. 3, 1986), chap. 5, pp. 99–172; B. E. Butterworth and T. J. Slaga, Eds., *Banbury Report 25. Non-Genotoxic Mechanisms in Carcinogenesis* (Cold Spring Harbor Laboratory, Cold Spring Harbor, NY, 1987).

16. The FDA has estimated the average U.S. dietary intake of 70 pesticides, herbicides, and industrial chemicals for 1981/1982 [M. J. Gartrell, J. C. Craun, D. S. Podrebarac, E. L. Gunderson, *J. Assoc. Off. Anal. Chem.* **69**, 146 (1986)]. The negative test on 2-ethylhexyl diphenyl phosphate is in J. Treon, F. Dutra, F. Cleveland, *Arch. Ind. Hyg. Occup. Med.* **8**, 170 (1953).

17. R. C. Beier *et al.*, *Food Chem. Toxicol.* **21**, 163 (1983).

18. L. Stoloff, M. Castegnaro, P. Scott, I. K. O'Neill, H. Bartsch, Eds., *Some Mycotoxins*, vol. 5 in *Environmental Carcinogens. Selected Methods of Analysis* (IARC Scientific Publ. No. 44, International Agency for Research on Cancer, Lyon, France, 1982); H. Mori *et al.*, *Cancer Res.* **44**, 2918 (1984); R. Röschenthaler, E. E. Creppy, G. Dirheimer, *J. Toxicol.-Toxin Rev.* **3**, 53 (1984); W. F. O. Marasas, N. P. J. Kriek, J. E. Fincham, S. J. van Rensburg, *Int. J. Cancer* **34**, 383 (1984); *Environmental Health Criteria 11: Mycotoxins* (World Health Organization, Geneva, Switzerland, 1979), pp. 21–85; W. F. Busby *et al.*, in *Chemical Carcinogens*, C. E. Searle, Ed. (ACS Monograph 182, American Chemical Society, Washington, DC, ed. 2, 1984), vol. 2, pp. 944–1136.

19. S. J. Van Rensburg *et al.*, *Br. J. Cancer* **51**, 713 (1985); S. N. Zaman *et al.*, *Lancet* **1985-I**, 1357 (1985); H. Austin *et al.*, *Cancer Res.* **46**, 962 (1986).

20. A. Takada, J. Nei, S. Takase, Y. Matsuda, *Hepatology* **6**, 65 (1986).

21. J. M. Elwood *et al.*, *Int. J. Cancer* **34**, 603 (1984).

22. Aldehydes and ketones are largely responsible for the aroma and flavor of bread [Y. Y. Linko, J. A. Johnson, B. S. Miller, *Cereal Chemistry* **39**, 468 (1962)]. In freshly baked bread, formaldehyde (370 µg per two slices of bread) accounts for 2.5% of the total carbonyl compounds [K. Lorenz and J. Maga, *J. Agric. Food Chem.* **20**, 211 (1972)]. Acetaldehyde, which is present in bread at about twice the level of formaldehyde, is a carcinogen in rats [R. A. Woutersen, L. M. Appelman, V. J. Feron, C. A. Vanderheijden, *Toxicology* **31**, 123 (1984)] and a DNA crosslinking agent in human cells [B. Lambert, Y. Chen, S.-M. He, M. Sten, *Mutat. Res.* **146**, 301 (1985)].

23. Ethyl alcohol contents of wine and beer were assumed to be 12% and 5%, respectively. The $TD_{50}$ calculation is based on M. J. Radike, K. L. Stemmer, E. Bingham, *Environ. Health Perspect.* **41**, 59 (1981). Rats exposed to 5% ethyl alcohol in drinking water for 30 months had increased incidences of endocrine and liver tumors.

24. C. S. Ough, *J. Agric. Food Chem.* **24**, 323 (1976). Urethane is also carcinogenic in hamsters and rhesus monkeys.

25. Y. Fujita, K. Wakabayashi, M. Nagao, T. Sugimura, *Mutat. Res.* **144**, 227 (1985); M. Nagao, Y. Fujita, T. Sugimura, in *IARC Workshop,* in press.

26. M. Petro-Turza and I. Szarfoldi-Szalma, *Acta Alimentaria* **11**, 75 (1982).

27. L. J. Marnett *et al., Mutat. Res.* **148**, 25 (1985).

28. Formaldehyde in air samples taken from all the mobile homes examined ranged from 50 to 660 ppb (mean, 167 ppb) [T. H. Connor, J. C. Theiss, H. A. Hanna, D. K. Monteith, T. S. Matney, *Toxicol. Lett.* **25**, 33 (1985)]. The important role of cell toxicity and cell proliferation in formaldehyde carcinogenesis is discussed in T. B. Starr and J. E. Gibson [*Annu. Rev. Pharmacol. Toxicol.* **25**, 745 (1985)].

29. J. A. Swenberg *et al., Carcinogenesis* **4**, 945 (1983).

30. G. Della Porta, M. I. Colnaghi, G. Parmiani, *Food Cosmet. Toxicol.* **6**, 707 (1968).

31. Formaldehyde develops postmortem in marine fish and crustaceans, probably through the metabolism of trimethylamine oxide. The average level found in shrimp from four U.S. markets was 94 mg/kg [T. Radford and D. E. Dalsis, *J. Agric. Food Chem.* **30**, 600 (1982)]. Formaldehyde is found in remarkably high concentrations (300 ppm, HERP = 29% per 100 g) in Japanese shrimp that have been bleached with a sulfite solution [A. Yoshida and M. Imaida, *J. Food Hygienic Soc. Japan* **21**, 288 (1980)].

32. J. F. Lawrence and J. R. Iyengar, *Int. J. Environ. Anal. Chem.* **15**, 47 (1983).

33. H. d'A. Heck *et al., Am. Ind. Hyg. Assoc. J.* **46**, 1 (1985).

34. T. Sugimura *et al.,* in *Genetic Toxicology of the Diet,* I. Knudsen, Ed. (Liss, New York, 1986), pp. 85–107; T. Sugimura, *Science* **233**, 312 (1986).

35. H. Ohgaki *et al., Cancer Lett.* **25**, 239 (1985).

36. T. Kinouchi, H. Tsutsui, Y. Ohnishi, *Mutat. Res.* **171**, 105 (1986).

37. T. Kawabata *et al.,* in *N-Nitroso Compounds: Analysis, Formation and Occurrence,* E. A. Walker, L. Griciute, M. Castegnaro, M. Borzsonyi, Eds. (IARC Scientific Publ. No. 31, International Agency for Research on Cancer, Lyon, France, 1980), pp. 481–490; T. Maki, Y. Tamura, Y. Shimamura, and Y. Naoi [*Bull. Environ. Contam. Toxicol.* **25**, 257 (1980)] have surveyed Japanese food for nitrosamines.

38. T. Fazio, D. C. Havery, J. W. Howard, in *N-Nitroso Compounds: Analysis, Formation and Occurrence,* E. A. Walker, L. Griciute, M. Castegnaro, M. Borzsonyi, Eds. (IARC Scientific Publ. No. 31, International Agency for Research on Cancer, Lyon, France, 1980), pp. 419–435; R. Preussmann and G. Eisenbrand, in *Chemical Carcinogenesis,* C. E. Searle, Ed. (ACS Monograph 182, American Chemical Society, Washington, DC, ed. 2, 1984), vol. 2, pp. 829–868; D. C. Havery, J. H. Hotchkiss, T. Fazio, *J. Food Sci.* **46**, 501 (1981).

39. W. A. Garland *et al., Cancer Res.* **46**, 5392 (1986).

40. E. A. Walker, L. Griciute, M. Castegnaro, M. Borzsonyi, Eds., *N-Nitroso Compounds: Analysis, Formation and Occurrence* (IARC Scientific Publ. No. 31, International Agency for Research on Cancer, Lyon, France, 1980), pp. 457–463; B. Spiegelhalder, G. Eisenbrand, R. Preussmann, *Oncology* **37**, 211 (1980); R. A. Scanlan and S. R. Tannenbaum, Eds., *N-Nitroso Compounds* (ACS Symposium Series No. 174, American Chemical Society, Washington, DC, 1981), pp. 165–180. Nitrosamines are formed in cured meats through reactions of secondary amines with nitrites added during the manufacturing process. One survey of bacon commercially available in Canada identified *N*-nitrosodimethylamine (DMN), *N*-nitrosodiethylamine (DEN), and *N*-nitrosopyrrolidine (NPYR) in most samples tested, with average levels of 3.4, 1.0, and 9.3 ppb, respectively. The cooked-out fat from the bacon samples contained DMN and NPYR at average levels of 6.4 and 21.9 ppb, respectively [N. P. Sen, S. Seaman, W. F. Miles, *J. Agric. Food Chem.* **27**, 1354 (1979); R. A. Scanlan, *Cancer Res.* **43**, 2435s (1983)]. The average levels of NPYR in cooked bacon have decreased since 1971 because of reduced levels of nitrite and increased levels of ascorbate used in bacon curing mixtures [D. C. Havery, T. Fazio, J. W. Howard, *J. Assoc. Off. Anal. Chem.* **61**, 1379 (1978)].

41. Y. Yamamoto *et al., Anal. Biochem.* **160**, 7 (1987).

42. B. N. Ames and R. L. Saul, in *Theories of Carcinogenesis,* O. H. Iversen, Ed. (Hemisphere, New York, in press); R. Cathcart, E. Schwiers, R. L. Saul, B. N. Ames, *Proc. Natl. Acad. Sci. U.S.A.* **81**, 5633 (1984).

43. B. P. Yu, E. J. Masoro, I. Murata, H. A. Bertrand, F. T. Lynd, *J. Gerontol.* **37**, 130 (1982); F. J. C. Roe, *Proc. Nutr. Soc.* **40**, 57 (1981); *Nature (London)* **303**, 657 (1983); M. J. Tucker, *Int. J. Cancer* **23**, 803 (1979).

44. Y. Tazima, *Environ. Health Perspect.* **29**, 183 (1979); M. Kinebuchi, T. Kawachi, N. Matsukura, T. Sugimura, *Food Cosmet. Toxicol.* **17**, 339 (1979).

45. F. W. Carlborg, *Food Chem. Toxicol.* **23**, 499 (1985).

46. T. H. Jukes, *Am. Stat.* **36**, 273 (1982); *J. Am. Med. Assoc.* **229**, 1920 (1974).

47. Allyl isothiocyanate (AITC) is the major flavor ingredient, and natural pesticide, of brown mustard and also occurs naturally in varying concentrations in cabbage, kale, broccoli, cauliflower, and horseradish [Y. M. Ioannou, L. T. Burka, H. B. Matthews, *Toxicol. Appl. Pharmacol.* 75, 173 (1984)]. It is present in the plant's volatile oil as the glucoside sinigrin. (The primary flavor ingredient of yellow mustard is *p*-hydroxybenzyl isothiocyanate.) The AITC yield from brown mustard is approximately 0.9% by weight, assuming all of the sinigrin is converted to AITC [A. Y. Leung, *Encyclopedia of Common Natural Ingredients Used in Food, Drugs and Cosmetics* (Wiley, New York, 1980), pp. 238–241]. Synthetic AITC is used in nonalcoholic beverages, candy, baked goods, meats, condiments, and syrups at average levels ranging from 0.02 to 88 ppm [T. E. Furia and B. Nicolo, Eds., *Fenaroli's Handbook of Flavor Ingredients* (CRC Press, Cleveland, OH, 2 ed., 1975), vol. 1, p. 19].

48. Estragole, one of numerous safrole-like compounds in plants, is present in the volatile oils of many edible plants, including basil, tarragon, bay, anise, and fennel, as well as in pine oil and turpentine [A. Y. Leung, *Encyclopedia of Common Natural Ingredients Used in Food, Drugs and Cosmetics* (Wiley, New York, 1980)]. Dried basil has a volatile oil content of about 1.5 to 3.0%, which contains (on average) 25% estragole [H. B. Heath, *Source Book of Flavors* (AVI, Westport, CT, 1981), pp. 222–223]. Estragole is used commercially in spice, anise, licorice, and fruit flavors. It is added to beverages, candy, baked goods, chewing gums, ice creams, and condiments at average levels ranging from 2 to 150 ppm [NAS/NRC Food Protection Committee, Food and Nutrition Board, *Chemicals Used in Food Processing* (NAS/NRC Publ. No. 1274, National Academy of Sciences, Washington, DC, 1965), p. 114].

49. The estimation of risk is from human data on uranium miners and estimates of intake. E. P. Radford, *Environ. Health Perspect.* 62, 281 (1985); A. V. Nero *et al.*, *Science* 234, 992 (1986); A. V. Nero, *Technol. Rev.* 89, 28 (1986); R. Hanley, *The New York Times*, 10 March 1986, p. 17.

50. The average daily adult dose of phenobarbital for sleep induction is 100 to 320 mg (HERP = 26 to 83%), though its use is declining [AMA Division of Drugs, *AMA Drug Evaluations* (American Medical Association, Chicago, IL, ed. 5, 1983), pp. 201–202]. The TD$_{50}$ data in the table is for phenobarbital, which, so far, has been shown to be carcinogenic only in mice; the sodium salt of phenobarbital is carcinogenic in both rats and mice. Human studies on phenobarbital and cancer are reviewed in A. E. M. McLean, H. E. Driver, D. Lowe, I. Sutherland, *Toxicol. Lett.* 31 (suppl.), 200 (1986).

51. Phenacetin use has gradually decreased following reports of urinary bladder and kidney tumors in heavy users [J. M. Piper, J. Tonascia, G. M. Matanoski, *N. Engl. J. Med.* 313, 292 (1985)]. Phenacetin also induces urinary bladder and kidney tumors in rats and mice.

52. The human dose of clofibrate is 2 g per day for many years [R. J. Havel and J. P. Kane, *Annu. Rev. Med.* 33, 417 (1982)]. The role of clofibrate as a peroxisome proliferator is reviewed in J. K. Reddy and N. D. Lalwani [*CRC Crit. Rev. Toxicol.* 12, 1 (1983)]. An epidemiologic study is in World Health Organization Report, *Lancet* 1984-II, 600 (1984).

53. L. S. Gold, G. Backman, N. K. Hooper, R. Peto, *Lawrence Berkeley Laboratory Report 23161* (1987); N. K. Hooper and L. S. Gold, in *Monitoring of Occupational Genotoxicants*, M. Sorsa and H. Norppa, Eds. (Liss, New York, 1986), pp. 217–228; K. Hooper and L. S. Gold, in *Cancer Prevention: Strategies in the Workplace*, C. Becker, Ed. (Hemisphere, Washington, DC, 1985), pp. 1–11.

54. California Department of Health Services, *EDB Criteria Document* (1985).

55. M. G. Ott, H. C. Scharnweber, R. R. Langner, *Br. J. Ind. Med.* 37, 163 (1980); J. C. Ramsey, C. N. Park, M. G. Ott, P. J. Gehring, *Toxicol. Appl. Pharmacol.* 47, 411 (1978). This has been disputed (*54*). The carcinogen dose reported in the table assumes a time-weighted average air concentration of 3 ppm and an 8-hour workday 5 days per week for 50 weeks per year for life.

56. R. Magaw, L. S. Gold, L. Bernstein, T. H. Slone, B. N. Ames, in preparation.

57. L. Bernstein, L. S. Gold, B. N. Ames, M. C. Pike, D. G. Hoel, *Fundam. Appl. Toxicol.* 5, 79 (1985); L. Bernstein, L. S. Gold, B. N. Ames, M. C. Pike, D. G. Hoel, *Risk Anal.* 5, 263 (1985).

58. D. B. Clayson, *Toxicol. Pathol.* 13, 119 (1985); D. B. Clayson, *Mutat. Res.*, in press.

59. R. Peto, S. E. Parish, R. G. Gray, in *Age-Related Factors in Carcinogenesis*. A. Likhachev, V. Anisimov, R. Montesano, Eds. (IARC Scientific Publ. No. 58, International Agency for Research on Cancer, Lyon, France, 1985), pp. 43–53.

60. D. M. Shankel, P. Hartman, T. Kada, A. Hollaender, Eds., *Antimutagenesis and Anticarcinogenesis: Mechanisms* (Plenum, New York, 1986).

61. B. N. Ames and J. McCann, *Cancer Res.* 41, 4192 (1981).

62. R. A. Weinberg, *Science* 230, 770 (1985).

63. A. G. Knudson, Jr., *Cancer Res.* 45, 1437 (1985).

64. J. E. Cleaver, in *Genes and Cancer*, J. M. Bishop, J. D. Rowley, M. Greaves, Eds. (Liss, New York, 1984), pp. 117–135.

65. A. D. Woodhead, C. J. Shellabarger, V. Pond, A. Hollaender, Eds., *Assessment of Risk from Low-Level*

*Exposure to Radiation and Chemicals: A Critical Overview* (Plenum, New York, 1985).

66. J. Cairns, *Nature (London)* **255**, 197 (1975); C. C. Harris and T. Sun, *Carcinogenesis* **5**, 697 (1984); A. M. Edwards and C. M. Lucas, *Biochem. Biophys. Res. Commun.* **131**, 103 (1985); H. Tsuda *et al.*, *Cancer Res.* **39**, 4491 (1979); W. H. Haese and E. Bueding, *J. Pharmacol. Exp. Ther.* **197**, 703 (1976).

67. J. E. Trosko and C. C. Chang, in *Methods for Estimating Risk of Chemical Injury: Human and Non-Human Biota and Ecosystems*, V. B. Vouk, G. C. Butler, D. G. Hoel, D. B. Peakall, Eds. (Wiley, New York, 1985), pp. 181–200; J. E. Trosko and C. C. Chang, in *Assessment of Risk from Low-Level Exposure to Radiation and Chemicals: A Critical Overview*, A. D. Woodhead, C. J. Shellabarger, V. Pond, A. Hollaender, Eds. (Plenum, New York, 1985), pp. 261–284; H. Yamasaki, *Toxicol. Pathol.* **14**, 363 (1986).

68. M. F. Rajewsky, in *Age-Related Factors in Carcinogenesis*, A. Likhachev, V. Anisimov, R. Montesano, Eds. (IARC Scientific Publ. No. 58, International Agency for Research on Cancer, Lyon, France, 1985), pp. 215–224; V. Kinsel, G. Furstenberger, H. Loehrke, F. Marks, *Carcinogenesis* **7**, 779 (1986).

69. E. Farber, *Cancer Res.* **44**, 5463 (1984); E. Farber, S. Parker, M. Gruenstein, *ibid.* **36**, 3879 (1976).

70. A. Denda, S. Inui, M. Sunagawa, S. Takahashi, Y. Konishi, *Gann* **69**, 633 (1978); R. Hasegawa and S. M. Cohen, *Cancer Lett.* **30**, 261 (1986); R. Hasegawa, S. M. Cohen, M. St. John, M. Cano, L. B. Ellwein, *Carcinogenesis* **7**, 633 (1986); B. I. Ghanayem, R. R. Maronpot, H. B. Matthews, *Toxicology* **6**, 189 (1986).

71. J. C. Mirsalis *et al.*, *Carcinogenesis* **6**, 1521 (1985); J. C. Mirsalis *et al.*, *Environ. Mutag.* **8** (suppl. 6), 55 (1986); J. Mirsalis *et al.*, Abstract for Fourth International Conference on Environmental Mutagens, held 24–28 June in Stockholm, Sweden (1985).

72. W. T. Stott, R. H. Reitz, A. M. Schumann, P. G. Watanabe, *Food Cosmet. Toxicol.* **19**, 567 (1981).

73. D. H. Moore, L. F. Chasseaud, S. K. Majeed, D. E. Prentice, F. J. C. Roe, *ibid.* **20**, 951 (1982).

74. J. K. Haseman, J. Huff, G. A. Boorman, *Toxicol. Pathol.* **12**, 126 (1984); R. E. Tarone, K. C. Chu, J. M. Ward, *J. Natl. Cancer Inst.* **66**, 1175 (1981).

75. S. H. Reynolds, S. J. Stowers, R. R. Maronpot, M. W. Anderson, S. A. Aaronson, *Proc. Natl. Acad. Sci. U.S.A.* **83**, 33 (1986); T. R. Fox and P. G. Watanabe, *Science* **228**, 596 (1985).

76. T. D. Jones, *Health Phys.* **4**, 533 (1984); J. B. Little, A. R. Kennedy, R. B. McGandy, *Radiat. Res.* **103**, 293 (1985).

77. T. W. Kensler and B. G. Taffe, *Adv. Free Radical Biol. Med.* **2**, 347 (1986); P. A. Cerutti, in UCLA *Symposium on Molecular and Biology Growth Factors, Tumor Promoters and Cancer Genes*, in press; P. A. Cerutti, in *Biochemical and Molecular Epidemiology of Cancer*, vol. 40 of UCLA Symposium on Molecular and Cellular Biology, C. Harris, Ed. (Liss, New York, 1986), p. 167; in *Theories of Carcinogenesis*, O. H. Iversen, Ed. (Hemisphere, New York, in press); H. C. Birnboim *Carcinogenesis* **7**, 1511 (1986); K. Frenkel and K. Chrzan, *ibid.* **8**, 455 (1987).

78. J. Rotstein, J. O. O'Connell, T. Slaga, *Proc. Assoc. Cancer Res.* **27**, 143 (1986); J. S. O'Connell, A. J. P. Klein-Szanto, J. DiGiovanni, J. W. Fries, T. J. Slaga, *Cancer Res.* **46**, 2863 (1986); J. S. O'Connell, J. B. Rotstein, T. J. Slaga, in *Banbury Report 25. Non-Genotoxic Mechanisms in Carcinogenesis*, B. E. Butterworth and T. J. Slaga, Eds. (Cold Spring Harbor Laboratory, Cold Spring Harbor, NY, 1987).

79. M. A. Trush, J. L. Seed, T. W. Kensler, *Proc. Natl. Acad. Sci. U.S.A.* **82**, 5194 (1985); A. I. Tauber and B. M. Babior, *Adv. Free-Radical Biol. Med.* **1**, 265 (1985); G. J. Chellman, J. S. Bus, P. K. Working, *Proc. Natl. Acad. Sci. U.S.A.* **83**, 8087 (1986).

80. I. U. Schraufstatter *et al.*, *Proc. Natl. Acad. Sci. U.S.A.* **83**, 4908 (1986); M. O. Bradley, in *Basic and Applied Mutagenesis*, A. Muhammed and R. C. von Borstel, Eds. (Plenum, New York, 1985), pp. 99–109.

81. L. Diamond, T. G. O'Brien, W. M. Baird, *Adv. Cancer Res.* **32**, 1 (1980); D. Schmahl, *J. Cancer Res. Clin. Oncol.* **109**, 260 (1985); O. H. Iversen and E. G. Astrup, *Cancer Invest.* **2**, 51 (1984); A Hagiwara and J. M. Ward, *Fundam. Appl. Toxicol.*, **7**, 376 (1986); J. M. Ward, in *Carcinogenesis and Mutagenesis Testing*, J. F. Douglas, Ed. (Humana, Clifton, NJ, 1984), pp. 97–100.

82. L. Zeise, R. Wilson, E. Crouch, *Risk Analysis* **4**, 187 (1984); L. Zeise, E. A. C. Crouch, R. Wilson, *ibid.* **5**, 265 (1985); L. Zeise, E. A. C. Crouch, R. Wilson, *J. Am. College Toxicol.* **5**, 137 (1986).

83. D. Hoel, personal communication.

84. N. Ito, S. Fukushima, A. Hagiwara, M. Shibata, T. Ogiso, *J. Natl. Cancer Inst.* **70**, 343 (1983).

85. F. W. Carlborg, *Food Chem. Toxic.* **20**, 219 (1982); *Food Cosmet. Toxicol.* **19**, 255 (1981).

86. K. G. Brown and D. G. Hoel, *Fundam. Appl. Toxicol.* **3**, 470 (1983); N. A. Littlefield and D. W. Gaylor, *J. Toxicol. Environ. Health* **15**, 545 (1985).

87. J. Ashby, *Mutagenesis* **1**, 3 (1986).

88. R. Doll, *Cancer Res.* **38**, 3573 (1978); —— and R. Peto, *J. Epidemiol. Community Health* **32**, 303 (1978).

89. J. E. Huff, E. E. McConnell, J. K. Haseman, *Environ. Mutagenesis* **7**, 427 (1985); H. S. Rosenkranz, *ibid.*, p. 428.

90. M. H. Harvey, B. A. Morris, M. McMillan, V. Marks, *Human Toxicol.* **4**, 503 (1985).

91. S. J. Jadhav, R. P. Sharma, D. K. Salunkhe, *CRC Crit. Rev. Toxicol.* **9**, 21 (1981).

92. Environmental Protection Agency, *Position Document 4* (Special Pesticide Review Division, Environmental Protection Agency, Arlington, VA, 1983).

93. R. Wilson and E. Crouch, *Risk/Benefit Analysis* (Ballinger, Cambridge, MA, 1982); W. F. Allman, *Science 85* **6**, 30 (1985).

94. P. Huber, *Regulation,* 33 (March/April 1984); C. Whipple, *ibid.* 9, 37 (1985).

95. B. A. Bridges, B. E. Butterworth, I. B. Weinstein, Eds., *Banbury Report 13. Indicators of Genotoxic Exposure.* (Cold Spring Harbor Laboratory, Cold Spring Harbor, NY, 1982); P. E. Enterline, Ed., Fifth Annual Symposium on Environmental Epidemiology, *Environ. Health Perspect.* **62**, 239 (1985).

96. A national survey of U.S. drinking water supplies identified the concentrations of about 20 organic compounds. The mean total trihalomethane concentration was 117 μg/liter, with the major component, chloroform, present at a mean concentration of 83 μg/liter (83 ppb). Raw water that is relatively free of organic matter results in drinking water relatively free of trihalomethanes after chlorination. These studies are reviewed in S. J. Williamson, *The Science of the Total Environment* 18, 187 (1981).

97. Public and private drinking water wells in Santa Clara Valley, California, have been found to be contaminated with a variety of halogenated hydrocarbons in small amounts. Among 19 public water system wells, the most commonly found contaminants were 1,1,1-trichloroethane (TCA), and 1,1,2-trichloro-1,2,2-trifluoroethane (Freon-113). TCA was found in 15 wells generally at concentrations of less than 30 ppb, though one well contained up to 8800 ppb, and Freon-113 was found in six wells at concentrations up to 12 ppb. Neither chemical has been adequately tested for carcinogenicity in long-term bioassays. In addition to these compounds, three wells also contained carcinogenic compounds at low concentrations. Water from public supply wells may be mixed with treated surface water before delivery, thus the concentrations of these compounds that people actually receive may be somewhat reduced. Thirty-five private drinking water supply wells were examined; the major contaminant was the carcinogen trichloroethylene (TCE), at levels up to 2800 ppb. TCA and Freon-113 were also found in some wells, at maximum levels of 24 ppb and 40 ppb, respectively. Though fewer people drink from private water wells, the contaminant concentrations may be higher because the water is not mixed with water from other sources [California Department of Health Services, California Regional Water Quality Control Board 2, Santa Clara County Public Health Department, Santa Clara Valley Water District, U.S. Environmental Protection Agency. *Ground Water and Drinking Water in the Santa Clara Valley: A White Paper*

(1984), table 8]. Trichloroethylene may not be a carcinogen in humans at low doses [R. D. Kimbrough, F. L. Mitchell, V. N. Houk, *J. Toxicol. Environ. Health* 15, 369 (1985)].

98. Contaminated drinking water in the area of Woburn, Massachusetts, was found to contain 267 ppb trichloroethylene, 21 ppb tetrachloroethylene, 12 ppb chloroform, 22 ppb trichlorotrifluoroethane, and 28 ppb 1,2-*trans*-dichloroethylene [S. W. Lagakos, B. J. Wessen, M. Zelen, *J. Am. Stat. Assoc.* **81**, 583 (1986)].

99. The amount of chloroform absorbed by a 6-year-old child in a chlorinated freshwater swimming pool has been estimated [J. A. Beech, *Med. Hypotheses* **6**, 303 (1980)]. Table 1 refers to the chloroform in an average pool (134 μg/liter) and for a 37-kg child. Three other trihalomethanes were identified in these freshwater pools: bromoform, bromodichloromethane and chlorodibromomethane. U. Lahl, J. Vondusze, B. Gabel, B. Stachel, W. Thiemann [*Water Res.* **15**, 803 (1981)] have estimated absorption in covered swimming pools.

100. J. McCann, L. Horn, J. Girman, A. V. Nero, in *Short-Term Bioassays in the Analysis of Complex Environmental Mixtures,* V. S. Sandhu, D. M. DeMarini, M. J. Mass, M. M. Moore, J. L. Mumford, Eds. (Plenum, New York, in press). This estimate (Table 1) for formaldehyde in conventional homes, excludes foam-insulated houses and mobile homes. The figure is a mean of the median or mean of the reported samples in each paper. For benzene, the figure is a mean of all reported median or mean samples. The level of benzene in Los Angeles outdoor air is similar (U.S. EPA Office of Air Quality Planning and Standards, EPA 450/4-86-012, 1986).

101. The average adult daily PCB intake from food estimated by the FDA in fiscal years 1981/1982 was 0.2 μg/day (*16*). Many slightly different PCB mixtures have been studied in long-term animal cancer bioassays; the calculation of $TD_{50}$ was from a test of Aroclor 1260 which was more potent than other PCBs (*14*).

102. The average consumption of EDB residues in grains has been estimated by the EPA for adults as 0.006 μg kg$^{-1}$ day$^{-1}$ and for children as 0.013 μg kg$^{-1}$ day$^{-1}$ [U.S. EPA Office of Pesticide Programs, *Ethylene Dibromide (EDB) Scientific Support and Decision Document for Grain and Grain Milling Fumigation Uses* (8 February 1984)].

103. The leaves and roots of Russian comfrey are widely sold in health food stores and are consumed as a medicinal herb or salad plant or are brewed as a tea. Comfrey leaf has been shown to contain 0.01 to 0.15%, by weight, total pyrrolizidine alkaloids, with an average level of 0.05% for intermediate size leaves [C. C. J. Culvenor, J. A. Edgar, J. L. Frahn, L. W. Smith, *Aust. J. Chem.* 33, 1105 (1980)]. The main pyrrolizidine alkaloids present in com-

frey leaves are echimidine and 7-acetyllycopsamine, neither of which has been tested for carcinogenicity. Almost all tested 1,2-unsaturated pyrrolizidine alkaloids have been shown to be genotoxic and carcinogenic [H. Mori *et al.*, *Cancer Res.* **45**, 3125 (1985)]. Symphytine accounts for 5% of the total alkaloid in the leaves and has been shown to be carcinogenic [C. C. J. Culvenor *et al.*, *Experientia* **36**, 377 (1980)]. We assume that 1.5 g of intermediate size leaves are used per cup of comfrey tea (Table 1). The primary alkaloids in comfrey root are symphytine (0.67 g per kilogram of root) and echimidine (0.5 g per kilogram of root) [T. Furuya and M. Hikichi, *Phytochemistry* **10**, 2217 (1971)]. Comfrey-pepsin tablets (300 mg of root per tablet) have a recommended dose of one to three tablets three times per day. Comfrey roots and leaves both induce liver tumors in rats [I. Hirono, H. Mori, M. Haga, *J. Natl. Cancer Inst.* **61**, 865 (1978)], and the $TD_{50}$ value is based on these results. Those pyrrolizidine alkaloids tested have been found to be at least as potent as carcinogens such as symphytine. If the other pyrrolizidine alkaloids in comfrey were as potent carcinogens as symphytine, the possible hazard of a daily cup of tea would be HERP = 0.6% and that of a daily nine tablets would be HERP = 7.3%.

104. *Agaricus bisporus* is the most commonly eaten mushroom in the United States with an estimated annual consumption of 340 million kilograms in 1984–85. Mushrooms contain various hydrazine compounds, some of which have been shown to cause tumors in mice. Raw mushrooms fed over a lifetime to male and female mice induced bone, forestomach, liver, and lung tumors [B. Toth and J. Erickson, *Cancer Res.* **46**, 4007 (1986)]. The 15-g raw mushroom is given as wet weight. The $TD_{50}$ value based on the above report is expressed as dry weight of mushrooms so as to be comparable to other values for $TD_{50}$ in Table 1; 90% of a mushroom is assumed to be water. A second mushroom, *Gyromitra esculenta*, has been similarly studied and found to contain a mixture of carcinogenic hydrazines [B. Toth, *J. Environ. Sci. Health* **C2**, 51 (1984)]. These mushrooms are eaten in considerable quantities in several countries, though less frequently in the United States.

105. Safrole is the main component (up to 90%) of oil of sassafras, formerly used as the main flavor ingredient in root beer [J. B. Wilson, *J. Assoc. Off. Anal. Chem.* **42**, 696 (1959); A. Y. Leung, *Encyclopedia of Common Natural Ingredients Used in Food, Drugs and Cosmetics* (Wiley, New York, 1980)]. In 1960, safrole and safrole-containing sassafras oils were banned from use in foods in the United States [*Fed. Regist.* **25**, 12412 (1960)]. Safrole is also naturally present in the oils of sweet basil, cinnamon leaf, nutmeg, and pepper.

106. Diet cola available in a local market contains 7.9 mg of sodium saccharin per fluid ounce.

107. Metronidazole is considered to be the drug of choice for trichomonal and *Gardnerella* infections [AMA Division of Drugs, *AMA Drug Evaluations* (American Medical Association, Chicago, IL, ed. 5, 1983), pp. 1717 and 1802].

108. Isoniazid is used both prophylactically and as a treatment for active tuberculosis. The adult prophylactic dose (300 mg daily) is continued for 1 year [AMA Division of Drugs, *AMA Drug Evaluations* (American Medical Association, Chicago, IL, ed. 5, 1983), pp. 1766–1777].

109. D. M. Siegal, V. H. Frankos, M. A. Schneiderman, *Reg. Toxicol. Pharmacol.* **3**, 355 (1983).

110. Supported by NCI Outstanding Investigator Grant CA39910 to B.N.A., NIEHS Center Grant ES01896, and NIEHS/DOE Interagency Agreement 222-Y01-ES-10066. We are indebted to numerous colleagues for criticisms, particularly W. Havender, R. Peto, J. Cairns, J. Miller, E. Miller, D. B. Clayson, J. McCann, and F. J. C. Roe.

# Technical Comment: Carcinogenic Risk Estimation

**SAMUEL S. EPSTEIN**
*University of Illinois Medical Center*
*Chicago*

**JOEL B. SWARTZ**
*University of Quebec*
*Montreal*

In their widely publicized and popularized article "Ranking possible carcinogenic hazard," Bruce N. Ames *et al.* (17 Apr. 1987, p. 271) conclude that "analysis on the levels of synthetic pollutants in drinking water and of synthetic pesticide residues in foods suggests that this pollution is likely to be a minimal carcinogenic hazard relative to the background of natural carcinogens" and thus that the "high costs of regulation" of such environmental carcinogens are unwarranted. These conclusions reflect both flawed science and public policy.

Although Ames *et al.* challenge the validity of animal carcinogenicity data for quantitative estimation of human risk, they nevertheless use such extrapolations, based on the percentage *H*uman *E*xposure dose/*R*odent *P*otency dose (HERP), for ranking carcinogenic hazards. Apart from the fact that HERP rankings are based on average population exposures excluding sensitive subgroups, such as pregnant women, the derived potencies of Ames *et al.*, doses inducing tumors in half the tumor-free animals, are misleading. Potencies for "synthetic pollutants," such as trichloroethylene, are derived from bioassays in which lowest doses are large fractions of the maximally tolerated dose (MTD), whereas potencies for more extensively studied "natural carcinogens," such as aflatoxins, are generally derived from titrated

doses, orders of magnitude below the MTD. Since dose-response curves are usually flattened near the MTD (*1*), potencies derived from high-dose testing yield artificially low risk estimates; HERPs for "synthetic" carcinogens are thus substantially underestimated compared with many "natural carcinogens."

Compounding this misconception, Ames *et al.* maintain that carcinogenic dose-response curves rise more steeply than linear curves and that tumor incidences increase more rapidly than proportional to dose. At high doses, dose-response curves are usually less steep than linear curves (*1*), as also recognized elsewhere by Ames and his colleagues (*2*). Thus at MTD doses, large further dose increases may induce only small increases in tumor incidence, perhaps reflecting competition between transformation and cytotoxicity (*3*); linear extrapolations from high-dose tests thus underestimate low-dose risks.

For Ames *et al.*, the term "carcinogen" heterogeneously includes direct and indirect influences, including promoting and modifying factors and mutagens. Caloric intake is considered "the most striking rodent carcinogen." However, no correlations have been established between food intake and tumor incidence among animals eating ad libitum, despite wide variations in caloric intake and body weight (*4*), nor have correlations been established between obesity and most human cancers. In the statement by Ames *et al.*, "at the MTD a high percentage of all chemicals might be classified as 'carcinogens'," toxicity and carcinogenicity are confused. However, among some 150 industrial chemicals selected as likely carcinogens and tested neonatally at MTD levels, fewer than 10% were carcinogenic (5). Many highly toxic chemicals are noncarcinogenic, and carcinogen doses in excess of the MTD often inhibit tumor yields. While Ames *et al.* revive the discredited theory that chronic irritation causes cancer, most irritants are noncarcinogenic, and there is no correlation between nonspecific cell injury and carcinogenic potency (6). . . .

While diffusely defining carcinogens, Ames *et al.* artificially categorize them as "natural" or "industrial," saying that the former hazards should somehow limit concerns on the latter. However, dietary levels of "natural carcinogens" such as aflatoxins and dimethylnitrosamine are influenced by harvesting and storage technologies and nitrite additives, respectively. Moreover, predominant exposure to other "natural carcinogens" results from industrial activity; examples include asbestos, heavy metals, uranium, and formaldehyde. While emphasizing "natural carcinogens" and "nature's pesticides" in food as major carcinogenic exposures, Ames *et al.* ignore natural dietary anticarcinogens and antimutagens, such as porphyrins, phenolics, and retinoids (7). Although risks from aflatoxin and alcohol, described as two most important and potent carcinogens, depend on synergism with hepatitis B virus and tobacco smoke, respectively, risk estimates for most synthetic carcinogens are based on single-agent exposures only. While "natural carcinogens" have long played a role in human cancer, concerns must also focus on recent incremental effects of increased production of and exposure to nonsynthetic carcinogens, such as asbestos and heavy metals, and on the novel and escalating production and exposure to "synthetic carcinogens" (8). Although some petrochemicals have been proved to be carcinogenic, most have not been tested; moreover, much industrial data is at best suspect or unavailable (9).

The National Institute for Occupational Safety and Health estimates that 11 million workers are exposed to ten high volume industrial carcinogens (10). Up to tenfold increases in organ-specific cancer rates are reported among those who work with asbestos, uranium, and arsenic and in coke plants and among those exposed to specific petrochemicals and to some 20 less well-defined processes, such as dry cleaning, spray painting, and plumbing (10); excess childhood leukemia is also associated with parental occupational exposures to organic solvents and related chemicals (11).

Just one of the few well-studied occupational carcinogens, asbestos, responsible for up to 10,000 annual cancer deaths (12), is second only to tobacco of all known causes of human cancer.

Growing evidence demonstrates that pervasive contamination of air, water, soil, and food with a wide range of industrial carcinogens, generally without public knowledge and consent, is important in causation of modern preventable cancer. Even if hazards posed by any industrial carcinogen are small, their cumulative, possibly synergistic, effects are likely substantial. Eating food contaminated with residues at maximum legal tolerances of only 28 of 53 known carcinogenic pesticides, excluding numerous other carcinogenic pesticides and incremental exposure in drinking water, is estimated to be potentially responsible for 1.5 million excess lifetime U.S. cancers (13). Trichloroethylene is a common contaminant of drinking water, generally resulting from improper disposal of industrial wastes; lifetime consumption levels of 250 parts per billion found in contaminated wells in Woburn, Massachusetts, together with other related carcinogens not considered by Ames, *et al.*, is associated with excess risks of cancer (14), childhood leukemia, perinatal deaths, and birth defects (15). Some 20 retrospective and case control studies have associated trihalomethane-contaminated water with gastrointestinal and urinary tract cancers (16). As only a few organic drinking water contaminants are characterized (17), and as inhalation and cutaneous exposures may be as important as ingestion (14), risk estimates, excluding possible interactive effects, are likely to be misleadingly low. Nevertheless, Ames *et al.* ignore these limitations and also the substantive epidemiologic data and assert that "the animal evidence provides no good reason to expect that chlorination of water or current levels of man-made pollution of water pose significant carcinogenic hazards," and that the risk from contaminated Woburn water is 1/10,000 that of a glass of wine.

Community air pollution from industrial emissions, and thus proximity of residence to certain industries, is a recognized cancer risk factor. Numerous studies, controlled or stratified for smoking, demonstrate associations between excess lung cancer rates and heavy metal and aromatic hydrocarbon emissions (18); exposure to benzo[*a*]pyrene, a conventional combustion index, increased lung cancer mortality by 5% per nanogram per cubic meter of air (19). Others estimate that "the proportion of lung cancer deaths in which air pollution is a factor is 21%" (20). Concerns have recently focused on defined industrial emissions, including arsenicals, benzene, chloroform, vinyl chloride, and acrylonitrile, which in both sexes are associated with excess overall and organ-specific, standardized community cancer rates; carcinogenic trace metals and volatile organic community air pollutants, have been incriminated in some 0.6 to 2.3 per 1000 excess lifetime cancers (21). Ames et al., however, trivialize risks from "general outdoor air pollution.". . .

Besides proper concerns about naturally occurring carcinogens and tobacco, prudent policy must reflect overwhelming data on incremental exposure to industrial carcinogens and their association with increasing cancer rates, besides reproductive, neurotoxic, and other toxic effects (22). The existence of natural hazards clearly does not absolve industry and government from the responsibility for controlling industrial hazards. From public health, ethical, and policy perspectives, the important distinction is not between "natural" and "synthetic" carcinogens, but between preventable and nonpreventable cancers.

---

Cosigners [of Epstein and Swartz, "Technical Comment: Carcinogenic Risk Estimation"]: John Bailar, McGill University, Montreal; Eula Bingham, University of Cincinnati Medical School; Donald L. Dahlsten, University of California, Berkeley; Peter Infante, Washington, DC; Philip Landrigan and William Nicholson, Mount Sinai School of Medicine, New York; Marc Lappé and Michael Moreno, University of Illinois Medical Center; Marvin Legator, University of Texas Medical Branch, Galveston; Franklin Mirer, Rafael Moure, and Michael Silverstein, United Auto Workers, Detroit; David Ozonoff, Boston University Medical School; Beverly Paigen, Oakland Children's Hospital, Oakland, CA; and Jacqueline Warren, Natural Resources Defense Council, New York.

---

## References and Notes

1. J. Swartz et al., Teratog. Carcinog. Mutagen. 2, 179 (1982); L. Davies, P. Lee, P. Rothwell, Brit. J. Cancer 30, 146 (1974); E. Hulse, K. Mole, D. Papworth, Int. J. Rad. Biol. 114, 437 (1978).

2. W. Hooper et al., Science 203, 602 (1979).

3. J. Marshall and P. Groer, Rad. Res. 71, 149 (1977).

4. A. Tannenbaum, Proc. Am. Ass. Cancer Res. 1, 56 (1953).

5. R. Innis et al., J. Natl. Cancer Inst., 42, 1101 (1969).

6. I. Berenblum, Carcinogenesis as a Biological Prob-

lem (American Elsevier, New York, 1974), chapter 5.

7. H. L. Newmark, Can. J. Physiol. Pharmacol. 65, 461 (1987).

8. D. Davis and B. McGee, Science 206, 1356 (1979).

9. S. S. Epstein, The Sciences 18, 16 (1978); S. S. Epstein, The Politics of Cancer (Anchor/Doubleday, New York, 1979); National Research Council. Toxicity Testing: Strategies to Determine Needs and Priorities (National Academy Press, Washington, DC, 1984).

10. D. Davis, Teratog. Carcinog. Mutagen. 2, 105 (1982).

11. R. A. Lowengart et al., J. Natl. Cancer Inst. 79, 39 (1987).

12. W. J. Nicholson, G. Perkel, I. J. Selikoff, Am. J. Ind. Med. 3, 259 (1982).

13. National Research Council, Regulating Pesticides in Food: The Delaney Paradox (National Academy Press, Washington, DC 1987).

14. C. R. Cothern, W. A. Coniglio, W. L. Marcus, Environ. Sci. Technol. 20, 111 (1986).

15. S. Lagakos et al., J. Am. Stat. Ass. 81, 580 (1986).

16. National Research Council, Drinking Water and Health (National Academy Press, Washington, DC, 1980), vol. 3; K. P. Cantor, Environ. Health Perspect. 46, 187 (1982); K. P. Cantor, J. Natl. Cancer Inst. 79, 1269 (1987).

17. National Research Council, Drinking Water and Health (National Academy Press, Washington, DC, 1977).

18. P. Stocks and J. M. Campbell, Brit. Med. J. 2, 923 (1955); S. Epstein and J. Swartz, Nature (London) 289, 127 (1981).

19. Committee on Biologic Effects of Atmospheric Pollutants, National Academy of Sciences, Particulate Polycyclic Organic Matter (National Academy Press, Washington, DC, 1972).

20. N. Karch and M. Schneiderman, Explaining the Urban Factor in Lung Cancer Mortality (report to the Natural Resources Defense Council, New York, 1981).

21. W. Hunt, R. Faoro, T. Curran, J. Muntz, Estimated Cancer Incidence Rates for Selected Toxic Air Pollutants Using Ambient Air Pollution Data (Office of Air Quality Planning and Standards, Environmental Protection Agency, Washington, DC, 1985). These estimates of excess cancer rates are supported by epidemiological studies on cancer clustering in highly urbanized and highly industrialized communities [for example, R. Hoover and J. F. Fraumeni, Environ. Res. 9, 196 (1975); W. J. Blot et al., Science 198, 51 (1977); M. S. Gottlieb et al., J. Natl. Cancer Inst. 63, 113 (1979); J. Kaldor et al., Environ. Health Perspect. 54, 319 (1984)].

22. S. S. Epstein, Congr. Rec. 133, E3449 (1987).

23. We thank M. Jacobson, T. Mancuso, M. Schneiderman, and A. Upton for their helpful comments.

# Response to Samuel S. Epstein *et al.*

BRUCE N. AMES
*University of California at Berkeley*

LOIS SWIRSKY GOLD
*Lawrence Berkeley Laboratory*
*Berkeley, California*

We agree with only the last two sentences of the letter of Epstein *et al.* Correcting each of their errors would require lengthy explanations and would duplicate previous detailed analyses (*1–3*), so here we cover only the main issues.

*Half the chemicals tested in animals are carcinogens.* Our exhaustive database of animal cancer tests listed 392 chemicals tested in *both* rats and mice at or near the maximum tolerated dose (MTD). Of these, 60% of the synthetic chemicals and 45% of the natural chemicals were carcinogens in at least one species (*1*). The finding that about half of tested chemicals are positive in rodents has been reported for many sets of data; we cited among others the studies of the National Toxicology Program (NTP). We concluded that the proportion of chemicals found to be carcinogens is strikingly high. Epstein *et al.* ignore our data and citations and cite the early Innes *et al.* study to support their conclusion that the proportion of carcinogens is low. This misrepresents the facts. The Innes tests (120 chemicals, not 150 as stated by Epstein *et al.*, 11 positive) used only one species and were much less thorough than modern tests: they therefore were less likely to detect a carcinogenic effect (*4*).

The proportion of carcinogens is about as high for natural chemicals as for industrial chemicals. Therefore, our diet is likely to be very high in natural carcinogens, since more than 99.99% of the pesticides we ingest are "nature's pesticides," chemicals that plants produce to defend themselves against insects, fungi, and other pests (*1, 2*). These are present in all plants and in enormous variety, and their concentration is commonly in parts per thousand (*1, 2, 5*) rather than the parts per billion level of synthetic pesticide residues or water pollution (*1, 2*). . . . Cooking food produces carcinogens (*1, 2*) and so does our normal metabolism (*2, 6*). A high proportion of the chemical elements tested are carcinogens. Epstein *et al.* do not address this problem. They do not acknowledge that at the MTD about one-third of all chemicals tested are teratogens (*1*), half of all chemicals are carcinogens, and many chemicals are mutagens; and these categories are not completely overlapping. Even when one considers that some chemicals are selected for testing because they are suspicious, these are strikingly high proportions (*1, 4*).

*Extrapolating rodent cancer test results to humans.* The key issue, given the above facts, is how to identify *significant* preventable exposures to carcinogens (*1, 7, 8*). It is reasonable to assume that if a chemical is a carcinogen in rats and mice it is likely to be a carcinogen in humans at the same (MTD) dose. However, until we understand more about mechanisms, knowing the shape of the dose response in the dose range tested in laboratory animals provides little scientific basis for predicting the risk to humans at low doses, often hundreds of thousands of times below the dose at which an effect is observed in rodents (*9*). Thus, quantitative risk assessment is currently not scientifically possible (*1, 7–10*).

Our HERP index uses the same toxicological information from animal bioassays that is generally used to estimate human risk, but is instead a relative ranking of the possible hazards of a variety of natural and synthetic chemical exposures to humans. We stated clearly that our HERP value should not be used to assess risks, because we do not know how to extrapolate to low doses. . . .

Epstein *et al.* have three erroneous objections to our comparisons.

(1) They say our HERP values are overestimates for natural chemicals relative to synthetic chemicals because (i) dose-response curves flatten out at high doses and therefore linear extrapolations underestimate low-dose risks, and (ii) natural chemicals are more thoroughly studied (at lower doses) than are synthetic chemicals. Neither (i) nor (ii) is true. As we discussed in our article, there is no way to calculate a low-dose risk from the two dose levels tested in an animal bioassay. In addition, our analysis of the animal dose-response curves indicated a better fit with a quadratic model (upward curving) than with a linear model, and that flat dose-response curves (supralinear) are a rarity. Synthetic chemicals are not less well studied than natural chemicals, as can be seen from our published database: 80% of the studies are on synthetic chemicals; most of the studies referred to were National Cancer Institute (NCI)—NTP tests done at the MTD and at half the MTD; the few chemicals tested at a wider range of doses are not biased toward natural chemicals.

(2) Epstein *et al.* say we ignore the fact that plants contain anticarcinogens. We do discuss this fact (*1, 2*), and it does not support their argument that this affects our comparisons: plant antioxidants, the major known type of ingested anticarcinogens, help to protect us against oxidant carcinogens *whether synthetic or natural in origin.*

(3) Epstein *et al.* say natural carcinogens can be synergistic with other substances. However, this is also true of synthetic chemicals, and it is also irrelevant to our argument that synthetic pesticide residues in food or water pollution appear to be a trivial increment over the background of natural carcinogens.

*Carcinogenesis mechanisms and the dose-response curve.* We discussed the rapidly developing field of mechanisms in carcinogenesis because this understanding is essential for rational risk assessment. Cell proliferation (promotion) and mutation are involved in carcinogenesis, with a basal spontaneous rate for each step (*6, 11, 12*). Thus, increasing either rate increases the chance of cancer. In addition, several mutations appear necessary, and we have many layers of defense against carcinogens. These considerations of mechanism suggest a sublinear dose-response relation, which is consistent with both the animal and human data (*1*). It also suggests that multiplicative relationships may be the norm in human cancer causation. Administering chemicals in cancer tests at near-toxic doses (the MTD) commonly causes cell proliferation (*9*). If a chemical is nonmutagenic, but is carcinogenic because of its toxicity, then it should have no effect at low doses. This is a major point (*1*). Epstein *et al.* raise two points concerning the above that we find erroneous.

(1) They say we should not call promoting agents carcinogens. However, well-studied promoting agents have been shown to cause cancer by themselves, as do those hormones that cause cell proliferation (*11*). In fact, this class of carcinogens may well include the most important risk factors for human cancer (*1, 8, 11, 12*).

(2) Chronic irritation as a risk factor for cancer is not "a discredited theory," but is supported by rodent and human evidence, and by recent evidence on cancer mechanisms indicating that cell-killing causes both cell proliferation and a mutagenic burst of oxygen radicals (*1*).

*Factors important in causing human cancer.* The major risk factors of tobacco (30% of U.S. cancer), dietary imbalances, hormones, and viruses appear to account for the bulk of human cancer (*1, 3, 7, 8, 11–13*). In our article we analyzed the evidence from animal cancer tests that was relevant to some of these risk factors and to occupational exposures and pollution.

Epstein *et al.* distort our discussion of the role of dietary fat and calories in cancer causation. Limiting calories in rats or mice (compared with ad libitum consumption) *reproducibly* extends life-span and decreases spontaneous tumor rates. Caloric intake is likely to be a significant risk factor in human cancer causation (*11, 14*). Excess saturated fat consumption is a clear risk factor for heart disease. Excess fat consumption is a plausible, but not proved, risk factor in several types of human cancer, a view supported by extensive animal evidence (*1, 3, 12–14*). However, disentangling the effect of excess fat from excess calories is difficult in both rodents and humans (*14*).

Alcohol consumption is certainly the major known chemical risk factor for birth defects and is thought to account for 3% of U.S. cancer (*15*). Epstein *et al.* discount the importance of alcohol because it is synergistic with smoking. They are inconsistent, because they do not discount the effects of radon, asbestos, or other occupational exposures that are also synergistic with smoking. For example, they attribute deaths to asbestos (exaggerated), but do not mention that the risk of lung cancer for asbestos workers would be an order of magnitude less if workers did not smoke. It is more reasonable to apportion, rather than to dismiss, these risks.

Occupational exposures to chemicals and possible hazards can be high, as we showed in our article. But the sweeping statements made by Epstein *et al.*, without a discussion of dose, do not clarify matters. In a separate analysis (*16*) we have ranked the potential carcinogenic hazards to U.S. workers using the PERP index (analogous to the HERP index except that Occupational Safety and Health Administration Permitted Exposure Levels replace actual exposures). The PERP values differ by more than 100,000-fold. For 12 substances, the permitted levels for workers are greater than 10% of the rodent $TD_{50}$ values. Priority should be given to reduction of the allowable worker exposures that appear most hazardous in the PERP ranking.

Epstein *et al.* misrepresent the conclusions of the NRC-NAS committee report on pesticides, which did not say there would be 1.5 million deaths from pesticide use; the report did not predict deaths from pesticide use at all (*17*). Our article showed that the actual levels of synthetic pesticide residues eaten in the United States are tiny relative to the background of natural pesticides in plants. The end result of disproportionate concern about tiny traces of synthetic pesticide residues, such as ethylene dibromide (*1*), is that plant breeders are breeding highly insect-resistant plants: this may create other risks (*18*).

Our conclusion that water pollution did not make toxicological sense as a significant cause of cancer (or birth defects) because the amounts involved were extremely small compared with the background levels, is not contradicted by the epidemiological studies cited by Epstein *et al.* It is almost always beyond the power of epidemiology to provide convincing evidence that clusters of cancer or birth defects are due to pollution or to chance, bias, or confounding variables (*7*). Epstein *et al.* discuss Woburn, Massachusetts, without mentioning severe criticisms of the study they cite (*19*). Our analysis showed that the polluted water in Woburn or in Silicon Valley was less of a possible hazard than the chloroform in average U.S. tap water, a minimal possible hazard itself compared with the background. Comparative toxicological analyses such as ours can help epidemiologists to set priorities in their efforts and to distinguish causal correlations from the myriad of chance correlations. For example, the intake of burnt material from outdoor air pollution is so tiny compared with that from smoking (or from cooking food) that it seems implausible as a major source of cancer, a view consistent with the epidemiology cited, and indicates that epidemiologists must rigorously control for smoking (*20*). . . .

## References and Notes

1. B. N. Ames, R. Magaw, L. S. Gold, *Science* **236**, 271 (1987); *ibid.* **237**, 235 (1987); *ibid.*, p. 1283; *ibid.*, p. 1399; B. N. Ames and L. S. Gold, *ibid.* **238**, 1633 (1987).

2. B. N. Ames, *ibid.* **221**, 1256 (1983).

3. R. Doll and R. Peto, *The Causes of Cancer* (Oxford Univ. Press, Oxford, England, 1981).

4. J. R. M. Innes *et al.* [*J. Natl. Cancer Inst.* **42**, 1101 (1969)] used two mouse strains (versus two species for NCI-NTP), had only 17 animals per group (versus 50 for NCI-NTP), had only one dose level (versus two for NCI-NTP), and were 18-month experiments (versus 24-month for most NCI-NTP); the dose was likely below the MTD (among 19 Innes chemicals also tested by another laboratory, the Innes dose was usually lower, sometimes by more than tenfold). We have discussed positivity in animal cancer tests in detail, including the Innes design (L. S. Gold *et al.*, *Environ. Health Perspect.*, in press).

5. For example, a recent analysis of lima beans showed an array of 23 natural alkaloids (those tested have biocidal activity) that ranged in concentrations in stressed plants from 0.2 to 33 parts per thousand fresh weight. None appear to have been tested for carcinogenicity or teratogenicity [J. B. Harborne, in *Natural Resistance of Plants to*

*Pests. Roles of Allelo-chemicals*, M. B. Green and P. A. Hedin, Eds. (ACS Symposium 296, American Chemical Society, Washington, DC, 1986), pp. 22–35].

6. R. Cathcart *et al.*, *Proc. Natl. Acad. Sci. U.S.A.* **81**, 5633 (1984); R. Adelman, R. L. Saul, B. N. Ames, *ibid.*, in press; C. Richter, J.-W. Park, B. N. Ames, *ibid.*, in press.

7. J. Higginson, *Cancer Res.* **48**, 1381 (1988).

8. R. Peto, in *Assessment of Risk from Low-Level Exposure to Radiation and Chemicals*, A. D. Woodhead, C. J. Shellabarger, V. Pond, A. Hollaender, Eds. (Plenum, New York, 1985), pp. 3–16.

9. J. A. Swenberg *et al.*, *Environ. Health Perspect.* **76**, 57 (1987).

10. D. A. Freedman and H. Zeisel, *Stat. Sci.*, in press. The discussion with several eminent practitioners on the subject of quantitative risk assessment is the best available discussion of this subject.

11. B. E. Henderson *et al.*, *Cancer Res.* **48**, 246 (1988).

12. M. Lipkin, *ibid.*, p. 235.

13. C. S. Yang and H. L. Newmark, *CRC Crit. Rev. Oncol.* **7**, 267 (1987); B. C. Pence and F. Buddingh, *Carcinogenesis* **9**, 187 (1988).

14. L. Kinlen, *Cancer Surv.* **6**, 585 (1987); D. M. Ingram, F. C. Bennett, D. Willcox, N. de Klerk, *J. Natl. Cancer Inst.* **79**, 1225 (1987); D. Albanes, D. Y. Jones, A. Schatzkin, M. S. Micozzi, P. R. Taylor, *Cancer Res.* **48**, 1658 (1988).

15. Ethyl alcohol itself (15 grams per drink) is the likely teratogen and carcinogen in alcoholic beverages [B. N. Ames, "Review of evidence for alcohol-related carcinogenesis" (report for Proposition 65 meeting, Sacramento, CA, 11 December 1987)]. The dose response in humans is of considerable interest: four drinks per day is associated with increased cancer and birth defects, yet one drink per day (the U.S. average) has not clearly been associated with increased risk. Both orange juice and bread naturally contain considerable amounts of alcohol.

16. L. S. Gold *et al.*, *Environ. Health Perspect.* **76**, 211 (1987).

17. At the press conference on the report, Arthur Upton, speaking for the committee, responded to a question by saying that the worse-case scenario possible might implicate pesticides in 400 cases of cancer per year at the present time. The worst-case scenario pictures every farmer using the maximum possible amount of every pesticide allowed, the public consuming this food for a lifetime, and a worst-case linearized multistage model for predicting risk. Since all of these worst cases appear much too pessimistic, an actual risk close to zero is more likely.

18. A recent case is instructive. A major grower introduced a new variety of highly insect-resistant celery into commerce. A flurry of complaints to the Centers for Disease Control from all over the country soon resulted when people who handled the celery developed a severe rash when they were exposed to sunlight. Some detective work uncovered that, instead of the normal level of 900 parts per billion of psoralens (light-activated carcinogens and mutagens), the pest-resistant variety contained 9000 ppb of psoralens. It is unclear whether other natural pesticides in the celery were increased as well. [S. F. Berkley *et al.*, *Ann. Intern. Med.* **105**, 351 (1986); P. J. Seligman *et al.*, *Arch. Dermatol.* **123**, 1478 (1987).]

19. W. Lagakos *et al.*, *Am. Stat. Assoc.* **81**, 583 (1986). The same issue contains articles by other epidemiologists critical of this study.

20. Epstein *et al.* cite an estimate that air pollution is a causal factor in 21% of lung cancer. This is not supported by other epidemiology. The cited study was not peer-reviewed and did not take into account important confounding variables [see (3)].

# Questions for Thought and Discussion

1. Given the views on aflatoxin and ethyl alcohol expressed in this paper, do you think that labels warning of cancer should be required on peanut butter jars and beer cans and bottles? Support your conclusion.

2. Show the calculations behind the HERP value of 2.1% for indoor air pollution attributed to formaldehyde in some mobile homes. Why do you suppose that the use of formaldehyde in the manufacture of mobile homes is still permitted, whereas the use of EDB as a pesticide in grain production—which leaves residues having a HERP value of only 0.0004%—has been banned.

3. Animal test results are used as the basis for most health risk assessments. What are the advantages and disadvantages of using such results to decide whether to ban products intended for public consumption on the grounds that they contain chemicals that cause cancer in animals? Suggest an alternative to using animal test data for this purpose.

4. Do you believe that the same standards that are used for making regulatory decisions about the use of man-made chemicals should also be applied to naturally occurring chemicals in popular consumer products? Explain.

# PART 3

# REGULATORY ISSUES

# Introduction ————————————————————

It seems that almost every day newspapers and broadcast media carry stories about government regulation of health, safety, and environmental risks. The reported issues range from regulation of carcinogens in foods, to setting standards for automobile safety, to restricting the application of pesticides that pollute the environment.

Governmental efforts to gauge and maintain control of such problems are complicated enormously by the inevitable conflicts among opposing interest groups. Frequently the industries whose activities or products are affected assert that a regulation is too strict, while citizens' groups and environmental advocates demand that the same regulation be made stricter.

Regulatory measures and the responses to them raise immediate questions about whether the government is employing the best available techniques to assess risks and whether it is establishing the most effective policies to manage them. Related doubts arise as to whether the most important risks are receiving the most attention and whether too much or too little money is being spent on risk reduction. The papers in part 3 include a review of regulatory risk assessment and risk management at the Environmental Protection Agency (EPA), an attack on government regulation by a political scientist who thinks that society has grown too concerned about risk, and a critique of the unbridled use of cost–benefit analysis in regulatory decision making.

William D. Ruckelshaus writes in "Risk, Science, and Democracy" that EPA instituted successful programs in the 1970s that reduced "touchable, visible, and malodorous pollution" in air and water, only to face later the far more difficult problems posed by toxic substances at exposure levels so low as to be undetectable by the senses. Ruckelshaus contends that regulators must strive to support the continued improvement of the science that underpins the risk assessment process. Further, they must strive to keep risk assessment separate from risk management, which considers risks in the light of related socioeconomic factors. He concludes that effective regulation requires that decisions be made at the local level, within broad bounds that are set at higher government levels.

In "No Risk Is the Highest Risk of All," Aaron Wildavsky asserts that the "richest, longest-lived, best-protected, most resourceful civilization . . . is on its way to becoming the most frightened." In his view the government has contributed to this process by taking responsibility for risk management away from individuals. In place of costly government programs that seek to anticipate and prevent a wide variety of risks, Wildavsky would prefer to see society organize to cope with the consequences when things go wrong, as new technologies are introduced and old ones expand. Such an approach, he says, would result in actions that seek to secure gains rather than just to prevent loss.

Steven Kelman, in "Cost–Benefit Analysis: An Ethical Critique," takes cost–benefit analysis to task for reducing all considerations—including risks to human life and the quality of the environment—to dollars and cents. He argues that putting price tags on risks contradicts our basic social tenets that personal and environmental well-being, like other special values such as free speech and the right to vote, have no price. Kelman maintains that certain regulations that cannot be justified by cost–benefit analysis are still the right thing to do if they reflect the values of the citizenry.

In their reply to Kelman's provocative views, economists Gerard Butters, John Calfee, and Pauline Ippolito take particular exception to his notion that an ethical system that balances rights and duties would provide a better basis for regulatory decision making than cost–benefit analysis. Who, they ask, would then make the necessary balancing judgments, and on whose behalf?

# Risk, Science, and Democracy

## WILLIAM D. RUCKELSHAUS

In the study of history, nothing is more fascinating than the emergence of those ideas that periodically galvanize mankind into urgent action. Such ideas leap onto the center stage of public awareness, stay for a time, and then effectively vanish — either discredited like witchcraft and the divine right of kings, or absorbed into the public consciousness to become part of the status quo. The most interesting moments in this process, of course, are those when the idea is on stage, when it engages the public in passionate debate, when people struggle to fit the idea into the existing order, and when, through their efforts, people inevitably change both the existing order and the character of the idea.

Environmentalism has been an idea "on stage" in this sense for nearly 20 years. Born in its modern form in the writings of such people as Rachel Carson, René Dubos, and Barry Commoner, environmentalism entered the world of action with startling speed. At all levels of government, dozens of important environmental laws were passed in a single decade. Government agencies were reorganized to administer the new legislation; at the

*Editors' note:* William D. Ruckelshaus is a former administrator of the U.S. Environmental Protection Agency. His current affiliation is Browning-Ferris Industries, Houston, Texas.

federal level this activity included the establishment of the Environmental Protection Agency (EPA) in December 1970. Hundreds of regulations were issued, all ensuring that the tenets of environmentalism would intrude into nearly every aspect of American life.

Environmentalism has changed the nation, but over the same period environmentalism itself has changed as well. During the past 15 years, there has been a shift in public emphasis from visible and demonstrable problems, such as smog from automobiles and raw sewage, to potential and largely invisible problems, such as the effects of low concentrations of toxic pollutants on human health. This shift is notable for two reasons. First, it has changed the way in which science is applied to practical questions of public health protection and environmental regulation. Second, it has raised difficult questions as to how to manage chronic risks within the context of free and democratic institutions. People are afraid of these environmental risks, and fearful people have too often traded freedom for the promise of security. Our current efforts to control environmental pollution represent, in a sense, an attempt by our society to deal with this immense issue of environmental risk. This essay is about how well these efforts have succeeded thus far and how certain problems inherent in the early statutory embodiment of environmentalism now make it difficult to move ahead with the effective management of risks from pollution.

**Figure 1. Major federal environmental and toxic substance laws.**

| STATUTE | AGENCY | COVERAGE |
|---|---|---|
| Clean Air Act | Environmental Protection Agency | Air pollutants |
| Federal Water Pollution Control Act | Environmental Protection Agency | Water pollutants |
| Safe Drinking Water Act | Environmental Protection Agency | Drinking water contaminants |
| Federal Insecticide, Fungicide and Rodenticide Act | Environmental Protection Agency | Pesticides |
| Marine Protection, Research and Sanctuaries Act | Environmental Protection Agency | Ocean dumping |
| Resource Conservation and Recovery Act | Environmental Protection Agency | Hazardous wastes |
| Toxic Substances Control Act | Environmental Protection Agency | All chemical hazards not covered by other laws |
| Comprehensive Environmental Response, Compensation, and Liability Act *(Superfund)* | Environmental Protection Agency | Hazardous substances, pollutants and contaminants |
| National Environmental Policy Act | Council on Environmental Quality | Environmental impacts of federal actions |
| Occupational Safety and Health Act | Occupational Safety and Health Administration | Workplace exposures |
| Consumer Product Safety Act | Consumer Product Safety Commission | Dangerous consumer products |
| Hazardous Materials Transportation Act | Department of Transportation | Transportation of hazardous materials |

*Source*: Environmental Protection Agency.

## II

The early years of EPA present a good example of what happens when a set of abstract ideas, in this case those of the environmental movement, are pushed into the world of action. The movement asserted, for example, that industrial society as a whole was injurious to the health of the natural systems on which life ultimately depends and did not recognize that there were limits to the capacity of the planet to provide resources and to absorb industrial wastes. Environmentalists argued that the protective functions of government were grossly inadequate, that industry had no interest in stopping pollution, and, absent government control, would not protect the public interest. They stated that because the planet and its life-forms are a seamless and interdependent web, efforts to control pollution must be comprehensive and coordinated with respect to all forms of pollution. Beyond these tenets, environmentalism in the late 1960s was characterized by the rhetoric of crisis and the pervading sense that unless we acted immediately, something irreplaceable and indispensable would be lost forever.

These tenets influenced Congress in the early 1970s and affected the initial character and charter of EPA. In addition, certain assumptions about the natural world and our knowledge of it were embodied in most early environmental laws. Those of us at EPA thought we knew what the so-called bad pollutants were, where they came from, and at what levels they caused harmful health and environmental effects. We were confident that we knew how to measure pollutants in the air and water and how to reduce pollution to acceptable levels at a reasonable cost. The only missing ingredient, we thought, was the setting of reasonable standards at the federal level and their strict enforcement. The states had long carried the responsibility for environmental protection, but it seemed that economic rivalry among them for industrial investment would always overwhelm their ability to enforce environmental laws.

The obvious solution—tough enforcement mandated by the new federal laws—was the proper one only if the assumptions listed above were correct. (As we shall see, however, many of them were not correct, and much of the trouble this country has had with its environmental legislation arises from this fact.) As pollution was the result of weak enforcement by the states, the new federal laws would be strong laws in that they would set tight deadlines for the achievement of highly specific environmental goals while providing little administrative flexibility in how they were to be attained. Congress in that era of Vietnam and general disillusionment with the existing order was in no mood to trust any administrative actors—state or federal. The Clean Air Act of 1970, for example, gave EPA 90 days from the date of enactment to propose national ambient air standards for the major pollutants, standards that would be fully protective of the public health, and told us we had five years to attain them. This was done in the face of evidence that the problem in such cities as Los Angeles would take 25 years to solve.

Yet EPA was organized around an idea that was manifestly true: the environmentalist concept of the unity of nature. This was expressed by President Richard M. Nixon in 1970 in his announcement of the reorganization of federal environmental responsibilities:

> Despite its complexity, for pollution control purposes the environment must be perceived as a single interrelated system. A single source may pollute the air with smoke and chemicals, the land with solid wastes, and a river or lake with chemical and other wastes. Control of air pollution may produce more solid wastes, which then would pollute the land or water. . . . A far more effective approach to pollution control would identify pollutants; trace them through the entire ecological chain, observing and recording changes in form as they occur; determine the total exposure of man and his environment; examine interactions among forms of pollution; [and] identify where on the ecological chain interdiction would be most appropriate.

### III

This whole complex of ideas went right out the window as far as practical attention at the nascent EPA was concerned, and it was not to be recovered until quite recently. The notion that we ought to look at what pollution does, how it actually affects the environment and public health as it moves through the world, collided both with the assumption that enforcement was the missing ingredient in cleaning up the environment and with the environmentalist tenet that there was not a moment to lose in the race with ecocatastrophe. From this collision, and from the prevailing tone of mistrust just mentioned, fol-

lowed the congressional prescriptions for progress expressed in unattainable goals, technological fixes, and unrealistic deadlines. Thus, in the early years, EPA was making progress but toward a standard of perfection that was impossible to achieve. This chasing of rainbows at EPA has declined somewhat, but many of the agency's initial congressional mandates still exist, and are still unattainable.

EPA's strict enforcement mandates were based on the belief that we knew our targets and how to hit them. It became clear very early in the history of EPA, however, that our scientific base was sorely inadequate, given the magnitude and potential effect of our regulatory mission. As late as 1975 the National Academy of Sciences (NAS) concluded: "The specific chemical species responsible for toxicity have not been identified, and the levels of pollutants necessary to cause toxic effects have not been determined." We were not prepared to "identify where on the ecological chain interdiction would be most appropriate," as Nixon had suggested. We barely had defendable standards for a handful of the most important air pollutants. (And "defendable" is relative; the original standard for carbon monoxide, for instance, was based on a single study involving 12 individuals.)

The situation was no better in water pollution control, where we had to contend with a scarcely identified soup of pollutants affecting an enormous variety of life forms in many different kinds of aquatic ecosystems. There was little hope of establishing reliable associations between pollutant discharges and ecosystem impacts, except in exceptional cases involving isolated discharges or dramatic fish kills. The focus of the Clean Water Act (passed in 1972) and its attendant regulation was therefore directed mainly at reduction in discharges of the most obvious, or conventional, pollutants, including such things as organic wastes, suspended particulates, and oil and grease. The federal government was to establish criteria for these substances—what levels were safe for human health and aquatic organisms—and the states were to establish the standards that lakes and streams had to meet to achieve the water quality necessary for particular uses, such as swimming or fishing.

The federal government was also required to set national effluent guidelines for each type of industrial plant. These were designed to control the bewildering variety of toxic substances that could appear in industrial waste streams. The guidelines were to be based on engineering and economic judgments, and they were to be put into effect by setting standards based on available technology rather than solely by reference to any specific level of water quality.

The environmentalists believed, and Congress agreed, that if quality-based standards were adopted, industry would use the uncertainties of the science to delay indefinitely any investments in controls. They pointed to the experience of state governments with water-quality standards, in which such delays had apparently happened, although the reluctance of states to get tough with some industries may have had a simpler origin: the fear that tough policies would lead to industrial flight.

Our efforts to establish the scientific base presupposed by the environmental laws were hindered by the difficulties of managing the six different scientific establishments that EPA had inherited. Our scientific resources were housed in 56 separate laboratories scattered across the country. From the first, it was extremely difficult to convey to EPA's scientific cadre the urgency of our need for authoritative findings to support the regulations we were obliged to turn out to the beat of those timetables in the legislation. Program officials would make lists of research "needs," which generally involved the connections between pollutants from particular sources and effects on the various values the law required them to protect. The scientists believed that it was difficult, if not impossible, to meet these needs within the generally recognized standards of scientific validity without significant improvements in the state of the science. Thus, a tension was established, which exists to this day, between the need to act and the availability of knowledge.

Even at the outset as the environmentalist idea moved, in Winston Churchill's phrase, "from the wonderful cloudland of aspiration to the ugly scaffolding of attempt and achievement," it confronted a problem that would plague EPA henceforth: the basic disjunction between the reality of science and the assumption, inherent in the mission of all governmental protective agencies, that it is always possible, and therefore obligatory, to provide swift and sure regulation of all substances and situations that threaten public health and the environment.

This disjunction is hardly ever acknowledged by the general public, so completely has science been accepted as the handmaiden of its material desires. The relationships among basic science, applied science, and improvements in daily life are usually regarded as simple, being much like those among growing trees, cutting lumber, and building houses. This concept feeds the notion that when we want something from science, we can order it, as we order lumber to build a house. If there is not enough lumber, we can grow and cut more trees. It follows that there is a way to "manage" this orderly process so as to make it more efficient, or more suitable to our current needs.

Even though a scientific explanation may appear to be a model of rational order, we should not infer from that order that the genesis of the explanation was itself orderly. Science is only orderly after the fact; in process, and especially at the advancing edge of some field, it is chaotic and fiercely controversial. Thus, the expectation built into environmental law, that science can provide definitive answers to the kinds of questions that policymakers are obliged to ask under the terms of that law, will be disappointed to the degree that such answers derive from the forward edge of research.

We can direct our resources in various directions with the expectation that talented researchers will work on questions that bear on important policy issues, but we cannot order answers as efficiently as the public would like. As Louis Pasteur said, "There is no such thing as applied science, there are only applications of science." Even in those scientific establishments specifically funded to explore policy questions for federal protective agencies, we find scientists behaving rather more like scientists in "basic" research than their budget justifications would suggest.

Nor can we order a consensus in the areas of greatest interest to environmental policy: pollutant exposure and effects. Policymakers, including me, have often deplored the tendency of scientific panels to engage in interminable debate rather than reach the agreement that was clearly indicated on the invitation. *Of course* scientists will disagree on issues involving the advancing edge of research; that is what they do for a living. And even if we could somehow get a group of scientists to endorse a consensus position, it would be, in the first place, only tentative and subject to revision with the arrival of new discoveries; and in the second place, it may be entirely wrong.

In science, the majority does not rule, as the history of science amply demonstrates. Everybody but Semmelweiss was wrong on childbed fever. Everybody but Wegener was wrong on continental drift. Scientists outside the consensus, who may not be averse to voicing their objections in public, may be cranks or they may be right. Public officials cannot make this distinction; only the slow mills of science can grind out the truth.

Public officials, though, do not have that kind of time. From its earliest days EPA was often compelled to *act under conditions of substantial scientific uncertainty*. The full implications of this problem were partially masked at the beginning because the kind of pollution we were trying to control was so blatant. Although scientists were often in the forefront of the early struggles against pollutants, most people did not need a scientific panel to tell them that air is not supposed to be brown, that streams are not supposed to ignite and stink, that beaches are not supposed to be covered with raw sewage. When I left EPA in 1973, the means for ending the worst forms of pollution were fairly well in hand. If emission controls were put on cars, if the requisite sewage treatment plants were built, and if industrial firms installed readily available control technologies, pollution would be essentially eliminated as a national problem. While I had some reservations about the scientific basis of some of the criteria we had established for health protection and about the ability of industry to meet congressional deadlines, and while we were probably not protecting as efficiently as we might have, all in all it did not seem to matter much. I was certain that Congress would adjust the more obvious infirmities in our environmental laws once it realized the social and economic distortion inherent in pursuing zero-risk environmental goals. We had made a start and had demonstrated that the federal government could move effectively against major polluters in such a way that the general public could literally see through its windows.

## IV

The events of the past dozen or so years have demonstrated that my opinion was in one sense quite correct and in another far too sanguine. Many of the grosser sorts of pollution are indeed under control. But the level of controversy about environmental protection has not diminished. On the evidence of my more recent experience as administrator of EPA, as indicated by what regulatory issues were of greatest concern, what the various congressional committees were interested in hearing about, press coverage, and so on, environmental controversy is now largely focused on the carcinogenic risk to human health from toxic chemicals, and on the removal of ever smaller increments of conventional pollution from the air and water.

"Risk" is the key concept here. It was hardly mentioned in the early years of EPA, and it does not have an important place in the Clean Air or Clean Water Acts passed in that period. Of the events that contributed to this change, the most important were the focus of public attention on PCBs and asbestos (two substances that are ubiquitous in the American environment and that are capable of causing cancer) and the realization that exposure to a very large number of unfamiliar and largely untested chemicals is universal. The discovery by cancer epidemiologists that cancer rates vary with environment suggested that pollution might play a role in causing this disease. And finally, the cancer risk was pushed to the forefront by the emergence of abandoned dumps of toxic chemicals as a consuming public issue. As a direct result of this shift in attention, the relation of EPA to its science base was altered; the problem of uncertainty was moved from the periphery to the center.

This shift occurred because the risks of effects from typical environmental exposures to toxic substances—unlike the touchable, visible, and malodorous pollution that stimulated the initial environmental revolution—are largely constructs or projections based on scientific findings. We would know nothing at all about chronic risk attributable to most toxic substances if scientists had not detected and evaluated them. Our response to such risks, therefore, must be based on a set of scientific findings. Science, however, is hardly ever unambiguous or unanimous, especially when the data on which definitive science must be founded scarcely exist. The toxic effects on health of many of the chemicals EPA considers for regulation fall into this class.

"Risk assessment" is the device that government agencies such as EPA have adopted to deal with this quandary. It is the attempt to quantify the degree of hazard that might result from human activities—for example, the risks to human health and the environment from industrial chemicals. Essentially, it is a kind of pretense; to avoid the paralysis of protective action that would result from waiting for "definitive" data, we assume that we have greater knowledge than scientists actually possess and make decisions based on those assumptions.

Of course, not all risk assessment is on the controversial outer edge of science. We have been looking at the phenomenon of toxic risk from environmental levels of chemicals for a number of years, and as evidence has accumulated for certain chemicals, controversy has diminished and consensus among scientists has become easier to obtain. For other substances—and these are the ones that naturally figure most prominently in public debate—the data remain ambiguous.

In such cases, risk assessment is something of an intellectual orphan. Scientists are uncomfortable with it when the method must use scientific information in a way that is outside the normal constraints of science. They are encroaching on political judgments and they know it. As Alvin Weinberg has written:

> Attempts to deal with social problems through the procedures of science hang on the answers to questions that can be asked of science and yet which cannot be answered by science. I propose the term *trans-scientific* for these questions. . . . Scientists have no monopoly on wisdom where this kind of trans-science is involved; they shall have to accommodate the will of the public and its representatives.

However, the representatives of the public, in this instance policy officials in protective agencies, have their problems with risk assessment as well. The very act of quantifying risk tends to reify dreaded outcomes in the public mind and may make it more difficult to gain public acceptance for policy decisions or push those decisions in unwise directions. It is hard to describe, say, one cancer case in 70 years among a population of a million as an "acceptable risk" when such a description may too easily summon up for any individual the image of some close relative on his deathbed. Also, the use of risk assessment as a policy basis inevitably provokes endless arguments about the validity of the estimates, which can seriously disrupt the regulatory timetables such officials must live by.

Despite this uneasiness, there appears to be no substitute for risk assessment, in that some sort of risk finding is what tells us that there is any basis for regulatory action in the first place. The alternative to not performing risk assessment is to adopt a policy of either reducing all *potentially* toxic emissions to the greatest degree technology allows (of which more later) or banning all substances for which there is any evidence of harmful effect, a policy that no technological society could long survive. Beyond that, risk assessment is an irreplaceable tool for setting priorities among the tens of thousands of substances that could be subjects of control actions—substances that vary enormously in their apparent potential for causing disease. In my view, therefore, we must use and improve risk assessment with full recognition of its current shortcomings.

This accommodation would be much easier from a public policy viewpoint were it possible to establish for all pollutants the environmental levels that present zero risk. This is prevented, however, by an important limitation of the current technique; the difficulty of establishing definitive no-effect levels for exposure to most carcinogens. Consequently, whenever there is any exposure to such substances, there is a calculable risk of disease. The environmentalist ethos, which is reflected in many of our environmental laws, and which requires that zero-risk levels of pollutant exposure be established, is thus shown to be an impossible goal for an industrial society, as long as we retain the no-threshold model for carcinogenesis.

## V

This situation has given rise to two conflicting viewpoints on protection. The first, usually proffered by the regulated community, argues that regulation ought not to be based on a set of unprovable assumptions, but only on connections between pollutants and health effects that can be demonstrated under the canons of science in the strict sense. It points out that for the vast majority of chemical species, we have no evidence at all that suggests effects on human health from exposures at environmental levels. Because many important risk assessments are based on assumptions that are scientifically untestable, the method is too susceptible to manipulation for political ends and, the regulated community contends, it has been so manipulated by environmentalists.

The second viewpoint, which has been adopted by some environmentalists, counters that waiting for firm evidence of human health effects amounts to using the nation's people as guinea pigs, and that is morally unacceptable. It proposes that far from overestimating the risks from toxic substances, conventional risk assessments underestimate them, for there may be effects from chemicals in combination that are greater than would be expected from the sum effects of all chemicals acting independently. While approving of risk assessment as a priority-setting tool, this viewpoint rejects the idea that we can use risk assessment to distinguish between "significant" and "insignificant" risks. Any identifiable risk ought to be eliminated up to the capacity of available technology to do so.

It is impossible to evaluate the merits of these positions without first drawing a distinction between the assessment of risk and the process of deciding what to do about it, which is "risk management." The arguments in the form sketched here are really directed at both these processes, a common confusion that has long stood in the way of sensible policymaking.

Risk assessment is an exercise that combines available data on a substance's potency in causing adverse health effects with information about likely human exposure, and through the use of plausible assumptions, it generates an estimate of human health risk. Risk management is the process by which a protective agency decides what action to take in the face of such estimates. Ideally the action is based on such factors as the goals of public health and environmental protection, relevant legislation, legal precedent, and application of social, economic, and political values. *Risk Assessment in the Federal Government*, a National Research Council (NRC) document, rec-

ommends that regulatory agencies establish a strict distinction between the two processes, to allay any confusion between them. In my view Congress should do the same in all statutes seeking to deal with risk.

Returning now to the opposing viewpoints we see that both reflect the fear that risk assessment may be imbued with values repugnant to one or more of the parties involved. That is, some people in the regulated community believe that the structure of risk assessment inherently exaggerates risk, while many environmentalists believe that it will not capture all the risk that may actually exist. As we have seen, this disagreement is not resolvable in the short run through recourse to science. Risk assessment is necessarily dependent on choices made among a host of assumptions, and these choices will inevitably be affected by the values of the choosers, whether they be scientists, civil servants, or politicians.

The NRC report suggests that this problem can be substantially alleviated by the establishment of formal public rules guiding the necessary inferences and assumptions. These rules should be based on the best available information concerning the underlying scientific mechanisms. Adoption of such guidelines reduces the possibility that an EPA administrator may manipulate the findings of some risk assessment so as to avoid making the difficult, and perhaps politically unpopular, choices involved in a risk-management decision. Both industry and environmentalists fear this manipulation—from different brands of administrator, needless to say. Although we cannot remove values from risk assessment, we can and should keep those values from shifting arbitrarily with the political winds.

The explicit and open codification suggested by the NRC will also ensure that the assumptions used in risk assessment will at least be uniform among all agencies that adopt them, will be plausible scientifically, and will reflect a predictable and relatively constant policy amid this complex and chaotic hybrid discipline. It also offers the possibility that one day all the protective agencies of government will speak with one voice when they address risks, so that estimates of risk will be comparable among agencies and the public at last will be able to make a fair comparison of the individual risk-management decisions of separate agencies.

The remaining points of both positions are really about risk management and on this issue both are flawed. At its extreme, the first position—that regulation should be based solely on scientifically provable connections between pollutants and health effects—would allow the release of unlimited quantities of substances that cause cancer in animals, on the assumption that there will be no analogous effect on people and that there must be thresholds for carcinogenesis. I expect that most Americans would reject that assumption as imprudent, given our current knowledge about carcinogenesis (for example, the similarity of cancer-causing genes across species). At some level we have to regard the possibility that we are controlling somewhat in excess of the true risk as a kind of insurance, with the cost of control as its premium. The effort to reduce apprehension, even so-called unreasonable apprehension, about the future results of current practices is a valid social function. Risk-management agencies such as EPA could be chartered to do precisely that. If so, we had better make clear what we are doing, and establish rules for doing it.

The weakness of the second viewpoint, that any identifiable risk ought to be eliminated up to the capacity of available technology to do so, lies in the concept of a best available technology that must invariably be applied where risk is discovered. "Best" and "available" are terms as infinitely debatable as the assumptions of risk assessment. There is always a technology conceivable that is an improvement on a previous one, and as the last increments of pollution are removed, the cost of each successive fix goes up very steeply. Because, according to the no-threshold assumption, even minute quantities of carcinogens can be projected out to cause cases of disease, arguments about technology reduce in the end to arguments about risk and cost: technology A allows a residual risk of $10^{-5}$ and costs $1 million; technology B allows a residual risk of $10^{-6}$ and costs $10 million, and so on ad infinitum. It is specious to pretend that costs do not matter, because it is always possible to show that at a certain level of removal, costs in fact do matter: technology Z allows residual risks of $10^{-15}$ and costs $1 trillion.

Once this is admitted, as it almost always is when we come down to debating actual regulations, the position is reduced to arguments about affordability.

This too is treacherous ground. Firms vary in their ability to pay, and what is affordable for one may bankrupt another. If requirements are adjusted so as not to cripple the poorest firms, the policy amounts to an environmental subsidy to the less efficient players in our economy. In the end, discussions about available technology and affordability develop into discussions of corporate management: "If they can't afford pollution control, how come their executives got such a big bonus?" This may be a fascinating topic to some, but it distracts from the main issue.

My point is that in confronting any risk there is no way to escape the question "Is controlling it worth it?" We must ask this question not only in terms of the relationship of the risk reduced and the cost to the economy but also as it applies to the resources of the agency involved. Policy attention is the most precious commodity in government, and a regulation that marginally protects only 20 people may take up as much attention as a regulation that surely protects a million.

"Is it worth it?" That this question must be asked and asked carefully is a token of how the main force of the environmental idea has been modified by the recent focus on toxic risk to human health. In truth this question should always have been asked, but because the early goals of environmentalism were so obviously good, the requirement to ask, "Is is worth it?" was not firmly built into all our environmental laws. Who would dare to question the worth of saving Lake Erie? Environmentalism at its inception was a grand vision, one that nearly all Americans willingly shared. Somehow that vision of the essential unity of nature and of the need for bringing industrial society into harmony with it has been lost among the parts per billion, and with it we have lost the capacity to reach social consensus on environmental policy.

## VI

Why has this happened? I believe it is because environmentalism, like many another social movement, is suffering from the excesses of its own youthful vigor. In the early legislation promises were made that could not reasonably be kept; expectations were raised that were bound to be disappointed. A case in point is the language of Section 112 of the Clean Air Act, which requires that EPA establish "an ample margin of safety" in the control of hazardous air pollutants. We can achieve this margin only when it is possible to determine a level at which there are no apparent effects on humans, as we may be able to with such substances as carbon monoxide and nitrogen oxides. It then becomes possible to roll that level back to one presumed to protect the most sensitive individuals in the population, which is, of course, the point of the "ample margin" language. Most of the air pollutants now under consideration, however, are suspected carcinogens. Because, as already noted here, we have assumed that it is impossible to establish no-effect levels for carcinogens, we cannot establish the margin of safety demanded by the law without banning these substances or the processes that produce them, and such action is nearly always unwise for social and economic reasons.

The Clean Air Act promised absolute protection from airborne carcinogens, but we cannot keep that promise. The American people were also promised under the clean air and water laws that the air would be pure and water pollution would be eliminated by a specific date; we have not kept those promises either. Few things are more corrupting to a free society such as ours than grand promises unfulfilled. Senator Daniel Patrick Moynihan (D-N.Y.) has written of this:

> The malaise of overpromising derives almost wholly, in my experience, from the failure of executives and legislators to understand what is *risked* when promises are made. . . . When things don't work out as promised it is all too easy to suspect that someone *intended* they should not.

Many of the promises made or implied by federal law as part of the triumph of environmentalism have indeed had this sad effect. Since the beginning of EPA, there has been insufficient appreciation of the difficulties involved in the successive tasks the agency has been assigned by Congress. As a result many people on Capitol Hill assume that if unrealistic promises have not been kept, some people are not doing their jobs—and that the way to ensure that they do is to write ever more stringent prescriptions into environmental law, with appropriate deadlines. Thus more promises are added to unrealistic promises, and the cycle of disappointment and growing mistrust is accelerated.

The foregoing should in no way be taken as implying that the control of toxic substances is unimportant or that EPA could not be better managed. My point is that EPA will not do an effective job in either case unless we realize that the debate about risk in its present form leads us ever deeper into a blind alley. We must revisit that crucial stage of problem solving in government at which one defines what kind of problem it is. We will emerge from the blind alley and reforge a practical consensus on the environment only when we are able to redefine the problem of environmental protection as being "the management of risk."

When we adopt this definition, we specifically will abandon the impossible goal of perfect security and accept the responsibility for making difficult and painful choices among competing goods. We can then move forward from the present impasse and begin to look at a more productive set of questions.

## VII

What then is risk management? As management in general can be defined as the distribution of current resources to shape some desirable future state, risk management in its broadest sense means adjusting our environmental policies to obtain the array of social goods—environmental, health-related, social, economic, and psychological—that forms our vision of how we want the world to be.

In practical terms, risk management means giving the protective agencies flexibility comparable to that which managers have traditionally exercised in other spheres. This suggestion, of course, runs counter to the inclination to further restrict the discretion accorded those agencies. Administrations do vary in their priorities and competence, so how can we ensure that this flexibility will not be misused? Obviously, flexibility cannot be taken to mean carte blanche for the agencies. Flexibility in risk management must be bounded by evidentiary rules of the type described in the NRC report, as well as by rules to assess the adequacy and competence of information. Most important, flexibility should be limited by broad public acceptance, including acceptance by individuals subject to risk, and also by more sensible and appropriate congressional oversight.

It may be argued that the administrator of EPA already has sufficient flexibility. Did we not, for example, attempt to get around the impossible standard in Section 112 of the Clean Air Act by adopting a best-available-technology approach that is not mentioned in the act? The answer is that this is the wrong kind of flexibility. Agency staffs can always finagle their way toward some workable solution out of almost any legislative language. This does not mean that we should not strive for clear and realistic language in our laws. Once the law is written, Congress has the responsibility to provide oversight to ensure that the reasonable goals of the law are carried out (an activity that Congress now performs very poorly). This is not the same as telling the EPA administrator what to do on specific issues. I believe that the tendency of Congress to give such direction will diminish over the long run if the protective agencies are able to redefine their missions and to accept the responsibility for showing what they are doing as clearly as possible; that is, what they are doing in terms of risk management. This approach requires much more flexibility to address specific pollution problems than many of the most important laws now permit.

Two characteristics of such problems make it imperative that we begin to address them in this way. The first of these has to do with the distribution of toxic substances; we typically find large variations in risk in different parts of the country or even in different parts of the same community. The second characteristic is that toxic risk engenders extraordinary fear, which may paralyze rational public policy and lead to unproductive or even perverse actions on the part of government. Let us deal with each of these characteristics in turn.

The distribution of pollutants is obviously not uniform across the country. We find abandoned hazardous waste dumps only in particular places, and we find the same sort of localization in connection with industrial-process releases of toxic substances. Receiving waters, whether on the surface or in the ground, differ in their vulnerability to pollution, as do air sheds. Further, public exposure, and hence risk, varies greatly depending on the location of pollution sources in relation to population centers. Finally, different populations may differ greatly in their opinions about what risks they are willing to endure in return for what benefits.

Current law makes it hard to recognize such distinctions. For example, the Clean Air Act requires EPA to impose expensive automobile inspection and maintenance programs in communities where certain pollutant criteria have been exceeded twice a year. EPA must do this even though the violations may be a consequence of the placement of the air-quality monitoring devices and may not reflect the general quality of the air, even though there is no discernible health effect, and even though the people in the community strongly oppose the action. The law does not allow the federal government to distinguish between (for example) Los Angeles and Spokane, Washington, in this regard—a restriction that defies common sense. In the same way, we cannot distinguish between a plant discharging pollutants into a highly stressed river in Connecticut and one discharging into Alaskan waters that bear no other pollutant burden. In other words, the law does not permit us to act sensibly.

## VIII

In my view, sound public policy would give EPA the flexibility to confront and deal with risks in the local context. This flexibility would entail, first, balancing risks against the local economic impacts of controlling them; second, ensuring that our national programs that attempt to deal with local risks operate according to risk-management principles; and third, involving the local public in a meaningful way in the decisionmaking process.

A typical example of the localization of risk is that in which an industrial plant imposes some local risks despite the installation of advanced pollution controls. Local economic interests then confront local health interests in reducing risk. A paradigm of such cases was the situation in Tacoma, Washington, where until recently a copper smelter processed arsenic-rich ore and released quantities of this carcinogen into the ambient air. Even after the plant was heavily controlled, it appeared impossible to eliminate the carcinogenic risk from this release; that is, eliminating the risk meant eliminating the plant. As EPA administrator, I believed EPA had a responsibility to explain to the people who would be most directly affected by the decision what we knew about the risks from the smelter.

True public involvement meant forcing the public to confront the tradeoffs involved in this risk-management decision. When we began to examine the

Tacoma situation in detail, we discovered that most of the local people were willing to view the problem in terms of the risk-management choice, although their positions on the arsenic risk depended (understandably) on where they lived in relation to the distribution pattern of the emissions, and on whether they had a personal economic stake in the plant remaining open. As the public discussions continued and we refined our data on both the risk and the means of reducing it, the citizens of Tacoma began to come up with helpful ideas about how we could minimize the actual impact of the arsenic emissions and keep the smelter operating.

Although the plant's owners eventually decided to close it (for reasons not directly connected to the pollution issue), we learned something valuable from the experience. Despite initial fears, it is possible for people subject to toxic risk to think rationally about it. It is possible for them to confront the hard truth that solutions to such problems necessarily involve an uneven distribution of risks and benefits. This rational thinking involves plunging into the uncertainties involved in the analyses we use to define the issues. How sure are we about exposure? How sure are we about the effects of the substance? How certain are the economic impacts?

Such rational thinking requires a kind of democratic citizenship that is willing to dig deeper than the glib headlines and the usual invitations to panic. It also requires much more from the regulatory agency providing the facts: The agency must be willing to explain and able to communicate, and most of all, it must admit the uncertainties buried in its calculations. Only then can the appropriate balancing decisions take place. Our environmental laws should specifically permit EPA to manage a localized risk in this manner.

Risk management thus provides us with the flexible approach needed to deal with the way toxic substances are actually distributed. However, it must also enable us to deal with the second characteristic of the current situation with respect to toxic substances: public fear and the mistrust often associated with it.

How can risk management, a logical and rational process, cope with emotionally based public reactions? While we do not yet have a perfect solution to this problem, we do know enough, I think, to identify some nonsolutions. Reliance on experts is not a solution. Psychologist Paul Slovic writes of this:

> Since even well-informed laypeople have difficulty judging risks accurately, it is tempting to conclude that the public should be removed from the risk assessment process. Such action would seem to be misguided on several counts. First, we have no assurance that experts' judgments are immune to biases once they are forced to go beyond hard data. . . . Second, in many if not most cases, effective hazard management requires the cooperation of a large body of laypeople. These people must agree to do without some things and accept substitutes for others; they must vote sensibly on ballot measures and for legislators who will serve them as surrogate hazard managers; they must obey safety rules and use the legal system responsibly. Even if the experts were much better judges of risk than laypeople, giving experts an exclusive franchise for hazard management would mean substituting short-term efficiency for the long-term effort needed to create an informed citizenry.

Burying quantitative estimates of risk within some procedural regulatory framework, an abstract scoring system perhaps, is also a nonsolution. In debates surrounding any environmental decision, however, the subject of risk is bound to arise, and then the agency will have to explain not only the risk itself but also why it was trying to "hide" it. The regulatory agencies must accept the fact that the public is part of the regulatory process. Given the atmosphere of mistrust that has characterized governmental relations with the public since Vietnam and Watergate, it is vital that the agencies be forthcoming with *all* the information involved in their decisions.

Furthermore, as we have seen, regulatory approaches not based on quantified risk must eventually confront the economic realities and become risk management in all but name. It is thus impossible for modern environmental policymakers to avoid dealing directly with named, quantified risks and the public fear these engender.

## IX

I believe that we can mitigate this fear only by adopting as policy the full disclosure of the risks involved in regulatory decisions. This belief is not universally shared within the environmentalist or the regulated communities or even at the protective agencies. It is not pleasant to tell people that some residual risk is nearly always associated with practical levels of pollution control. It becomes less and less pleasant as one gets closer to the most affected locality. In a way, the great national debates on environmental policy are resolvable only in high school auditoriums across the country and in terms of what public officials are able to say to worried citizens. The questions that these audiences naturally ask (Can we drink the water? Will my child get cancer?) can hardly ever be answered unequivocally. The uncertainties involved in risk assessment must be explained, and people must be brought to understand that "safety" is a social construct, the definition of which is, or ought to be, a part of each citizen's duty in our society.

Our skill at developing such constructs for new apparent dangers is not yet what it should be. Communication and public education must become a much more important part of the work of EPA and the other protective agencies, on a level with their scientific, engineering, policymaking, and legal functions. Risk management is not merely a set of techniques for arriving at the correct answers. It must include communication to the public about how we arrive at environmental protection decisions. The values and assumptions that underlie all such decisions must be made manifest. Transparency is the object of the whole process, and public trust is the ultimate goal.

But it is not easy, for we must confront one of the inevitable dilemmas of leadership in a democracy, that of having to choose between telling people what they want to hear and telling them what they ought to know. A good deal of our environmental legislation has bent toward the former, which has made things more difficult for the officials who have to stand up on the stages of those high school auditoriums. If people have been told in statutes that risk can be made to vanish through governmental action, they will be satisfied with nothing less, and they will not treat kindly the messenger who has to tell them it cannot be done.

If risk management is to be used as a means of coming to decisions at special locales, the principle must be specifically established in environmental law. Our statutes should reflect the reality that environmental protection in an imperfect world means the assessment of risks and the ability to manage them as particular situations warrant. Note that this should go beyond health risk in the strict sense. We should learn to look at all the impacts of pollution from the perspective of balancing some definable improvement against our always finite resources, whether control expenditures or governmental attention.

This is always a hard case to make, especially to Congress, where the prizes seem to go for new programs and ever more stringent ones. It sometimes seems that those who write our protective statutes would rather have EPA pretend to be doing a thousand tasks than have it select the hundred most important tasks and do them well. I am not saying that our present laws have not worked. There is no question that we have mounted a successful national effort against the more gross sorts of pollution. Those laws, however, will become increasingly ineffective if they force us to deal with toxic risk, or even with the final remaining increments of conventional pollutants, in the wrong way. Our laws need less pious hope and more determination to strike at the sources of actual harm. The management of EPA would surely benefit from such legislative changes.

For example, in the air program we would allocate resources not to the inevitably slow process of listing and controlling a handful of the immense number of chemical species that could cause trouble but rather to those actual situations in which some unacceptable risk can be found. We would work with industry and the affected communities to develop acceptable levels, which would naturally vary with conditions specific to different areas. Where appropriate, we would have the flexibility to establish ambient, source-specific, or technology-based standards tailored to the particularities of each risk situation.

In the water program we would pay more attention to the impact of pollutants on aquatic ecosystems, and invest time and attention where it promises to do the most good. We would question whether it pays to go from 95 to 99 percent removal of toxic substances and conventional pollutants at industrial sources where we know that the bulk of the pollution load on the water comes from urban and agricultural

*Our societies are beginning to realize that the very social and technological practices that have made them economically prosperous and politically powerful, increasingly damage human and environmental health. Slowly and grudgingly they are developing palliative measures to deal with the most obvious threats; but this piecemeal approach will not be sufficient to solve the ecological crisis and improve the quality of life. Technological fixes amount to little more than putting a finger in the bursting dike, whereas what is needed is a sociotechnological philosophy of man in his environment.*

Source: *Reason Awake: Science for Man,* by René Dubos. New York: Columbia University Press, 1970, p. 178. [Copyright © 1970 Columbia University Press. Used by permission.]

nonpoint runoff. By now it is clear that we are not going to eliminate all pollutant discharges into our waterways by the date specified in the Clean Water Act. Should we therefore wring our hands and cry "failure"? No, we should decide what kind of water quality we can live with, in which bodies of water, and what we are really prepared to pay to get it, and then we should put that kind of flexibility into law.

In groundwater protection we should start relating our concern for this resource to its presumptive use. Groundwater is a resource like any other. Obviously we cannot allow our groundwater to be carelessly fouled, but let us cultivate a sense of proportion as well as a sense of outrage. There is a great difference between the careless and casual destruction of a supply of drinking water and the demand that a vast pool of groundwater lying beneath an industrial metropolis be scrubbed as clean as it was when a dozen farmers lived above it.

We accept risks from chlorinated drinking water in most places in the United States right now, risk levels that many communities have declared to be unacceptable when presented by way of groundwater. Most chlorinated drinking waters contain chloroform, a product of the reaction between chlorine and the humic acids occurring naturally in many water sources. We have established an acceptable level for chloroform in drinking water, based on a balancing of the carcinogenic risk of this substance against the risk of pathogenic diseases that could result from insufficient chlorination. We ought to eventually establish the same kind of acceptable levels for the potential carcinogens that may show up in groundwater.

The hazardous waste problem boils down to specifying where we want the waste to go. In the last analysis, after we have reduced production and have recycled and treated as much as we can of our waste, we have three choices as to where to put it: the land, the air, or the water. Congress has never analyzed the total risk associated with a certain hazardous waste policy and the total cost of eliminating it and then used the analysis as the basis for legislation. However, that is what we will increasingly have to do if our national environmental policies are to retain any contact at all with day-to-day realities they are supposed to influence.

In the control of pesticides, we have relied heavily on the banning of chemicals that entail unacceptable risks. We must continue to do so, of course, but we must go beyond the single-substance ban as our major weapon. The industrial laboratory will always outpace the regulatory agency in providing substitutes for banned chemicals, and some of those substitutes in field use may prove as troublesome as the ones they replace. We must start looking at the long-term risks posed by whole systems of agriculture—land management, tillage practices, and herbicide and pesticide applications—and not just at the health risks associated with chemical residues on food or applicator contact.

What is the impact of all this chemical loading over the years on the ecological systems in which human culture is embedded? After decades of so-called pesticide control, we have not even begun to ask this question. Indeed, it is odd how little time is spent at the upper levels of EPA thinking about such things and how much time is spent worrying about tiny increases in the risk of a single human disease. We need a return to the vision of environmentalism embodied in the documents that created EPA.

## X

Some years ago, when pollution control seemed a far more overwhelming task than it does now, René Dubos said that the way to cope with such massive problems was to "think globally and act locally." It is still good advice, and may serve as a general prescription for successful management of technological risk in a democratic society as big and diverse as ours. Global thinking in the present case means dealing explicitly with the central questions of risk management: how to reconcile technological systems with social values; how to develop the consensus about potentially dangerous technologies that is necessary for continued growth; and how to establish and maintain trust in our protective institutions. We do not yet know how to deal very well with any of these questions, although I have tried to suggest some promising directions to follow.

This is why local action—a diversity of local actions—is necessary as well. I suspect that the most efficient way for our society to learn how to cope with risk is to enable hundreds or thousands of locally based risk-management endeavors to take place. Local risk-taking preferences could then be expressed, under broad limits set by higher levels of government. This will inevitably change public perceptions of risk, for in such perceptions familiarity breeds, not contempt, but the ability to discriminate between trivial and important risks. Fear is, after all, what has tended to paralyze public policy on these issues, and all our research shows that familiar risks engender less fear than unfamiliar ones. As our people begin to assess and manage risks at the local level, they will be preparing themselves to cope as citizens of a democratic society moving into a future dominated by barely imaginable technologies and fraught with unfamiliar risks.

# Questions for Thought and Discussion

1. Where do your sympathies lie with regard to the major risk assessment controversy described in this paper: with the side that believes the regulation of toxic substances should not be based on a set of unprovable assumptions, or with the side that holds that it is morally unacceptable to wait for firm evidence of human health effects? What arguments would you make to support your position?

2. In 1983 the National Research Council recommended the explicit and open codification of guidelines for making the inferences and assumptions that are a necessary part of carcinogenic risk assessment. Suppose that EPA adopts new guidelines in response to this recommendation and the following problems arise as a result: (1) a chemical that was formerly considered to be safe for a particular use is declared to be unsafe under the new rules, and (2) a chemical that was formerly declared unsafe for some use is now declared to be safe. If you were the head of EPA, how would you handle these problems?

3. What are the arguments for and against encouraging local authorities to do more risk assessment and having the federal government do less? Cast your response in the context of the problem of municipal waste incineration.

4. Ruckelshaus points up the contrast between the public's general acceptance of the cancer risks associated with chloroform in chlorinated drinking water and its refusal to accept smaller carcinogenic risks from other chemicals in drinking water. Do you think these two attitudes are inconsistent? Why or why not?

5. One proposed solution for siting hazardous waste disposal facilities is to hold an auction in which communities would be invited to submit closed bids stating the amount of monetary compensation they would accept in return for agreeing to be the host community for such a facility. What are the pros and cons of using such a device for risk management? Who would the stakeholders be? How would you expect them to react to such a proposal?

# No Risk Is the Highest Risk of All

## AARON WILDAVSKY
*University of California at Berkeley*

What to believe? Whom to trust? How to decide what to believe and whom to trust? These are not new questions. When they are applied to the birth of societies or the origins of religions, or even to the remnants of seemingly strange tribes, these questions do not jar our contemporary consciousness. But in a technological age, where the secrets of heredity appear within our grasp and space travel is more than science fiction, it does seem strange to hear them asked about ordinary aspects of everyday life.

Will you and I be able to breathe amidst the noxious fumes of industrial pollution? Will we be able to eat with poisonous substances entering our orifices? How can we sleep knowing that the light of the life-giving sun may be converted into death rays that burn our bodies to a crisp? How do we know that our mother's milk does not contain radiation or our meats putrefaction or our water cancer-causing chemicals? Should people be allowed to take risks?

Or should cars and factories be engineered so that risks can't be taken? Seat belts do increase driver and passenger safety; is it enough to tell people about them or must cars be required to have them or, since belts can be circumvented, should air bags be in-

stalled that inflate upon impact, whether the driver wants them or not? Deodorant commercials may urge, "Don't be half-safe!" but is it possible or desirable to be all safe? What sort of society raises these questions insistently everywhere one turns?

How extraordinary! The richest, longest-lived, best-protected, most resourceful civilization, with the highest degree of insight into its own technology, is on its way to becoming the most frightened. Has there ever been, one wonders, a society that produced more uncertainty more often about everyday life? It isn't much, really, in dispute—only the land we live on, the water we drink, the air we breathe, the food we eat, the energy that supports us. Chicken Little is alive and well in America. Evidently, a mechanism is at work ringing alarms faster than most of us can keep track of them. The great question is this: Is there something new in our environment or is there something new in our social relations?

Is it our environment or ourselves that have changed? Would people like us have had this sort of concern in the past? Imagine our reaction if most of modern technology were being introduced today. Anyone aware of the ambience of our times must be sensitive to the strong possibility that many risks, such as endless automotive engine explosions, would be postulated that need never occur, or, if they did, would be found bearable. Wouldn't airliners crash into skyscrapers that would fall on others, killing tens of thousands? Who could prove otherwise? Even today there are risks from numerous small dams far exceeding those from nuclear reactors. Why is the one feared and not the other? Is it just that we are used to the old or are some of us looking differently at essentially the same sorts of experience?

The usual way to resolve differences about the effects of technology is to rely on expert opinion—but the experts don't agree. Experts are used to disagreeing but they are not so used to failing to understand why they disagree. The frustration of scientists at the perpetuation rather than the resolution of disputes over the risks of technology is a characteristic feature of our time. So is the complaint of citizens that they no longer feel they are in touch with a recognizable universe. Unable or unwilling to depend on governmental officials or acknowledged experts, beset by the "carcinogen-of-the-month club," people see their eternal verities turn into mere matters of opinion. Is it the external environment that creates uncertainty over values, or is it a crisis in the culture that creates the values?

[Harm to health caused by accident or food or physical environment was far more prevalent in the past] than it is now. Each generation is richer in human ingenuity and purchasing power than its predecessor. People at the same ages are healthier than they once were—except possibly at the oldest ages, because so few made it that far in the good old days. Industrial accidents occurred more often and with greater severity and were attended to less frequently and with little success. Anyone who thinks otherwise knows little of the disease-ridden, callous, and technically inept past. As Aharoni has said (1):

> The hazards to health in the last century, or even a few decades ago, were many times greater than they are today; the probability of a deep depression fifty years ago was much greater than that of a much smaller recession today; water pollution in some lakes at the beginning of the century was at such a high level that no fish could exist in the water, and pollution and sanitary conditions in the cities often caused epidemics. Pneumonia, influenza and tuberculosis were fatal diseases at the beginning of the century and the majority of the population suffered many hazards almost unknown today, such as spoiled food, nonhygienic streets, recurring waves of diphtheria, danger-

ously exposed machinery in factories, a variety of hazards in the rural areas with almost no doctors around, and so on.

Why is it that so much effort is put into removing the last few percent of pollution or the last little bit of risk? Why do laws insist on the best available technology rather than the most cost-effective? Is it that we are richer and more knowledgeable and can therefore afford to do more? Why, then, if accidents or health rates are mainly dependent on personal behavior, does the vast bulk of governmental resources go into engineering safety in the environment rather than inculcating it into the individual? Perhaps the answer is that government can compel a limited number of companies, whereas it cannot coerce a multitude of individuals.

This answer is its own question: Why aren't individuals encouraged or allowed to face or mitigate their own risks? The great question asked here might be rephrased to inquire why risk is no longer individualized but is socialized, requiring a collective governmental response. Put this way, the question turns into an anomaly: since another important answer to the expansion of concern about risk is the decline of trust in government to give appropriate warning, why is risk being collectivized at the same time as the collective response mechanism is thought defective?

Even if the theory of increasing risk were reasonable, so many evils appear to lurk about us that it would, in any event, be necessary to choose among them. Since the awful things that *do* occur must be vastly exceeded by the virtually infinite number that *may* occur, societies might well exhaust their energies in preventive measures. Risks as well as resources must be allocated.

## Reducing Risks Is Not Riskless

Risk cannot be reduced for everyone. The sheer supply of resources that have to be devoted to minimizing risk means that surpluses are less likely to be available for unforeseen shocks. What is spent to lessen risk cannot be used to increase productivity. Since there is a scheme of progressive taxation (and perhaps regressive tax avoidance), the costs are borne disproportionately by classes of people whose ability to meet other contingencies must be proportionately reduced.

Rather than rely on a priori argument, however, it is desirable to distinguish various ways in which attempting to lower risks actually raises them or, alternatively, displaces them on to other objects.

Risk displacement has, in fact, become routine. When the conditions for certifying drugs are made difficult, a consequence is that far more tests are done on foreigners, who are generally poorer and possess far fewer resources with which to alleviate any unfortunate consequences. Resistance to nuclear power and other forms of energy does not necessarily mean that the demand for energy decreases but may well signify that energy must be imported, thereby, for example, flooding Indian lands to create dams in Canada. Eliminating certain pesticides and fertilizers in the United States may result in less food and higher food prices in countries that import from us. Alternatively, food might be favored at the expense of flying, as a National Academy of Sciences *News Report* expressed it (2):

> Clearly, food production is more vital to society than stratospheric aviation . . . , and limiting the use of nitrogen in food production [if both nitrogen fertilizer and jet fuel damage the atmosphere] would not appear to be an acceptable [control] strategy.

In one area after another the possible personal risk must be weighed against the probable personal harm. Nitrates might cause cancer, but they are also preservatives that guard against botulism. Early and frequent sexual intercourse with plenty of partners appears related to cervical cancer (hence the low reported rate among nuns), but the same phenomenon may be linked to lesser rates of breast cancer (hence the higher rates among nuns). Women who suffer frequent hot flashes relieved by estrogens must choose between probable relief now or possible cancer later.

The trade-offs become apparent when different government agencies, each in charge of reducing a single risk, insist on opposing policies. In order to reduce accidents, for instance, the Occupational Health and Safety Administration ordered the meat industry to place guardrails between carcasses and the people who slaughtered them; but the Food and Drug Administration forbade guardrails because they carry disease to the workers. Risk management would be easy if all you had to do was reduce one risk without worrying about others.

If it were merely a matter of being cautious, all of us might do well to err on the safe side. However, not only do we not know or agree on how safe is safe, but by trying to make ourselves super-safe, for all we know, we may end up super-sorry. Not only are there risks in reducing risks, but safety features themselves, as with catalytic converters on cars, may become a hazard. The more safeguards on line in an industrial plant, the greater the chance that something will go wrong.

Safety may become its own defect, and the new "safe" technologies may turn out to be more harmful than the ones they replaced. The superior safety of solar versus nuclear energy, for example, has become virtually an article of faith. Yet, as a Canadian study shows, if the entire production process (not merely the day-to-day workings) are considered, nuclear is considerably safer than solar, with natural gas the safest of all. How could that be? Solar energy is inefficient compared to nuclear. Since so much more effort per kilowatt generated is involved in producing solar panels and piping, the normal relative accident rate in industry is far greater for solar than for nuclear power production—and the evidence for the number of people falling off roofs or chemicals leaking into heated water is not in yet (3)!

The dilemma of risk is the inability to distinguish between risk reduction and risk escalation. Whether there is risk of harm, how great it is, what to do about it, whether protective measures will not in fact be counterproductive are all in contention. So far the main product of the debate over risk appears to be uncertainty.

Analyzing the argument in the abstract, divorced from confusing content, may help us understand its central tendencies. Planet Earth has a fragile life-support system whose elements are tightly linked and explosive and whose outcomes are distant and uncertain. Because the effects of current actions have long-lasting future consequences, the interests of future generations have to be considered. Because the outcomes are uncertain, calculations have to be conservative: when in doubt, don't. If the moral is "Don't start what you can't stop," the conclusion will probably be "This trip isn't strictly necessary." Indeed, this is the central tendency: under most circumstances most of the time doing little is safer than risking much. The consequence for scientists is that you can't do anything that won't have unanticipated effects. If required to give a guarantee that future generations will be better off in regard to every individual action, the scientist (or the businessman or the politician) cannot so certify.

By what right, one may ask, does anyone enrich himself by endangering future generations? By what right, it may be asked in return, does anyone impoverish the future by denying it choice? What would you have to believe about the future to sacrifice the present to it? It is not necessary to postulate a benevolent social intelligence that is a guarantee of good no matter what happens in order to believe that a capable future will be able to take care of itself. The future can acquire new strengths to compensate for old weaknesses; it can repair breakdowns, dampen oscillations, divert difficulties to productive paths. Not to be able to do this implies that consequences which current actions create are both unique and irreversible, so that acceptable alternatives are no longer available and cannot be created. Without more knowledge than anyone possesses about most phenomena, it is difficult to see how this claim could be made with confidence.

Rather, the reverse appears plausible: the future is being sacrificed for the present. Fear of the future is manifested by preempting in the present what would otherwise be future choices. The future won't be allowed to make mistakes because the present will use up its excess resources in prevention. One way of looking at this is that cocoonlike concern for the future leads to sacrifice of present possibilities; another way is that the purveyors of prevention want not only to control the present but, by preempting its choices, the future as well.

## The Politics of Prevention

Reducing risk for some people someplace requires increasing it for others elsewhere. For if risk is not in infinite supply, it also has to be allocated, and less for me may mean more for you. The distribution of risk, like the allocation of uncertainty, becomes an ordinary object of political contention (4). Insofar as the argument over risk is about power (who will bear the costs of change?), the observation that it occasions political passions is less surprising.

Most accidents occur at home and most industrial accidents are traceable to individual negligence. Yet remedies are sought in governmental regulation of industry, and the onus for adverse experience is increasingly placed on manufacturers, not on users of products. Indeed, product liability insurance has become such a major impediment to doing business, especially for small firms, that Congress is considering subsidizing the cost. The risks of carelessness in use and/or in manufacture are thus spread among the population of taxpayers.

But is a political approach mandated when individuals do not choose to follow healthy habits? If individuals eat regularly, exercise moderately, drink and smoke little, sleep six to eight rather than four to fourteen hours a night, worry a little but not too much, keep their blood pressure down, drive more slowly, wear seat belts, and otherwise take care of themselves, they would (on the average) live longer with far fewer illnesses.

Evidently, they don't do these things, but they are increasingly asking government to constrain their behavior (mandating seat belts, imposing speed limits) or to insure them against the consequences (by paying medical bills or providing drug rehabilitation programs). Politicization follows socialization, and politicization is not inadvertent but is a direct consequence of attempting to shift the burden of risk.

Most of the time most decisions are reactions to what has already happened. Within changing circumstances, security depends on our collective capacity to cope—after the fact—with various disasters. Suppose, however, that collective confidence wanes, either because of decreasing trust in internal social relations or increasing menace from without. The institutions through which people relate may be in disrepute, or the environment perceived as precarious, so that the slightest error spreads, causing catastrophe throughout the system.

The evils that do crop up are limited, if by nothing else, by our capacity to recognize them. But the evils that might come about are potentially limitless. Shall we err by omission, then, taking the chance that avoidable evils if anticipated might overwhelm us? Or shall we err by exhaustion, using up our resources to forestall the threat of possible evils that might never have occurred in order to anticipate unperceived but actual disasters that might do us in?

The centerpiece of the call to swift, centralized action is the domino theory—generalized to include the evils of domestic policy. Once we feared that an unfriendly regime in Laos would bring the inevitable "fall" of Vietnam and Cambodia, which in turn would undermine Thailand, and then the Philippines, then even Japan or West Berlin. Our alliances appeared a tightly linked, explosive system, in which every action carried a cumulative impact, so that if the initial evil were not countered, the international system would explode.

As in the past we saw the inevitability of the balance of power, so now we see each domestic action linked inextricably to the next. When all systems are interrelated, a single mistake opens the door to cumulative evils. The variety of areas to which this approach might be applied can be seen from the numerous calls for impact statements. Inflation, energy, health, research, safety, and now war impact statements have permeated the smallest segments of governmental activity. Understanding the impacts in any one of these areas could be effective. But identifying impacts, in essence, is a claim for special priority. To include all possible impacts makes for a world of all constants and no variables, with no room for trade-offs, bargaining, or judgment.

If all policy areas could be subject to interrelated impact statements, we would achieve an *anticipatory democracy*. In that world we would expect each policy area to take on those same characteristics prominent at the height of the cold war: secrecy, "cooperation" with or deference to executive leadership, and an insistence on the moral authority of the president to make these (unending) crucial choices. A preventive policy, in fact, could succeed only if it were invested with enormous faith (action in the absence of things seen), for that is what is required to trust leadership when it acts to control dangers that cannot yet be observed by the rest of us.

This pattern might be called the collectivization of risk: citizens may not choose what dangers they will face. Individual judgment and choice is replaced by governmental action that wards off potential evils automatically before they materialize. Yet if leaders act to avoid every imagined evil, where will their control end—or begin?

What will the fight be over? Money, naturally. The politics of prevention is expensive. The cost of anticipating evils, whether or not they materialize, is enormous. (Biologists tell of organisms that die not so much because of invasion by lethal foreign bodies but rather because the attack arouses all defense mechanisms of the organism at once, thereby leading to annihilation by exhaustion.) Ways must be found, therefore, either to raise huge sums by taxation or to get citizens to bear the cost personally.

## Market Socialism

Who benefits from a preventive polity—warding off evils, as yet unseen, by spreading the risk around? Who gains from this collectivization of risk? To answer, we must find out how costs are assessed when risks are collectivized. We might call the process "market socialism." Costs, instead of appearing directly in the government budget (to be supported by taxes and borrowing), are passed on to consumers indirectly through higher charges to producers. Indeed, costs even may appear as government revenues—taxes on business—that also are ultimately passed on to consumers.

When new safety standards are imposed, for example, governmental expenditures appear relatively small—usually just for inspectors—but larger costs are incurred by industry for more equipment, additional safety supervisors, and new tests that become part of production costs. Naturally, the consumer identifies the company rather than the government with higher prices. Think of regulations about housing density, insulation, safety features, and the like: each raises the cost of housing, diffusing it among the population, with only a few inspectors charged up as governmental expenditure.

And consider effluent charges: by placing a tax on pollution, government hopes to make it worthwhile for producers to reduce levels of noxious substances without direct regulation. Producers who prefer not to spend money clearing up streams shrug their shoulders at government imperatives and, in effect, become tax collectors by passing on charges to private citizens. Market socialism is the name of this reverse-tax process.

The chief concomitants of market socialism are low visibility, capital intensity, and social selectivity. Consider, first, low visibility: market socialism blurs cause and effect between public and private sectors. Which sector, for instance, is responsible for increased housing costs? Which sector inflates the cost of safety features in manufacturing? Suppose, to make the contrast vivid, government closes down a plant because it is unsafe for workers or neighbors. Everyone can see what is happening. Should workers and neighbors prefer existing hazards to threatened unemployment, they might be able to exert their will.

Suppose, however, that environmental and safety regulations prevent a projected plant from locating in a low-income area, shifting it instead to a more affluent suburb. No one would know that the location had been passed over, thus imposing costs on residents who wish employment. It is harder to mobilize support for something that did not happen than for something about to be undone.

Allied with low visibility is market socialism's widespread displacement of costly effects onto unsuspecting (and unconsenting) adults. Take the well-known problem of drug lag: the call for countless trials and procedures, as well as the demand for levels of safety difficult to prove, means that drugs which may be on the horizon may not be developed at all, are produced later, or come at a higher cost.

These sorts of considerations suggest that a preventive polity (collectivizing risk through market socialism) does not have the same impact on people in different social and economic strata. The lower classes—with no cars or smaller ones, and less income with which to absorb price increases—will not do so well. Predictably, the very people who are supposed to benefit from the collectivization of risk will turn out to be the losers. Why this should be so, and how the underlying mechanism works, should be clarified by investigating the other side of market socialism—its capital intensity and social selectivity.

When government seeks to anticipate a wide variety of risks or to mitigate those that do exist, less money is available for investment to create future jobs, or for expenditure to support existing jobs. To some extent, there is a substitution effect, as jobs devoted to production give way to jobs supporting safety. (Whatever happens, the price goes up because consumers pay for safety whether they want it or not.) These jobs, however, go to middle-class professionals and technicians, lawyers, inspectors, clerks, etc., who displace production workers.

In addition, companies make intensive use of labor-saving machinery, wherever possible, in order to economize, and thereby displace manual workers. But companies have no choice, according to government regulations, other than to hire people to perform necessary regulatory functions. Predictably, there has been a sudden surge in demand for specialists in industrial medicine. In this there is a certain symmetry: the preventive polity is a style of government suitable for people who already have more than enough and want to safeguard it.

A politics of prevention reflects a lack of confidence in existing social relations. People who trust each other, and the institutions through which they interrelate, will have sufficient self-confidence to deal with most difficulties and the faith to follow leaders in trying to head off a few evils before they occur. When trust declines, self-confidence declines, and the need for leadership increases, just as willingness to follow it decreases.

In a proverbial world—nothing ventured, nothing gained; a bird in hand is worth two in the bush; look before you leap; but he who hesitates is lost—each adage is countered by its opposite. Folk wisdom degenerates into indecision. When one person can't make up his mind, that is his personal problem. But when a society suffers from the same disease, that is a severe social problem. Once again we are back to the social roots of risk.

## Social Roots of Risk

Most of the decisions that affect our lives are not made directly. They are the results of innumerable other choices made on other grounds in other places for other purposes, whose cumulative impact may still be considerable. They are, so to speak, indirect resultants rather than direct decisions. The distribution of income, for example, or the size of the population are not decisions, strictly speaking, but end up as resultants of innumerable individual and collective choices whose cumulative consequences add up.

When a resultant becomes a decision, taken on a given day and formalized as legislation and regulation, as happened where governments were given responsibility for unemployment, the transition marks a watershed in collective consciousness. The conversion of individual reward-seeking into safety-conscious collective choices about how much risk is acceptable is just such a turning point. It has special interest because it appears divorced from (or at least not necessarily required by) environmental evidence.

It is not my intention, I hasten to say, to allege that risk seekers are rational and risk avoiders are not. On the contrary, I wish to rationalize, to make explicable in terms of a different set of values, beliefs, and social relations, these otherwise unreasonable actions. People think by way of actions and objects as well as through categories of scientific thought. If we hear that chemicals are carcinogens and nuclear waste is industrial excrement, if we hear that nature is natural, good, and golden, pure and perfect, that sunshine is a disinfectant in politics and sunsets close the curtain on the big bad bureaucracy, these symbols speak to us in strange tongues.

Society is symbolized not only by what we take into our bodies but by the environment we create for ourselves on the outside. At first blush it seems strange to say that the physical environment is man-made. God or nature or geologic time have created the atmosphere and the mountains and the lakes and all that, apart from human intervention. No doubt. No doubt, too, that mankind cannot change its physical form at will. But how people feel about their environment, whether it is friend or enemy, a place of sanctuary or a sign for sacrifice, is man-made.

The environment as it enters man's moral universe is part of the rationale for why he lives the way he does. The environment is a cultural construct that is part of the cosmology that justifies the structure of society. If the parts of man's life are to fit together, the physical and the social must be in harmony.

So Mary Douglas declares in her seminal paper on "Environments at Risk" (5):

> Once there is a consensus on a moral order, this determines, within wide limits, the way a community experiences physical conditions: this perception is then enforced by methods of social control. Among verbal weapons of control, time is one of the four final arbiters. Time, money, God, and nature, usually in that order, are universal trump cards plunked down to win an argument.

Restriction is coercion. Society cannot go on as it was before. Its people must live differently. Doubt it? Ask an environmentalist about the consequences of running out of energy (the very word suggests losing strength): immediately you will hear about getting rid of "gas-guzzling" automobiles, walking or riding a bicycle to work, residing in closer and more self-contained communities, and otherwise doing what comes naturally. Before your eyes, an entirely new way of life will unfold out of a seemingly small restriction in fossil fuels.

Changes in the acceptable level of risk would be both harbingers of things to come and consequences of a new way of life. What to favor and what to fear are cultural constructs that enable us to walk right past snarling monsters and run away from little-bitty things. If we want to know why we are fearful about what and whether we should be, this is equivalent to asking the cultural question: How should we live?

The extent of cultural change in our times, roughly the last quarter century, has been astounding—phenomenal—remarkable. Choose whatever superlative you like, values, beliefs, and social structure have been changing fantastically fast. There has been an enormous expansion in the types of transactions permitted, even encouraged, in society. Sex roles—what is expected and accepted by whom toward whom—are being transformed. Family authority and the old acceptance of racial superiority are being undermined. Heroes and villains, spenders versus savers, conservation versus exploitation, artist versus entrepreneur are changing places.

Answer these questions for yourself: Is there anything most people could do twenty-five years ago that they can't do today? Hardly. Is there behavior acceptable today that was frowned upon back then? No doubt. And the few exceptions are instructive. Though individuals must follow far fewer rules, business must follow many more. As marital vows become less holy, nature becomes more sanctified.

All this—vast and speedy cultural change, the decline in scope and acceptance of authority and expertise, disagreements over the relative risks of technology—represents not random occurrences but interconnected phenomena. The decline in acceptable physical risk and the increasing desire to shift [responsibility for] that risk to government are caused by cultural change. How might we conceive of risk as social phenomenon?

Assume that the level of social risks varies with the form of society. If social risks may be perceived as loss of possible insurance cover for the social effects of physical risks, there is a relation between perceived ability to deliver social insurance and perceived danger from physical risks. (Define physical risks as physical harm caused by accident or malicious attack. Define social risks as loss of social standing, dishonor, unemployment, loss of social initiative for close family relationships, or general lack of ability to make a go of it in terms of prevailing social status.) Social risks are constituted by the inability of the social system to deliver compensation for physical risks. Were individual social units with a given distribution of social and physical risks unable to underwrite their own risks, so to speak, they might seek to pass them on to others through government.

Now assume, with good reason, a tendency to keep the risk load tailored to the same size and distribution of resources so that the regular underwriting of risks seems reasonable. Partitioning of resources through the community results in de facto partitioning of risks and creation of risk subunits. The smaller the risk unit, naturally, the smaller its perceived appropriate risk load. Anything that reduces ability of individuals to deliver social insurance (e.g., bureaucratization), therefore, will increase perception of risk. If the risks we fear are social, not physical, I conclude that answers to questions should not be sought in the world outside but inside society (6).

The formal literature on risk, as well as public discussion, is preoccupied with the period of calculation before decision. But there is not only a degree of risk that may be associated with a particular act, there is ability to cope with the consequences *after* they occur. The period "after" (in systems that are are social and therefore interactive) has a strong effect on the decision "before."

Knowing that they will be taken care of afterward, that others "near and dear" will share their risks, enables individuals to take many risks habitually without profound assurance or with no assurance at all. Few demands need be placed on future knowledge because the consequences—for their position or their family—are being cared for. Action may then be taken to secure gain, not only to prevent loss. Then public policy would not ape the great aircraft carrier, most of whose effect is devoted to defending itself.

## Reprise

The vision of a no-risk society may be considered a caricature. What about risks of the middle range? No one wants to get hurt by a product that contains hidden dangers. Thalidomide that deforms babies is a case in point. Nor would anyone relish the thought of delayed effects, like black lung diseases from coal mining or radiation from plutonium piling, debilitating our bodies. Error, ignorance, and avarice may combine to cause harm. Identifying and eliminating or modifying these effects is evidently in order. When harmful effects are immediate, obvious, and traceable to the substance, regulatory action can be relatively straightforward. But when hazards are hidden, or caused by unforeseen combinations, or delayed over long periods of time, correction involves compromise between safety and stultification.

When most substances are assumed safe and only a few are judged harmful, the line may be drawn against risking contact. But when the relationship is reversed, when most substances are suspect and few are safe, refusing risk may become ruinous. It is not, in other words, the individual instance but the across-the-board concern with contamination that is of cultural significance.

Theodore Roosevelt was fond of dividing people into those who pointed with pride or viewed with alarm. On an a priori basis, without understanding, there should at any time be ample reason to do both. What I want to know is why the most resourceful society of all times does so little pointing and so much viewing, for asking others to take over risks may be tantamount to inviting them to take away our freedom.

## References

1. Yair Aharoni, pers. comm.

2. NAS *News Report,* October 1978, No. 10, p. 1.

3. Herbert Inhaber. May 1978. *Risk of Energy Production,* 2nd ed. Ottawa, Ontario: Atomic Energy Control Board, AECB 1119/REV-1.

4. This section is based on my article with Sanford Weiner, "The Prophylactic Presidency," Russell Sage Foundation, for the Hoover Institute.

5. Mary Douglas. *Times Literary Supplement,* 30 Oct. 1970, pp. 1273–75.

6. For further discussion see Mary Douglas, *Cosmology, An Inquiry into Cultural Bias,* The Frazer Lecture, given at Cambridge in 1976, and Aaron Wildavsky, "Economy and the Environment/Rationality and Ritual: A Review of the Uncertain Search for Environmental Quality." *Stanford Law Review,* Issue I, Volume 29, Sept. 1976.

# Questions for Thought and Discussion

1. What are the pros and cons of simply educating people about air bags as an option, instead of requiring that they be installed?

2. The author observes that in the past, hazards to health were generally far greater than they are now. Can you identify five specific exceptions to this observation, that is, five risks that you believe are *worse* now than in the past?

3. The nuclear accidents at Three Mile Island and at Chernobyl, which happened after this paper was published, have undoubtedly increased societal concerns about the risks of nuclear power. Yet, assuming it is true that the risks from numerous small dams far exceed those from nuclear reactors, why do you suppose, as Wildavsky contends, that there is more fear about the latter?

4. How would you respond to the following questions posed by the author?

- "Is it the external environment that creates uncertainty over values, or is it a crisis in the culture that creates the values?"
- "Why is risk being collectivized at the same time as the collective responsive mechanism is thought defective?"

5. Wildavsky cites thalidomide and black lung disease as examples in which the "compromise between safety and stultification" came down heavily on the side of safety. List three other risks in which you think safety should be weighed heavily. Explain your choices. Some individuals consider motorcycle helmet requirements to be stultifying. Do you agree, and why or why not?

# Cost-Benefit Analysis:
# An Ethical Critique

## STEVEN KELMAN
*Harvard University*
*Cambridge, Massachusetts*

At the broadest and vaguest level, cost-benefit analysis may be regarded simply as systematic thinking about decision-making. Who can oppose, economists sometimes ask, efforts to think in a systematic way about the consequences of different courses of action? The alternative, it would appear, is unexamined decision-making. But defining cost-benefit analysis so simply leaves it with few implications for actual regulatory decision-making. Presumably, therefore, those who urge regulators to make greater use of the technique have a more extensive prescription in mind. I assume here that their prescription includes the following views:

(1) There exists a strong presumption that an act should not be undertaken unless its benefits outweigh its costs.

(2) In order to determine whether benefits outweigh costs, it is desirable to attempt to express all benefits and costs in a common scale or denomina-

tor, so that they can be compared with each other, even when some benefits and costs are not traded on markets and hence have no established dollar values.

(3) Getting decision-makers to make more use of cost-benefit techniques is important enough to warrant both the expense required to gather the data for improved cost-benefit estimation and the political efforts needed to give the activity higher priority compared to other activities, also valuable in and of themselves.

My focus is on cost-benefit analysis as applied to environmental, safety, and health regulation. In that context, I examine each of the above propositions from the perspective of formal ethical theory, that is, the study of what actions it is morally right to undertake. My conclusions are:

(1) In areas of environmental, safety, and health regulation, there may be many instances where a certain decision might be right even though its benefits do not outweigh its costs.

(2) There are good reasons to oppose efforts to put dollar values on non-marketed benefits and costs.

(3) Given the relative frequency of occasions in the areas of environmental, safety, and health regulation where one would not wish to use a benefits-

outweigh-costs test as a decision rule, and given the reasons to oppose the monetizing of non-marketed benefits or costs that is a prerequisite for cost-benefit analysis, it is not justifiable to devote major resources to the generation of data for cost-benefit calculations or to undertake efforts to "spread the gospel" of cost-benefit analysis further.

## I

How do we decide whether a given action is morally right or wrong and hence, assuming the desire to act morally, why it should be undertaken or refrained from? Like the Molière character who spoke prose without knowing it, economists who advocate use of cost-benefit analysis for public decisions are philosophers without knowing it: the answer given by cost-benefit analysis, that actions should be undertaken so as to maximize net benefits, represents one of the classic answers given by moral philosophers—that given by utilitarians. To determine whether an action is right or wrong, utilitarians tote up all the positive consequences of the action in terms of human satisfaction. The act that maximizes attainment of satisfaction under the circumstances is the right act. That the economists' answer is also the answer of one school of philosophers should not be surprising. Early on, economics was a branch of moral philosophy, and only later did it become an independent discipline.

Before proceeding further, the subtlety of the utilitarian position should be noted. The positive and negative consequences of an act for satisfaction may go beyond that act's immediate consequences. A facile version of utilitarianism would give moral sanction to a lie, for instance, if the satisfaction of an individual attained by telling the lie was greater than the suffering imposed on the lie's victim. Few utilitarians would agree. Most of them would add to the list of negative consequences the effect of the one lie on the tendency of the person who lies to tell other lies, even in instances when the lying produced less satisfaction for him than dissatisfaction for others. They would also add the negative effects of the lie on the general level of social regard for truth-telling, which has many consequences for future utility. A further consequence may be added as well. It is sometimes said that we should include in a utilitarian calculation the feeling of dissatisfaction produced in

the liar (and perhaps in others) because, by telling a lie, one has "done the wrong thing." Correspondingly, in this view, among the positive consequences to be weighed into a utilitarian calculation of truth-telling is satisfaction arising from "doing the right thing." This view rests on an error, however, because it *assumes* what it is the purpose of the calculation to *determine*—that telling the truth in the instance in question is indeed the right thing to do. Economists are likely to object to this point, arguing that no feeling ought "arbitrarily" to be excluded from a complete cost-benefit calculation, including a feeling of dissatisfaction at doing the wrong thing. Indeed, the economists' cost-benefit calculations would, at least ideally, include such feelings. Note the difference between the economist's and the philosopher's cost-benefit calculations, however. The economist may choose to include feelings of dissatisfaction in his cost-benefit calculation, but what happens if somebody asks the economist, "Why is it right to evaluate an action on the basis of a cost-benefit test?" If an answer is to be given to that question (which does not normally preoccupy economists but which does concern both philosophers and the rest of us who need to be persuaded that cost-benefit analysis is right), then the circularity problem reemerges. And there is also another difficulty with counting feelings of dissatisfaction at doing the wrong thing in a cost-benefit calculation. It leads to the perverse result that under certain circumstances a lie, for example, might be morally right if the individual contemplating the lie felt no compunction about lying and morally wrong only if the individual felt such a compunction!

This error is revealing, however, because it begins to suggest a critique of utilitarianism. Utilitarianism is an important and powerful moral doctrine. But it is probably a minority position among contemporary moral philosophers. It is amazing that economists can proceed in unanimous endorsement of cost-benefit analysis as if unaware that their conceptual framework is highly controversial in the discipline from which it arose—moral philosophy.

Let us explore the critique of utilitarianism. The logical error discussed before appears to suggest that we have a notion of certain things being right or wrong that *predates* our calculation of costs and benefits. Imagine the case of an old man in Nazi

Germany who is hostile to the regime. He is wondering whether he should speak out against Hitler. If he speaks out, he will lose his pension. And his action will have done nothing to increase the chances that the Nazi regime will be overthrown: he is regarded as somewhat eccentric by those around him, and nobody has ever consulted his views on political questions. Recall that one cannot add to the benefits of speaking out any satisfaction from doing "the right thing," because the purpose of the exercise is to determine whether speaking out *is* the right thing. How would the utilitarian calculation go? The benefits of the old man's speaking out would, as the example is presented, be nil, while the costs would be his loss of his pension. So the costs of the action would outweigh the benefits. By the utilitarians' cost-benefit calculation, it would be *morally wrong* for the man to speak out.

Another example: two very close friends are on an Arctic expedition together. One of them falls very sick in the snow and bitter cold, and sinks quickly before anything can be done to help him. As he is dying, he asks his friend one thing, "Please, make me a solemn promise that ten years from today you will come back to this spot and place a lighted candle here to remember me." The friend solemnly promises to do so, but does not tell a soul. Now, ten years later, the friend must decide whether to keep his promise. It would be inconvenient for him to make the long trip. Since he told nobody, his failure to go will not affect the general social faith in promise-keeping. And the incident was unique enough so that it is safe to assume that his failure to go will not encourage him to break other promises. Again, the costs of the act outweigh the benefits. A utilitarian would need to believe that it would be *morally wrong* to travel to the Arctic to light the candle.

A third example: a wave of thefts has hit a city and the police are having trouble finding any of the thieves. But they believe, correctly, that punishing someone for theft will have some deterrent effect and will decrease the number of crimes. Unable to arrest any actual perpetrator, the police chief and the prosecutor arrest a person whom they know to be innocent and, in cahoots with each other, fabricate a convincing case against him. The police chief and the prosecutor are about to retire, so the act has no effect on any future actions of theirs. The fabrication is perfectly executed, so nobody finds out about it. Is the *only* question involved in judging the act of framing the innocent man that of whether his suffering from conviction and imprisonment will be greater than the suffering avoided among potential crime victims when some crimes are deterred? A utilitarian would need to believe that it is *morally right to punish the innocent man* as long as it can be demonstrated that the suffering prevented outweighs his suffering.

And a final example: imagine two worlds, each containing the same sum total of happiness. In the first world, this total of happiness came about from a series of acts that included a number of lies and injustices (that is, the total consisted of the immediate gross sum of happiness created by certain acts, minus any long-term unhappiness occasioned by the lies and injustices). In the second world the same amount of happiness was produced by a different series of acts, none of which involved lies or injustices. Do we have any reason to prefer the one world to the other? A utilitarian would need to believe that the choice between the two worlds is a *matter of indifference*.

To those who believe that it would not be morally wrong for the old man to speak out in Nazi Germany or for the explorer to return to the Arctic to light a candle for his deceased friend, that it would not be morally right to convict the innocent man, or that the choice between the two worlds is not a matter of indifference—to those of us who believe these things, utilitarianism is insufficient as a moral view. We believe that some acts whose costs are greater than their benefits may be morally right and, contrariwise, some acts whose benefits are greater than their costs may be morally wrong.

This does not mean that the question whether benefits are greater than costs is morally irrelevant. Few would claim such. Indeed, for a broad range of individual and social decisions, whether an act's benefits outweigh its costs is a sufficient question to ask. But not for all such decisions. These may involve situations where certain duties—duties not to lie, break promises, or kill, for example—make an act wrong, even if it would result in an excess of benefits over costs. Or they may involve instances where people's rights are at stake. We would not permit rape even if it could be demonstrated that the rapist derived enormous happiness from his act, while the victim experienced only minor displeasure. We do not do cost-benefit analyses of freedom of speech or trial by jury. The Bill of Rights was not RARGed.* As the United Steelworkers noted in a comment on the Occupational Safety and Health Administration's economic analysis of its proposed rule to reduce worker exposure to carcinogenic coke-oven emissions, the Emancipation Proclamation was not subjected to an inflationary impact statement. The notion of human rights involves the idea that people may make certain claims to be allowed to act in certain ways or to be treated in certain ways, even if the sum of benefits achieved thereby does not outweigh the sum of costs. It is this view that underlies the statement that "workers have a right to a safe and healthy work place" and the expectation that OSHA's decisions will reflect that judgment.

In the most convincing versions of non-utilitarian ethics, various duties or rights are not absolute. But each has a *prima facie* moral validity so that, if duties or rights do not conflict, the morally right act is the act that reflects a duty or respects a right. If duties or rights do conflict, a moral judgment, based on conscious deliberation, must be made. Since one of the duties non-utilitarian philosophers enumerate is the duty of beneficence (the duty to maximize happiness), which in effect incorporates all of utilitarianism by reference, a non-utilitarian who is faced with conflicts between the results of cost-benefit analysis and non-utility-based considerations will need to undertake such deliberation. But in that deliberation, additional elements, which cannot be reduced to a question of whether benefits outweigh costs, have been introduced. Indeed, depending on the moral importance we attach to the right or duty involved, cost-benefit questions may, within wide ranges, become irrelevant to the outcome of the moral judgment.

In addition to questions involving duties and rights, there is a final sort of question where, in my view, the issue of whether benefits outweigh costs should not govern moral judgment. I noted earlier that, for the common run of questions facing individuals and societies, it is possible to begin and end our judgment simply by finding out if the benefits of the contemplated act outweigh the costs. This very fact means that one way to show the great importance, or value, attached to an area is to say that decisions involving the area should not be determined by cost-benefit calculations. This applies, I think, to the view many environmentalists have of decisions involving our natural environment. When officials are deciding what level of pollution will harm certain vulnerable people—such as asthmatics or the elderly—while not harming others, one issue involved may be the right of those people not to be sacrificed on the altar of somewhat higher living standards for the rest of us. But more broadly than this, many environmentalists fear that subjecting decisions about clean air or water to the cost-benefit tests that determine the general run of decisions removes those matters from the realm of specially valued things.

## II

In order for cost-benefit calculations to be performed the way they are supposed to be, all costs and benefits must be expressed in a common measure, typically dollars, including things not normally bought and sold on markets, and to which dollar prices are therefore not attached. The most dramatic example of such things is human life itself; but many of the other benefits achieved or preserved by environmental policy—such as peace and quiet, fresh-smelling air, swimmable rivers, spectacular vistas—are not traded on markets either.

Economists who do cost-benefit analysis regard the quest after dollar values for non-market things as a difficult challenge—but one to be met with relish. They have tried to develop methods for imputing a person's "willingness to pay" for such things, their approach generally involving a search for bundled goods that *are* traded on markets and that vary as to whether they include a feature that is, *by itself*, not marketed. Thus, fresh air is not marketed, but houses in different parts of Los Angeles that are similar except for the degree of smog are. Peace and quiet is not marketed, but similar houses inside and outside airport flight paths are. The risk of death is not marketed, but similar jobs that have different levels of risk are. Economists have produced many often ingenious efforts to impute dollar prices to non-marketed things by observing the premiums accorded

*Editors' note:* RARG stands for Regulatory Analysis Review Group.

homes in clean air areas over similar homes in dirty areas or the premiums paid for risky jobs over similar nonrisky jobs.

These ingenious efforts are subject to criticism on a number of technical grounds. It may be difficult to control for all the dimensions of quality other than the presence or absence of the non-marketed thing. More important, in a world where people have different preferences and are subject to different constraints as they make their choices, the dollar value imputed to the non-market things that most people would wish to avoid will be lower than otherwise, because people with unusually weak aversion to those things or unusually strong constraints on their choices will be willing to take the bundled good in question at less of a discount than the average person. Thus, to use the property value discount of homes near airports as a measure of people's willingness to pay for quiet means to accept as a proxy for the rest of us the behavior of those least sensitive to noise, of airport employees (who value the convenience of a near-airport location) or of others who are susceptible to an agent's assurances that "it's not so bad." To use the wage premiums accorded hazardous work as a measure of the value of life means to accept as proxies for the rest of us the choices of people who do not have many choices or who are exceptional risk-seekers.

A second problem is that the attempts of economists to measure people's willingness to pay for non-marketed things assume that there is no difference between the price a person would require for *giving up* something to which he has a preexisting right and the price he would pay to *gain* something to which he enjoys no right. Thus, the analysis assumes no difference between how much a homeowner would need to be paid in order to give up an unobstructed mountain view that he already enjoys and how much he would be willing to pay to get an obstruction moved once it is already in place. Available evidence suggests that most people would insist on being paid far more to assent to a worsening of their situation than they would be willing to pay to improve their situation. The difference arises from such factors as being accustomed to and psychologically attached to that which one believes one enjoys by right. But this creates a circularity problem for any attempt to use cost-benefit analysis to determine

*whether* to assign to, say, the homeowner the right to an unobstructed mountain view. For willingness to pay will be different depending on whether the right is assigned initially or not. The value judgment about whether to assign the right must thus be made first. (In order to set an upper bound on the value of the benefit, one might hypothetically assign the right to the person and determine how much he would need to be paid to give it up.)

Third, the efforts of economists to impute willingness to pay invariably involve bundled goods exchanged in *private* transactions. Those who use figures garnered from such analysis to provide guidance for *public* decisions assume no difference between how people value certain things in private individual transactions and how they would wish those same things to be valued in public collective decisions. In making such assumptions, economists insidiously slip into their analysis an important and controversial value judgment, growing naturally out of the highly individualistic microeconomic tradition—namely, the view that there should be no difference between private behavior and the behavior we display in public social life. An alternative view—one that enjoys, I would suggest, wide resonance among citizens—would be that public, social decisions provide an opportunity to give certain things a higher valuation than we choose, for one reason or another, to give them in our private activities.

Thus, opponents of stricter regulation of health risks often argue that we show by our daily risk-taking behavior that we do not value life infinitely, and therefore our public decisions should not reflect the high value of life that proponents of strict regulation propose. However, an alternative view is equally plausible. Precisely because we fail, for whatever reasons, to give life-saving the value in everyday personal decisions that we in some general terms believe we should give it, we may wish our social decisions to provide us the occasion to display the reverence for life that we espouse but do not always show. By this view, people do not have fixed unambiguous "preferences" to which they give expression through private activities and which therefore should be given expression in public decisions. Rather, they may have what they themselves regard as "higher" and "lower" preferences. The latter may come to the fore in private decisions, but people may want the former to come to the fore in public decisions. They may sometimes display racial prejudice, but support antidiscrimination laws. They may buy a certain product after seeing a seductive ad, but be skeptical enough of advertising to want the government to keep a close eye on it. In such cases, the use of private behavior to impute the values that should be entered for public decisions, as is done by using willingness to pay in private transactions, commits grievous offense against a view of the behavior of the citizen that is deeply engrained in our democratic tradition. It is a view that denudes politics of any independent role in society, reducing it to a mechanistic, mimicking recalculation based on private behavior.

Finally, one may oppose the effort to place prices on a non-market thing and hence in effect incorporate it into the market system out of a fear that the very act of doing so will reduce the thing's perceived value. To place a price on the benefit may, in other words, reduce the value of that benefit. Cost-benefit analysis thus may be like the thermometer that, when placed in a liquid to be measured, itself changes the liquid's temperature.

Examples of the perceived cheapening of a thing's value by the very act of buying and selling it abound in everyday life and language. The disgust that accompanies the idea of buying and selling human beings is based on the sense that this would dramatically diminish human worth. Epithets such as "he prostituted himself," applied as linguistic analogies to people who have sold something, reflect the view that certain things should not be sold because doing so diminishes their value. Praise that is bought is worth little, even to the person buying it. A true anecdote is told of an economist who retired to another university community and complained that he was having difficulty making friends. The laconic response of a critical colleague—"If you want a friend why don't you buy yourself one"—illustrates in a pithy way the intuition that, for some things, the very act of placing a price on them reduces their perceived value.

The first reason that pricing something decreases its perceived value is that, in many circumstances, non-market exchange is associated with the production of certain values not associated with market exchange. These may include spontaneity and various other feelings that come from personal relation-

ships. If a good becomes less associated with the production of positively valued feelings because of market exchange, the perceived value of the good declines to the extent that those feelings are valued. This can be seen clearly in instances where a thing may be transferred both by market and by non-market mechanisms. The willingness to pay for sex bought from a prostitute is less than the perceived value of the sex consummating love. (Imagine the reaction if a practitioner of cost-benefit analysis computed the benefits of sex based on the price of prostitute services.)

Furthermore, if one values in a general sense the existence of a non-market sector because of its connection with the production of certain valued feelings, then one ascribes added value to any non-marketed good simply as a repository of values represented by the non-market sector one wishes to preserve. This seems certainly to be the case for things in nature, such as pristine streams or undisturbed forests: for many people who value them, part of their value comes from their position as repositories of values the non-market sector represents.

The second way in which placing a market price on a thing decreases its perceived value is by removing the possibility of proclaiming that the thing is "not for sale," since things on the market by definition are for sale. The very statement that something is not for sale affirms, enhances, and protects a thing's value in a number of ways. To begin with, the statement is a way of showing that a thing is valued for its own sake, whereas selling a thing for money demonstrates that it was valued only instrumentally. Furthermore, to say that something cannot be transferred in that way places it in the exceptional category—which requires the person interested in obtaining that thing to be able to offer something else that is exceptional, rather than allowing him the easier alternative of obtaining the thing for money that could have been obtained in an infinity of ways. This enhances its value. If I am willing to say "You're a really kind person" to whoever pays me to do so, my praise loses the value that attaches to it from being exchangeable only for an act of kindness.

In addition, if we have already decided we value something highly, one way of stamping it with a cachet affirming its high value is to announce that it is "not for sale." Such an announcement does more, however, than just reflect a preexisting high valuation. It signals a thing's distinctive value to others and helps us persuade them to value the thing more highly than they otherwise might. It also expresses our resolution to safeguard that distinctive value. To state that something is not for sale is thus also a source of value for that thing, since if a thing's value is easy to affirm or protect, it will be worth more than an otherwise similar thing without such attributes.

If we proclaim that something is not for sale, we make a once-and-for-all judgment of its special value. When something is priced, the issue of its perceived value is constantly coming up, as a standing invitation to reconsider that original judgment. Were people constantly faced with questions such as "how much money could get you to give up your freedom of speech?" or "how much would you sell your vote for if you could?", the perceived value of the freedom to speak or the right to vote would soon become devastated as, in moments of weakness, people started saying "maybe it's not worth *so much* after all." Better not to be faced with the constant questioning in the first place. Something similar did in fact occur when the slogan "better red than dead" was launched by some pacifists during the Cold War. Critics pointed out that the very posing of this stark choice—in effect, "would you *really* be willing to give up your life in exchange for not living under communism?"—reduced the value people attached to freedom and thus diminished resistance to attacks on freedom.

Finally, of some things valued very highly it is stated that they are "priceless" or that they have "infinite value." Such expressions are reserved for a subset of things not for sale, such as life or health. Economists tend to scoff at talk of pricelessness. For them, saying that something is priceless is to state a willingness to trade off an infinite quantity of all other goods for one unit of the priceless good, a situation that empirically appears highly unlikely. For most people, however, the word priceless is pregnant with meaning. Its value-affirming and value-protecting functions cannot be bestowed on expressions that merely denote a determinate, albeit high, valuation. John Kennedy in his inaugural address proclaimed that the nation was ready to "pay any price [and] bear any burden . . . to assure the survival and the success

of liberty." Had he said instead that we were willing to "pay a high price" or "bear a large burden" for liberty, the statement would have rung hollow.

## III

An objection that advocates of cost-benefit analysis might well make to the preceding argument should be considered. I noted earlier that, in cases where various non-utility-based duties or rights conflict with the maximization of utility, it is necessary to make a deliberative judgment about what act is finally right. I also argued earlier that the search for commensurability might not always be a desirable one, that the attempt to go beyond expressing benefits in terms of (say) lives saved and costs in terms of dollars is not something devoutly to be wished.

In situations involving things that are not expressed in a common measure, advocates of cost-benefit analysis argue that people making judgments "in effect" perform cost-benefit calculations anyway. If government regulators promulgate a regulation that saves 100 lives at a cost of $1 billion, they are "in effect" valuing a life at (a minimum of ) $10 million, whether or not they say that they are willing to place a dollar value on a human life. Since, in this view, cost-benefit analysis "in effect" is inevitable, it might as well be made specific.

This argument misconstrues the real difference in the reasoning processes involved. In cost-benefit analysis, equivalencies are established *in advance* as one of the raw materials for the calculation. One determines costs and benefits, one determines equivalencies (to be able to put various costs and benefits into a common measure), and then one sets to toting things up— waiting, as it were, with bated breath for the results of the calculation to come out. The outcome is determined by the arithmetic; if the outcome is a close call or if one is not good at long division, one does not know how it will turn out until the calculation is finished. In the kind of deliberative judgment that is performed without a common measure, no establishment of equivalencies occurs in advance. Equivalencies are not aids to the decision process. In fact, the decision-maker might not even be aware of what the "in effect" equivalencies were, at least before they are revealed to him afterwards by someone pointing out what he had "in effect" done. The decision-maker would see himself as simply having made a deliberative judgment; the "in effect" equivalency number did not play a causal role in the decision but at most merely reflects it. Given this, the argument against making the process explicit is the one discussed earlier in the discussion of problems with putting specific quantified values on things that are not normally quantified—that the very act of doing so may serve to reduce the value of those things.

My own judgment is that modest efforts to assess levels of benefits and costs are justified, although I do not believe that government agencies ought to spon-sor efforts to put dollar prices on non-market things. I also do not believe that the cry for more cost-benefit analysis in regulation is, on the whole, justified. If regulatory officials were so insensitive about regulatory costs that they did not provide acceptable raw material for deliberative judgments (even if not of a strictly cost-benefit nature), my conclusion might be different. But a good deal of research into costs and benefits already occurs—actually, far more in the U.S. regulatory process than in that of any other industrial society. The danger now would seem to come more from the other side.

# Reply to Steven Kelman

## GERARD BUTTERS, JOHN CALFEE, and PAULINE IPPOLITO
*Federal Trade Commission, Washington, D.C.*

In his article, Steve Kelman argues against the increased use of cost-benefit analysis for regulatory decisions involving health, safety, and the environment. His basic contention is that these decisions are moral ones, and that cost-benefit analysis is therefore inappropriate because it requires the adoption of an unsatisfactory moral system. He supports his argument with a series of examples, most of which involve private decisions. In these situations, he asserts, cost-benefit advocates must renounce any moral qualms about lies, broken promises, and violations of human rights.

We disagree (and in doing so, we speak for ourselves, not for the Federal Trade Commission or its staff). Cost-benefit analysis is not a means for judging private decisions. It is a guide for decision making involving others, especially when the welfare of many individuals must be balanced. It is designed not to dictate individual values, but to take them into account when decisions must be made collectively. Its use is grounded on the principle that, in a democracy, government must act as an agent of the citizens.

We see no reason to abandon this principle when health and safety are involved. Consider, for example, a proposal to raise the existing federal standards on automobile safety. Higher standards will raise the costs, and hence the price, of cars. From our point of view, the appropriate policy judgment rests on whether customers will value the increased safety sufficiently to warrant the costs. Any violation of a cost-benefit criterion would require that consumers purchase something they would not voluntarily purchase or prevent them from purchasing something they want. One might argue, in the spirit of Kelman's analysis, that many consumers would want the government to impose a more stringent standard than they would choose for themselves. If so, how is the cost-safety trade-off that consumers really want to be determined? Any objective way of doing this would be a natural part of cost-benefit analysis.

Kelman also argues that the process of assigning a dollar value to things not traded in the marketplace is rife with indignities, flaws, and biases. Up to a point, we agree. It *is* difficult to place objective dollar values on certain intangible costs and benefits. Even with regard to intangibles which have been systematically studied, such as the "value of life," we know of no cost-benefit advocate who believes that regulatory staff economists should reduce every consideration to dollar terms and simply supply the decision maker with the bottom line. Our main concerns are two-fold: (1) to make the major costs and benefits explicit so that the decision maker makes the trade-offs consciously and with the prospect of being held accountable, and (2) to encourage the move toward a more consistent set of standards.

The gains from adopting consistent regulatory standards can be dramatic. If costs and benefits are not balanced in making decisions, it is likely that the returns per dollar in terms of health and safety will be small for some programs and large for others. Such programs present opportunities for saving lives, and cost-benefit analysis will reveal them. Perhaps, as Kelman argues, there is something repugnant about assigning dollar values to lives. But the alternative can be to sacrifice lives needlessly by failing to carry out the calculations that would have revealed the means for saving them. It should be kept in mind that the avoidance of cost-benefit analysis has its own cost, which can be gauged in lives as well as in dollars.

Nonetheless, we do not dispute that cost-benefit analysis is highly imperfect. We would welcome a better guide to public policy, a guide that would be efficient, morally attractive, and certain to ensure that governments follow the dictates of the governed. Kelman's proposal is to adopt an ethical system that balances conflicts between certain unspecified "duties" and "rights" according to "deliberate reflection." But who is to do the reflecting, and on whose behalf? His guide places no clear limits on the actions of regulatory agencies. Rather than enhancing the connections between individual values and state decisions, such a vague guideline threatens to sever them. Is there a common moral standard that every regulator will magically and independently arrive at through "deliberate reflection"? We doubt it. Far more likely is a system in which bureaucratic decisions reflect the preferences, not of the citizens, but of those in a peculiar position to influence decisions. What concessions to special interests cannot be disguised by claiming that it is degrading to make explicit the trade-offs reflected in the decision? What individual crusade cannot be rationalized by an appeal to "public values" that "rise above" values revealed by individual choices?

# Questions for Thought and Discussion

1. In one of his illustrations, Kelman refers to OSHA's regulation of the exposure of workers to carcinogenic coke-oven emissions, which should reflect the principle that "workers have a right to a safe and healthy work place." Characterize the kinds of costs and benefits that are associated with this issue, and explain the role that cost–benefit analysis might play in resolving those issues. Explain how this role would change if the principle were interpreted to mean that coke-oven workers are entitled to a totally risk-free environment.

2. In critiquing the ethics of cost–benefit analysis, the author cites the moral problem associated with regulations which permit a level of air pollution that will harm asthmatics and the elderly so that others may continue to enjoy a high standard of living. What is the connection between air pollution regulation and standard of living? What could be done to resolve the moral problem referred to here? If stricter regulation of air pollution were also to reduce the number of jobs available to family breadwinners, how would the complexion of this problem change?

3. The difficulties that arise in cost–benefit analysis over converting benefits to monetary terms can sometimes be avoided by settling for cost-effectiveness measures, such as the number of diseases prevented or lives saved per dollar spent. Describe two health or safety situations, one in which the trade-offs allow such measures to be used, and another in which the need for monetary conversion is unavoidable.

4. The author's position is that cost–benefit analysis is used too much for deciding how to regulate health and safety. What problems do you think would arise if cost–benefit analysis were to be replaced exclusively by the "deliberative judgments" of government regulators? If cost–benefit analysis continues to be used for ranking health and safety programs on the basis of expected benefit per dollar spent, what kinds of deliberative judgments should be part of the process? How might they be incorporated into the process? To what extent should they be permitted to influence the rankings?

# PART 4

# HEALTH RISK ASSESSMENT

# Introduction ————————————————————————

As hundreds of new chemicals find their way into the workplace and market-place each year, suspicions abound as to whether some will eventually prove to be hazardous to our health, as has happened in the past with mercury, lead, and other substances that we now know to be harmful even at low doses. Heightened awareness of the health risks of chemicals in our food and in the environment has led to public demands for more information about the toxic effects of both new and existing chemicals.

The process of health risk assessment has emerged as a systematic (albeit far from perfect) approach to responding to the questions being raised. It has four stages:

1. The identification of chemicals that, based on toxicological and epidemiological evidence, are health hazards;

2. The measurement or, more often, the estimation of the amounts of these substances to which people are exposed;

3. The calculation, using biological and mathematical models, of the risks associated with doses that are usually far below the doses that produced the effects observed in laboratory tests on animals or in epidemiological studies of human beings; and

4. The risk assessor's characterization of the risks, which includes an explanation of the underlying assumptions and uncertainties in the quantitative estimates.

The three papers in part 4 provide a general description of the methods used to assess health risks from chemicals, a criticism of some methods and their application, and an illustration of a risk assessment that became the basis for public health policy.

Joseph Rodricks and Michael R. Taylor's paper entitled "Application of Risk Assessment to Food Safety Decision Making" highlights the differences between assessing risks from noncarcinogens and carcinogens. Most noncarcinogenic effects are treated as "threshold" phenomena; that is, there is assumed to be a dose below which no toxic response will occur. In contrast, except in rare instances, risks from carcinogens are assessed on the basis that any exposure, no matter how small, may be harmful. The authors draw particular attention to the fact that data for assessing risk are often limited and that the assessment

141

techniques tend to be conservative in the sense that they intentionally avoid understating the risks.

The shortcomings of health risk assessment are the focus of "Assessing Risks from Health Hazards: An Imperfect Science." Authors Dale Hattis and David Kennedy point out that dose-response relationships and human exposure levels are hard to estimate, and that it is difficult, if not impossible, to validate the models used. While conceding that our primitive ability to assess risks may mean that some risks go underregulated, they conclude that society must nevertheless rely on risk assessment for guidance about managing chemical risks. But instead of a single "bottom-line" number for risk, they argue, the results should be presented in the form of a range of the most plausible estimates.

In 1980 the Supreme Court overturned an OSHA regulation restricting occupational exposure to benzene, ruling that OSHA first had to demonstrate that the existing level of exposure presented a significant risk to workers' health. To meet that requirement, scientists Mary C. White, Peter F. Infante, and Kenneth C. Chu derived a formula relating the rate of leukemia to the level of benzene exposure. Their health risk assessment, described in "A Quantitative Estimate of Leukemia Mortality Associated with Occupational Exposure to Benzene," provided the basis for OSHA's lowering of the occupational exposure level for benzene and for EPA's estimation of the health risks of benzene outside the workplace.

The one-hit model for carcinogenicity, used by White, Infante, and Chu, posits a linear relationship between the intensity of exposure to carcinogens and the excess risk of cancer. Although this model is now part of most cancer risk assessments, the exchange of letters, included here, between scientist Jerry L. R. Chandler and the authors following publication of their paper demonstrates the controversy surrounding the model's use.

# Application of Risk Assessment to Food Safety Decision Making[1]

JOSEPH RODRICKS
*Environ Corporation*
*Washington, D.C.*

MICHAEL R. TAYLOR
*King and Spalding*
*Washington, D.C.*

## Summary

The purpose of this report is to present the scientific basis of risk assessment and to demonstrate that risk assessment can be used to make decisions about the safety of our food supply. The report has been prepared as a resource for those involved in the current discussions in Congress and elsewhere over the policies by which the safety of the American food supply is evaluated and assured. The document is not intended to present an argument for any particular food safety policy, but rather to address a key scientific issue that arises in the review of policy options. It distinguishes risk assessment—the scientific process of identifying and evaluating potential risks—from risk management, the separate policy decision regarding what constitutes "safety" or an acceptable degree of risk.

[1]Prepared in cooperation with the members of the Risk Assessment Committee of the International Life Sciences Institute, Washington, D.C.

## Scientific and Policy Background

Scientists have recognized for many years that substances in the food supply are extremely complex and diverse. If every one were thoroughly evaluated, some safety questions, however minor, could be raised about many. Based on this understanding, Congress established in 1958 the still-existing safety standard applicable to most food additives: an additive is "safe" if there is a "reasonable certainty" that "no harm" will result under its intended conditions of use. However, at the same time Congress also enacted the Delaney Clause, which absolutely prohibits the use of food additives found to induce cancer in man or animal. In contrast to the general safety standard applicable to noncarcinogens, which clearly recognizes the impossibility of assuring the complete absence of risk, the Delaney Clause has been interpreted as taking a "zero risk" approach to substances implicated as carcinogens. This policy was based on the uncertainty in 1958 that an acceptable level of human exposure to carcinogens could be defined, on the assumption that only a very small number of substances were capable of inducing cancer, and on the belief that the identification of carcinogens was a straightforward scientific exercise.

The discussion about food safety policy is a result of

significant developments in science since the present law was enacted. Advances in analytical chemistry and toxicology, including new information about the varying mechanisms of action and potencies of carcinogens, have shown that the identification and evaluation of carcinogens is more complicated than once thought and requires the exercise of scientific judgment. It is also now recognized that conditions of exposure can be defined for carcinogens under which the residual risk of harm to humans, if any, is extremely low, similar to the unspecified risk that has been tolerated historically for all substances other than carcinogens. Indeed, the Food and Drug Administration (FDA) and other health agencies in the United States have already begun applying the new scientific knowledge to establish acceptable or "safe" levels of exposure to animal carcinogens to the extent permitted by law.

## The Current Issue Concerning Risk Assessment

An important impetus to the current review of food safety policies is the recognition that the law should not be a barrier to the use of all available scientific knowledge and, if necessary, should be changed to accommodate new scientific findings. Consequently, several proposals have been made recently to define safety more explicitly in terms of "negligible" or "insignificant" risk for all additives, including animal carcinogens. If such proposals are adopted, risk assessment would be used to assist policymakers in determining whether a particular risk meets the standard of "negligible" or "insignificant" risk. There is a justifiable concern, however, about whether the methods of risk assessment used to implement such a standard are indeed adequate to protect the public health.

The thesis of this report is that risk assessment is adequate for this purpose, and most of the report is devoted to explaining the scientific rationale for this conclusion. This document is directed to those involved in the current deliberations on policy alternatives who may lack direct experience with the science of risk assessment.

## Definition of Risk Assessment

Risk assessment is the scientific process by which the toxic properties of a substance are identified and evaluated. This process determines the likelihood

that humans exposed to the substance will be adversely affected and characterizes the nature of the effects they may experience. Using this information, the regulatory agency can then make a judgment about whether exposure to the substance under certain defined conditions can be considered acceptable or "safe." The ultimate conclusion about safety—a part of risk management—remains a policy judgment about the degree of risk to be tolerated, and what level of exposure results in residual risks so small as to be of no public health concern.

## The Practice of Risk Assessment

For carcinogens and noncarcinogens alike, the first step in risk assessment is the collection of toxicity information about the substance. These data include basic knowledge about the properties of the substance and its effects in various biological systems. The necessary information comes from a variety of sources, but much of it is derived from studies in animals.

If the effects noted in animals at various dose levels do not include cancer, but are deemed relevant to what might occur in humans, the next steps generally involve identifying the highest dosage level at which no adverse effects are observed in the animals, and dividing that dose by a safety factor (usually 100). This calculation yields an exposure level—called the "acceptable daily intake" (ADI)—that is considered to be a safe level of exposure for humans. If this level would not be exceeded under the intended conditions of use, the substance is approved as a "safe" food additive. While this approach assumes there is a threshold dose below which no toxic effects occur, it does not assure this. Rather, it is thought to assure that if there are any residual risks at the ADI, they are at most trivial and of no public health significance. This system, with its underlying assumptions and uncertainty, has been used successfully for many years to ensure the safety of noncarcinogenic food additives, not only in the United States but in most other Western countries.

For substances which are carcinogens, the process of risk assessment is more complicated, because some types of carcinogens may pose some risk even at very low doses. The first step in the process is to determine whether the substance represents a carcinogenic risk that is relevant to humans. Ideally, many factors should be considered in determining relevance to humans, such as the metabolism of the substance, its effects on DNA, the nature of its effects in different species, and whether these effects occur at doses which do not grossly disrupt the normal homeostasis of the animal. Detailed knowledge of the physiological processes or mechanisms resulting in an effect in the animal model would allow an unambiguous determination of relevance to humans. For some substances associated with cancer in experimental animals, the establishment of relevance is straightforward, but for most, significant uncertainty exists.

Assuming relevance to humans is established or cannot be discounted, the next step involves an evaluation of the degree of risk the substance might pose to humans. Animal toxicity studies generally include dosage levels far in excess of what humans would ever consume. Scientists recognize, however, that the likelihood of a carcinogenic effect is related to the dose, declining as dose declines. Therefore, in order to determine what the risks might be at the dosage levels comparable to those experienced by humans, it is necessary to extrapolate downward from the dosage levels used in the animal study. Mathematical models are used to perform this extrapolation, which yields a hypothetical estimate of the risks animals would experience if dosed at the much lower levels to which humans are exposed.

The final step in a risk assessment involves interspecies extrapolation—that is, using the animal risk estimates to make some assessment of the potential risks to humans. This procedure includes a comparison of the actual human and animal doses, because the manner of expressing dose, the route of administration, and the dosing schedule may differ between the animal study and the pattern of human exposure.

## The Adequacy of Risk Assessment

Scientists do not yet have a complete understanding of the process of carcinogenesis, and some uncertainties remain about how results seen in animals relate to human risk. Thus, at each step of the risk assessment process, decisions or assumptions must be made to deal with that uncertainty. These decisions are made today by regulatory scientists on the basis of a clear policy that pervades the current practice of risk assessment: whenever uncertainty exists at any point in a risk assessment, the decision or assumption is made that will most likely avoid understating the risk.

For example, in determining first whether a substance is a carcinogen and is likely to pose any risk to humans, a single positive observation in a valid animal study is considered sufficient; tumors of uncertain biological significance are assumed to be related to cancer; negative evidence is rarely allowed to overcome a positive finding; humans are assumed to be as susceptible as experimental animals, although this is not always the case; and, without very strong evidence to the contrary, all carcinogens are assumed to act by the mechanism (genotoxicity) that presents the greatest risk at low dose. Similarly, in extrapolating from high dose to low dose, the data that will yield the highest estimate of low dose risk are selected from all available data; the statistical upper limit on observed cancer incidence in the animals is used in extrapolation rather than the observed incidence itself; and the mathematical models used for high to low dose extrapolation are the ones that yield the highest prediction of risk at low doses. Finally, in extrapolating from animals to humans, "worst case" assumptions are made about the equivalence of the animal and human dose measurements and about the total amount of human exposure to the substance.

The effect of these conservative policies is to produce carcinogenic risk assessments that consistently overstate the true risk, sometimes substantially.

## Uses and Advantages of Risk Assessment

While current approaches to risk assessment overstate risk, they nevertheless provide a reliable tool to public health policymakers for establishing the relative risks posed by various substances in a systematic manner. Risk assessments do not tell the policymaker what level of risk should be considered safe; however, they do provide a high degree of assurance that the actual risk will not exceed the estimate of risk. These assessments can therefore be used to assist in making regulatory decisions about individual substances and in establishing priorities for regulating various categories of substances.

Risk assessment has the further advantage of increasing the contribution science can make to food safety decisionmaking. It can take advantage of all available evidence about a substance—such as its dose-response characteristics, mechanism of action, and human exposure—rather than relying solely on the observation that the compound induces cancer in one species of animal. It encourages scientists to develop more information about specific substances, because as additional knowledge becomes available it can be substituted for one or more of the conservative assumptions that now tend so strongly to overstate risk.

These advantages all reflect the basic premise of risk assessment, which is that it is possible to utilize the vast amount of information scientists generate to make intelligent decisions about managing potential risks. The cautiousness built into current methods of risk assessment ensures that those decisions can be made in a way that is fully protective of the public health.

## Introduction

### A. Purpose and Thesis of the Report

Developments in science over recent years have placed great pressure on the legal standards, administrative policies, and scientific methods by which the safety of the American food supply is evaluated and assured. Scientists have responded to these developments by devising more sophisticated methods for assessing potential risks to humans posed by substances used or found in food, especially those determined or suspected to cause cancer in animals. As a matter of administrative policy, the U.S. Food and Drug Administration (FDA) has recognized that, while a high standard of safety is appropriate and can be achieved, it is scientifically unrealistic to expect a food supply that is absolutely risk free. Now Congress is considering legislation that would acknowledge these scientific developments and reaffirm that pursuing the fundamental goal of a safe food supply cannot mean pursuing a policy of absolute zero risk.

A principal feature of the pending legislative proposals is the definition of "safe" in terms of "insignificant" or "negligible" risk.* An important element of the debate over such proposals concerns the adequacy of current risk assessment methods for

*Editors' note:* As of January 1990, no such legislation has passed, but "negligible risk" bills are being considered in both the House and the Senate.

making sound safety decisions under such a standard, especially in cases involving known or suspected animal carcinogens. This report does not argue that any particular form of food safety legislation should be adopted, although from a scientific standpoint the general direction of the legislative proposals—including the approach to defining "safe"—seems desirable. Instead, the report addresses the scientific issue of whether current methods of assessing risk would be adequate in the event one of the proposed definitions of "safe" is adopted. Defining the term "safe" in legislation is fundamentally a policy matter, but the debate over such a policy should be based upon a clear understanding of its scientific rationale and implications.

The thesis of this report is that current methods of assessing potential health risks from exposures to substances in food are conceptually and scientifically well founded and, if properly applied, can form the basis for sound decisions under an "insignificant" or "negligible" risk safety standard. There is general, if not complete, agreement on this point insofar as the evaluation of most noncarcinogens is concerned.

The current policy and scientific debate centers instead on carcinogens—in particular, on the human health and regulatory implications of a finding that a substance affects the development of cancer in one or more species of laboratory animals. Such a finding traditionally has triggered close regulatory scrutiny of the substance, which is appropriate. The question increasingly asked by scientists and regulatory officials, however, is whether such a finding should automatically preclude the use of a substance in food without any further scientific consideration of whether the finding in animals is relevant or significant from a human health standpoint. The various legislative proposals would allow that further scientific consideration. They thus raise the question of whether available methods of assessing the human health significance of animal carcinogens would be adequate for making sound public health decisions under an "insignificant" risk standard.

This report will show that risk assessment can be used to protect the public health because it approaches the assessment of potential human risk in a manner designed to err on the side of overstating the potential risk—the actual risk, if any, is likely to be substantially less than that projected.

**Table 1. Annual Risk of Death from Selected Common Human Activities[a]**

| | Number of deaths in representative year | Individual risk/year |
|---|---|---|
| Coal mining | | |
|     Accident | 180 | $1.3 \times 10^{-3}$ or 1/770 |
|     Black lung disease | 1,135 | $8 \times 10^{-3}$ or 1/125 |
| Fire fighting | — | $8 \times 10^{-4}$ or 1/1,250 |
| Motor vehicle | 46,000 | $2.2 \times 10^{-4}$ or 1/4,500 |
| Truck driving | 400 | $10^{-4}$ or 1/10,000 |
| Falls | 16,339 | $7.7 \times 10^{-5}$ or 1/13,000 |
| Football (averaged over participants) | | $4.10^{-5}$ or 1/25,000 |
| Home accidents | 25,000 | $1.2 \times 10^{-5}$ or 1/83,000 |
| Bicycling (assuming one person per bicycle) | 1,000 | $10^{-5}$ or 1/100,000 |
| Air travel: one transcontinental trip/year | | $2 \times 10^{-6}$ or 1/500,000 |

[a]Selected from Hutt (1978, *Food, Drug, Cosmetic Law J.* **33**, 558–589).

Risk assessment is sometimes questioned because scientists lack a complete understanding of the process of carcinogenesis. The strength of current methods rests, however, on the fact that at each point in the decision process where uncertainty exists about how to apply the animal results to humans, conservative assumptions or choices are made. As gaps in current knowledge are filled, risk estimates for some chemicals will come closer to being projections of true risk rather than projections of very conservative upper limits on risk. But in the meantime, regulatory officials can make decisions based on risk assessment, confident of the degree of public health protection being provided. . . .

## B. Basic Concepts: Risk, Safety, and Risk Assessment

Risk is the measured or estimated probability of injury, disease, or death. It is usually expressed in quantitative terms, taking values from zero (certainty that injury, disease, or death will not occur) to one (certainty that injury, disease, or death will occur). In some cases it is not possible to describe risks in quantitative terms, so qualitative expressions are used, such as "high," "low," or "negligible."

All human activities carry some degree of risk. The risk of many activities, such as driving a car, working in a coal mine, and playing professional football, can be assessed readily and with considerable accuracy. Some of these are displayed in table 1. Cause and effect relationships are clear and there is direct empirical evidence concerning the nature and frequency of adverse effects.

In general, it is much more difficult to assess the risk that might result from human exposure to chemical substances. Statistical data on the risks of some chemical exposures are available, but most such data concern immediately detectable injury or death resulting from overdoses or accidental exposure to drugs, household products, pesticidal chemicals, solvents, etc. In these cases, the effects occur because the substances are usually quite toxic and exposure to them is very high. The more difficult problem is to evaluate potential risks resulting from low-level exposures to compounds that do not cause immediately detectable injury or disease. An enormous amount of scientific information and knowledge has been developed, however, that enables scientists to identify and assess such risks.

In its common usage, the term "safe" is usually taken to mean "without risk." One of the first principles in evaluating the "safety" of substances, however, is that substances cannot be classified simply as either "safe" or "unsafe." The risk associated with a substance is a function of both its toxic properties and the conditions of human exposure to it.[2] Moreover, safety is a negative condition—that is, it entails the

---

[2]It is common, even among scientists, to use the term "toxic chemical." This implies that there are "nontoxic" chemicals. Chemicals should not be grouped in such a simplistic fashion. To do so yields a false view of potential risk, which must take into account both toxicity and exposure. The use of the phrase "toxic chemical" is based on a somewhat arbitrary division, in which chemicals of high toxicity (e.g., cyanide, mercury, lead) are separated from those of very low toxicity (e.g., sugar, carbon dioxide, polyethylene). All chemicals can be made to produce some form of toxicity under some conditions of exposure.

absence of risk. While certain risks are measurable, the limitations of science make it impossible to identify with absolute accuracy the conditions under which risk becomes zero. Thus, for the purpose of making public health policy decisions, it has become a practical necessity to define safety as a condition of very low risk. This has been the usual approach taken by regulatory agencies in defining acceptable human exposure limits for chemical substances. By careful analysis of both toxicity and exposure data, it is possible to define the conditions under which a substance is almost certainly dangerous as well as those under which its risks are sufficiently low to protect the health of the exposed population.

As defined by the National Academy of Sciences, *risk assessment* is:

> The scientific activity of evaluating the toxic properties of a chemical and the conditions of human exposure to it in order both to ascertain the likelihood that exposed humans will be adversely affected, and to characterize the nature of the effects they may experience.

Present knowledge does not permit strictly quantitative expressions of the actual risk from chemical exposures, but a proper risk assessment provides substantial information—both quantitative and qualitative—about the likelihood a substance will cause harm.

The ultimate conclusion about safety—risk management—is a policy judgment about the degree of risk to be tolerated or, stated in more common, functional terms, what level of exposure results in residual risks so small as to be of no public health concern, and what, if any, controls on the substance are needed to ensure safety. *Risk management* is an important topic for policymakers, but it is beyond the scope of this report. The purpose here is to set forth the scientific foundations and adequacy of current methods of risk assessment, the products of which may be used to make risk management decisions.

## C. Traditional Approaches to Food Safety Evaluation and the Problem of Carcinogens

Food is probably the most chemically complex part of the environment to which humans are directly exposed. Most of the substances in food are those that occur naturally. These include the essential micro- and macronutrients, as well as other natural substances which impart flavor or color, or that result from such traditional processes as cooking, fermenting, and pickling. The number of naturally occurring substances present in food is far greater than the number added by food processors or present through contamination. Most of these natural substances have not been thoroughly studied, but human experience with them has been extensive and provides considerable assurance that the natural background of risk is, at most, slight.

With advances in food technology and the increased use of food additives, the need arose in the 1950s for a systematic approach to assuring the safety of substances added to food. Scientists responded by developing the concept of acceptable daily intake (ADI), which remains in use today as the basic tool for regulating substances intentionally added to food. The ADI is the amount of a specific substance that it is believed can be ingested daily, for a full lifetime, without producing adverse health effects. The ADI is established by examining the toxic properties of a chemical, usually in experimental animals, and determining the maximum amount that produces no observable toxicity. This amount is termed the "no-observed-effect level" (NOEL); the NOEL is then divided by a safety factor to calculate an ADI.

The ADI approach has been used to implement the general safety standard for food additives that Congress established in 1958. This standard provides that a food additive, unless found to be a carcinogen, will be deemed "safe" if there is a reasonable certainty that no harm to human health will result from the intended use of the additive. Because the ADI is calculated on the basis of a no-observed-effect level and incorporates a safety factor, it can be said to provide a "reasonable certainty" that "no harm" will result.

The ADI clearly has a strong element of public health caution built into it, but it is not possible to claim that consumption of a substance at the ADI level is totally without risk; and FDA does not regard the ADI approach as one that assures zero risk. Instead, the ADI approach is a conservative means of assuring that any risks presented are at most very low. Experience shows that such an approach provides adequate assurance of a safe food supply.

Significantly, the ADI system has not been applied to food additives that display carcinogenic properties. This is due in part to the special treatment given carcinogens in the law governing the intentional addition of substances to food. When Congress enacted the general safety standard in 1958, it included a provision—the Delaney Clause—that any additive found to induce cancer when ingested by animals is prohibited from use in food, without any further consideration of whether (or to what extent) it poses any risk of harm to humans. The scientific rationale for the Delaney Clause was that it was impossible to define an exposure level that could be said with reasonable certainty to pose no risk of cancer to humans. That is, it was believed that there was no "threshold effect" for carcinogens. These views were based in 1958 upon then-current understandings about the biological mechanism by which carcinogens produce their effect. While the ADI approach to evaluating noncarcinogens recognizes at least implicitly the possibility that some residual risk remains even at the "safe" level of exposure, the Delaney Clause applies a "zero-risk" approach to carcinogens: once the carcinogenic effect has been observed the inquiry ends, and the substance is prohibited from use.

Since the Delaney Clause was enacted, this "zero-risk" approach has been challenged by developments in toxicology and analytical chemistry. For example, investigators are uncovering compounds having carcinogenic potential among the natural components and contaminants of food, and even as by-products of the cooking of food. Potentially carcinogenic substances also are being detected in food packaging materials and as trace constituents of food and color additives, albeit at levels in the parts-per-billion or parts-per-trillion range. Substances present in these amounts would have gone undetected with the analytical methods available in 1958, but they are now being found with increasing frequency using today's extraordinarily sensitive detection methods. If the full force of the Delaney Clause were applied in each such case, the food supply would be substantially disrupted.

The "zero-risk" approach of the Delaney Clause is also being challenged by the increasing recognition

that distinctions can be made among carcinogens based upon the differing mechanisms by which they act. Some substances directly initiate cancer and others are only secondarily involved in the process of carcinogenesis. The secondary carcinogens may involve an indirect method of action that requires special exposure levels or conditions, or may act only on tissues previously affected by direct-acting carcinogens. With this new information, scientists are recognizing that may substances found to induce or enhance cancer in animal studies may be having that effect through processes for which a threshold can be identified. Thus, for some carcinogens, as for noncarcinogens, there may well be levels of exposure for which the possibility of harm to humans can be ruled out with reasonable certainty and for which an ADI-type approach to safety evaluation might then be appropriate.

FDA itself has found that, in some cases, substances found to induce cancer in animals may be deemed safe under the "reasonable certainty of no harm" safety standard, depending on such factors as potency of the animal carcinogen and level of human exposure. This conclusion is reflected in several FDA decisions on specific substances and in its recently proposed "constituents policy" for regulating certain carcinogenic constituents of food and color additives and migrants from packaging materials. As a scientific and policy matter, the FDA's position in this area is based principally on the concept that the risks of carcinogens decline with dose and that at some sufficiently low level of human exposure the risk posed by the substance becomes so low as to be simply inconsequential.[3]

As a legal matter, of course, FDA is constrained by the Delaney Clause. Thus, in order to make a comprehensive evaluation of an animal carcinogen and a judgment about whether it is safe for human use, FDA must in each case find a sufficient legal or factual basis for concluding that the Delaney Clause does not apply. When FDA can avoid the Delaney

---

[3]FDA also seems to have concluded that the potential risk posed by a carcinogen may be outweighed in some cases by other public health considerations. For example, there is evidence that many essential nutrients, such as selenium, dietary fat, and vitamin D, may cause cancer under some conditions, yet FDA has understandably not taken action to prohibit or restrict their use.

Clause, it is using risk assessment to make judgments about the safety of substances. . . .

## The Use of Risk Assessment for Public Health Decision Making

### A. Introduction

. . . It has been seen that there are important similarities between carcinogens and noncarcinogens in the nature of the safety decisions made about them. In the case of noncarcinogens, it has long been believed that an ADI approach adequately protects the public health. As pointed out earlier, this method cannot provide absolute assurance of safety—no method could do so—but there is great confidence that any residual risks at an ADI are very low.

One can have similar confidence that, under certain conditions, risks from carcinogens will be very low or non-existent. For certain types of carcinogens—directly acting, genotoxic agents—there may be a finite risk at all dose levels, but there is certainly a range of doses at which the probability of cancer occurring is extremely small, and as indicated by recent regulatory trends, of questionable public health significance. For carcinogens that do not act directly, it may even be possible to assure with all but absolute certainty, in a manner similar to setting an ADI for a noncarcinogen, that there is no human cancer risk.

This concluding section summarizes the scientific reasons for the adequacy of risk assessment and suggests how risk assessment might be applied in making public health decisions.

### B. The Adequacy of Risk Assessment

Current methods of carcinogenic risk assessment as used by the FDA and other regulatory agencies can be relied upon for making public health decisions because they provide a consistent, reliable basis for distinguishing among the widely varying levels of risk posed by different carcinogens. Current methods cannot be claimed, however, to provide accurate estimates of risk because they incorporate a series of assumptions and policy choices that are designed to overstate the degree of risk posed by carcinogens. . . .

(1) Although evidence that a substance is likely to be a human carcinogen can vary greatly in quality and quantity, agencies ordinarily conclude that a substance is carcinogenic on the basis of minimum evidence, usually a single observation in an animal experiment. Moreover, uncertainties regarding the biological meaning of certain types of tumors are usually resolved by assuming the worst plausible interpretation.

(2) Agencies assume that a chemical that displays carcinogenic properties in experimental animals will also do so in humans.

(3) Agencies assume humans and animals to be equally susceptible. Although the available data are limited, there is no case in which data are available for both humans and animals that reveals humans to be more susceptible to a carcinogen than are animals, and some suggest that humans are less susceptible.

(4) Agencies assume all carcinogens act by the same mechanism (genotoxicity), which is the mechanism that predicts the greatest risk at low dose. This may be correct for some carcinogens, but is incorrect for others. This assumption will thus lead in the aggregate to an overestimation of low dose risks.

(5) In selecting sets of data from different experiments for high-to-low extrapolation, agencies will choose that set which will yield the highest estimate of low dose risk.

(6) FDA ordinarily uses the statistical upper confidence limit on the observed cancer incidence for high-to-low dose extrapolation, rather than the observed incidence.

(7) Agencies select mathematical models for high-to-low dose extrapolations that yield the highest prediction of risk at low doses. These models may be approximately correct for some carcinogens, but are likely to overstate risk for others.

(8) The models used by agencies do not assume the existence of a population threshold for any carcinogen.

(9) Quantitative extrapolation from animals to humans is performed by the agencies using assumptions about dose equivalence that will either approximate or, more likely, contribute further to overestimation of human risk relative to animal risk.

(10) Methods used to estimate human consumption involve the application of "worst plausible case" assumptions and are likely to overstate dose for many, if not most, members of exposed populations.

The cumulative effect of these choices is to make carcinogenic risk assessment an exceedingly cautious activity. The degree to which risk is likely to be overstated cannot be calculated, but it is certain to be substantial in many cases.

It is important to note that, as new scientific knowledge and information are acquired, the assumptions now used can be replaced or modified, and scientists will be able to make more accurate predictions of risk. There are several substances for which information regarding mechanisms of carcinogenic action has already been incorporated in risk assessment to produce a more accurate picture of the risk they present. Such progress toward more accurate risk assessments is obviously in the public interest. It will improve risk assessment as a scientific tool and give the results of risk assessment even greater credibility and usefulness in public health decision making. It should be emphasized again, however, that this room for improvement in risk assessment does not affect its present adequacy as a tool for making sound public health decisions. Because the true risk will be less than the predicted risk (perhaps substantially), policymakers can have great confidence in the degree of protection afforded by a decision based upon a risk assessment.

## C. The Notion of Insignificant Risk

This report has addressed the subject of risk assessment as a scientific tool for implementing a safety standard stated in terms of negligible or insignificant risk. As noted, risk assessment does not itself answer the question of what is safe—what risk is significant or insignificant. Such decisions have a qualitative component and thus cannot be reduced to a simple numerical quotient. They are decisions for public health policy officials to make. This section illustrates, however, how risk assessment can assist policymakers in determining when a risk is insignificant.

FDA and other regulatory agencies have already introduced the notion of insignificant carcinogenic risk for certain classes of substances, such as migrants from food-packaging materials, animal drug residues, trace constituents of food additives, and food contaminants. For animal drug residues that may be carcinogenic, for example, FDA has stated that a lifetime cancer risk of one in one million,

estimated by the cautious methods already described, is sufficiently low to protect public health. The selection of this risk level as acceptably low was based in large part upon the agency's recognition that for most carcinogens the actual risk is probably very much lower than that predicted using current methods of risk assessment. Thus, although a one in one million lifetime risk of cancer translates to approximately 3.6 extra cancer cases per year in a population of 250 million people,[4] the FDA recognized that any such prediction is subject to great uncertainty and that for most carcinogens the true number of cases would be much less than this figure. In fact, even without considering the highly cautious nature of the several assumptions used to estimate risk, the figure of 3.6 would be predicted only if it were assumed that all 250 million people were exposed, every day of their lives, to an amount of drug residue that posed a risk of one in one million. Thus, if there were a means to predict risk accurately, it would almost certainly reveal that true lifetime risk from such a drug residue was well below one in one million.

A properly prepared and reported risk assessment can provide considerable guidance on the question of how closely the risk predicted using the current cautious methods approximates the actual risk. In general, some judgment must be made about the degree to which risk is likely to be overstated by each of the conservative assumptions used in the assessment. Suppose, for example, that there was in a particular case fairly strong evidence that a carcinogen is genotoxic and that it does not have to be metabolized to do its damage. Moreover, the carcinogenic effects of the chemical occur in several species of test animals and in several different tissues. Under these circumstances, low dose risks predicted from one-hit or multistage models may be close approximations to actual risk. This conclusion might be further strengthened if an assessor had applied the high-to-low dose extrapolation models not only to the animal data set showing highest risk, but to all the available data sets, and observed that the predicted risks fell into a fairly narrow range.

On the other hand, in other situations the available

---

[4]Calculated as follows: (250 million) × (1/one million) = 250 ÷ 70 (average lifespan) = 3.6 per year.

data will demonstrate that a risk is highly overstated. If, for example, it is shown that the evidence for carcinogenicity is relatively weak, that there are considerable conflicting animal and epidemiological data, that the chemical is probably not genotoxic, that models probably greatly overstate low dose risk, and that the estimate of human exposure is probably overestimated, then it may be safely concluded that the quantitative estimate of risk greatly exceeds the actual risk, which may in fact be zero.

If two chemicals with the sets of properties described in these examples yielded quantitatively identical risks as estimated by current methods, it is clear that the actual risks to humans would be quite different. One of the goals of risk assessment is to incorporate the various types of chemical and biological data that may be available on specific substances into the risk assessment process, where they would replace or modify the ordinary, cautious assumptions. If this were successfully accomplished in the situations described above, it would be possible to document, ideally in a quantitative fashion, the fact that these two chemicals would pose quite different risks. However, even without a quantitative incorporation of the pertinent biological data, decision makers could take into account the qualitative differences between these two substances in deciding whether any risks they pose should be considered insignificant.

In sum, although risk assessment cannot answer by itself the question of whether a risk is insignificant, it can provide public health officials with the information needed to make such decisions on a sound, scientifically rigorous basis.

## D. Maintaining the Integrity of the Risk Assessment Process

Throughout the discussion of risk assessment . . ., it was noted that several important assumptions must be made to complete an assessment and that these could not be based entirely on scientific evidence. In no case was the decision seen to be without a scientific basis. Rather, it was seen that science could narrow the range of choices to a few, but could not reduce it to a single choice. These decisions are reflected in the ten points listed in Section B [The Adequacy of Risk Assessment]. In every case, it was seen that agencies have made the policy decision to select that scientific approach or assumption that leads to the highest estimate of risk. In the face of scientific uncertainty, this may be prudent public health policy, but it needs to be recognized by decision makers as a policy choice, not as a strictly scientific decision.

If risk assessment is to form the basis for sound public health decisions, care must be taken to ensure that its scientific and policy basis is not distorted. Distortion can arise in specific cases if one or more of the basic operating assumptions used in risk assessment are altered to accommodate some predetermined decision about how a chemical should be regulated. Certain safeguards can be developed to prevent such distortions.

First, it seems essential that risk assessors make explicit all their assumptions, and the uncertainties in them, when presenting an assessment. Only if the scientific assumptions are explicitly described can it be readily determined if and why the usual assumptions were not adopted.

This is not to suggest that the ordinary, conservative, scientific assumptions should never be abandoned. In many cases, there will be sufficient data available on a specific chemical to replace assumption with fact. But an explicit statement about what was done in a specific case and the justification for it must be made.

Peer review is a second method for ensuring that any significant departure from the usual assumptions employed in risk assessment is scientifically justified. Most EPA risk assessments undergo peer review, and FDA appears to be heading in a similar direction. Peer review is important to ensure the overall scientific quality and integrity of agency risk assessments.

Finally, it is important that agencies separate to the extent feasible the scientific procedures of risk assessment from those policy questions bearing on the management (or regulation) of a chemical. Ideally, a full risk assessment, including peer review, would be completed before an agency considers whether a risk is important and whether and to what extent a chemical needs to be controlled. Such an ideal separation (which is recognized not to be fully achievable in practice) can do much to prevent the tailoring of a risk assessment to fit some predetermined regulatory position. . . .

# References

## Risk Assessment

Anderson, M. W., Hoel, D. G., and Kaplan, N. L. (1980). A general scheme for the incorporation of pharmacokinetics in low-dose risk estimation for chemical carcinogenesis: Example-vinyl chloride. *Toxicol. Appl. Pharmacol.* 55, 154–161.

Armitage, P. (1982). The assessment of low-dose carcinogenicity. *Biometrics Suppl.: Current Topics in Biostatistics and Epidemiology*, pp. 119–129.

Crouch, E., and Wilson, R. (1979). Interspecies comparison of carcinogenic potency. *J. Toxicol. Environ. Health* 5, 1095–1118.

Crump, K. S., Hoel, D. G., Langley, C. H., and Peto, R. (1976). Fundamental carcinogenic processes and their implications for low-dose risk assessment. *Cancer Res.* 36, 2973–2979.

Food Safety Council (1980). *Proposed System for Food Safety Assessment.* Final Report of the Scientific Committee of the Food Safety Council, Washington, D. C.

Interagency Regulatory Liaison Group, Work Group on Risk Assessment (1979). Scientific bases for identification of potential carcinogens and estimation of risks. *J. Nat. Cancer Inst.* 63, 241–268.

Krewski, D., and Brown, C. (1981). Carcinogenic risk assessment: A guide to the literature. *Biometrics* 37, 353–366.

National Research Council, Safe Drinking Water Committee (1977). *Drinking Water and Health.* National Academy Press, Washington, D. C.

National Research Council, Safe Drinking Water Committee (1980). *Drinking Water and Health,* Vol. 3. National Academy Press, Washington, D. C.

National Research Council, Committee on the Institutional Means for Assessment of Risks to Public Health. (1983). *Risk Assessment in the Federal Government: Managing the Process.* National Academy Press, Washington, D. C.

Office of Technology Assessment. (1981). *Assessment of Technologies for Determining Cancer Risks from the Environment.* OTA-H-138. Washington, D. C.

Whittemore, A. S. (1980). Mathematical models of cancer and their use in risk assessment. *J. Environ. Pathol. Toxicol.* 3, 353–362.

## Use of Risk Assessment in Regulation

Allera, E. J. (1978). An overview of how the Food and Drug Administration regulates the carcinogens under the Federal Food, Drug and Cosmetic Act. *Food Drug Cosmet. Law J.* 33, 59–70.

Hutt, P. B. (1982). Legal considerations in risk assessment under federal regulatory statutes. Paper presented at symposium on Assessing Health Risks from Chemicals, Annual Meeting of the American Chemical Society, Kansas City. September 16, 1982.

Leape, J. P. (1980). Quantitative risk assessment in regulation of environmental carcinogens. *Harvard Environ. Law Rev.* 4, 86–116.

U. S. Environmental Protection Agency (1980). Water Quality Criteria Documents; Availability. Appendix C—Guidelines and Methodology used in the Preparation of Health Effect Assessment Chapters on the Consent Decree Water Criteria Documents. *Fed. Reg.* 45, 79347–79379.

U. S. Food and Drug Administration (1978). *Assessment of Estimated Risk Resulting from Aflatoxins in Consumer Peanut Products and Other Food Commodities.* Bureau of Foods, FDA, Washington, D. C.

U. S. Food and Drug Administration (1979). Chemical Compounds in Food-producing Animals; Criteria and Procedures for Evaluating Assays for Carcinogenic Residues. *Fed. Reg.* 44, 17070–17114.

U. S. Food and Drug Administration (1980). Lead Acetate; Listing As a Color Additive in Cosmetics that Color the Hair on the Scalp. *Fed. Reg.* 45, 72112–72117.

U. S. Food and Drug Administration (1982). D & C Green No. 6; Listing As a Color Additive in Externally Applied Drugs and Cosmetics. *Fed. Reg.* 47, 14138–14147.

U. S. Food and Drug Administration (1982). Policy for Regulating Carcinogenic Chemicals in Food and Color Additives; Advance Notice of Proposed Rulemaking. *Fed. Reg.* 47, 14463–14470.

# Questions for Thought and Discussion

1. The Delaney Clause, we are told in this paper, "absolutely prohibits the use of food additives found to induce cancer in man or animal." Suppose that a particular ingredient used in making chocolate was found to cause cancer in rats at consumption levels equivalent to a person eating at least 10 pounds of chocolate a day for a lifetime, but that no link had been found to cancer in humans. Further, suppose that life without chocolate would have little meaning for you, and that without this additive chocolate would hold no appeal. Would you support the Delaney Clause in this case? Defend your position.

2. Risk assessment, as described in this paper, does not necessarily provide accurate estimates of risk because the methods are based on assumptions and choices that tend to overstate the risk when scientific uncertainty exists. The authors state that it does, however, provide estimates that are useful for regulatory decision making. What are these assumptions and choices, and why are they made? In what ways might the use of overly conservative risk estimates seriously distort the decision-making process?

3. The results of laboratory tests demonstrate that not every member of a population of inbred, genetically homogeneous animals will respond in the same way to a given dose of a toxic substance. These results tend to reinforce the concern that the health effects of toxic substances place a higher burden on the most sensitive humans, especially when those persons also happen to be in the most highly exposed segments of the population. Do you believe that regulatory programs should be designed to protect the most sensitive individuals? If so, how would you identify those individuals? Should protection depend on *why* they are sensitive? If not, what criterion would you suggest instead? Support your conclusions.

4. Suppose the following information is available about two chemicals:

*Case 1.* There is strong evidence that a carcinogenic chemical is genotoxic, does not have to be metabolized to be harmful, and that its carcinogenic effects occur in several different species of test animals.

*Case 2.* The chemical is probably not genotoxic and the evidence for carcinogenicity is fairly weak, since an increase in cancer was observed in only one of the species tested.

Also suppose that the calculated carcinogenic potencies of the two chemicals are quite similar. If the federal government were to regulate the chemicals, do you think that the permissible exposures ought to be the same or different? Explain.

# Assessing Risks from Health Hazards: An Imperfect Science

DALE HATTIS
*Massachusetts Institute of Technology*
*Cambridge, Massachusetts*

DAVID KENNEDY
*Harvard University*
*Cambridge, Massachusetts*

When William Ruckelshaus came back to run the embattled Environmental Protection Agency (EPA) in 1983, one of his primary goals was to separate the "science" of assessing health hazard risks from the "policy" of managing those risks. During the previous administration of Anne Burford Gorsuch, risk assessment had become so entangled with politics that many public observers felt that the EPA was acting as an advocate for the very industries it was supposed to regulate. For example, the EPA had concluded that there was no significant health risk to workers from exposure to formaldehyde, a chemical used to make particle board, plywood, and some permanent-press fabrics. The actual evidence on the health risks from formaldehyde was less reassuring than the EPA's position indicated.

In an effort to make risk assessment more impartial, Ruckelshaus argued that scientists should first make an "objective" study of the extent of risk from exposure to a particular hazardous chemical or situation. How many people, for example, will die from cancer after 20 years of exposure to airborne arsenic emitted from a copper smelter in Tacoma, Wash.? Only after that assessment is made should governmental agencies move into the political realm and decide what to do about that risk.

Moreover, Ruckelshaus thought that his agency would better serve the public by being more explicit about the extent of risk from various environmental hazards. Potent carcinogens such as dioxin are very frightening, but just how dangerous is a very low concentration of the chemical? Should the EPA take great pains to reduce dioxin in waste sites to very low levels, or might the money spent on such cleanups save more lives if channeled toward controlling other kinds of pollution? Ruckelshaus believed that risk assessments comparing the threat from different hazards would help him, and the public, resolve such issues.

Ruckelshaus' intent was to create a special authority and credibility for risk assessment. He wanted to build a strong scientific foundation upon which EPA and society at large could balance social, economic, and political concerns and reach sensible decisions about managing environmental risks.

Other policymakers shared his views. The National Research Council recommended in 1983 that

"Assessing Risks from Health Hazards: An Imperfect Science" by Dale Hattis and David Kennedy is reprinted, with permission, from *Technology Review*, copyright 1986, vol. 89, no. 4 (May/June 1986), pp. 60–71.

regulatory agencies take steps to make the process of assessing risks more formal and scientific, and to maintain a clear conceptual distinction between risk assessment and risk management. Industry groups such as the American Industrial Health Council also insisted that "the scientific determination should be made separately from the regulatory determinations." Such industry groups have often suggested that environmental controversies be resolved by experts capable of critically evaluating specific facts.

There is only one problem with this call for authoritative, scientific risk assessment: such a commodity does not exist. In classical times, there was a great demand for the skills of soothsayers in reading entrails, and there is a similar amount of wishful thinking going on today. The fact is that the science behind risk assessment is not up to the challenge of consistently providing accurate answers about the degree of risk individuals or populations face from health hazards.

Scientific uncertainties remain in the process of assessing risks—uncertainties that some people find devastating. Vernon Houk, a senior official of the Centers for Disease Control (CDC), has said that the difference between risk assessment and a five-year weather forecast is that at least with the weather forecast, if you wait five years you find out whether you were right. Many risk assessments project numbers of cancers or deaths that, while large enough to arouse public concern, are too small to be definitively separated from those occurring normally in a given society. Furthermore, one usually has to wait decades—the length of time cancer often takes to emerge—before an unusually high incidence of disease can be confirmed.

There is also no way risk analysts at the EPA or other agencies can escape making value-laden choices in the course of their work—choices that render their results far less "scientific" and objective than Ruckelshaus envisioned. Even apparently neutral reports of what is known and not known generally reflect value-based assumptions about what matters and what does not.

Consider, for example, two recent analyses of daminozide, a chemical that regulates growth (apple growers use it to control the time when their crops are ready). At a hearing in Massachusetts on daminozide, Ian Nisbet, a consultant for the Massachu-

setts Department of Public Health, presented an analysis of the special risks the substance might pose to infants and young children. Young children generally eat more food relative to their body weight than adults, and they also seem to eat more processed apple products such as applesauce. Furthermore, there is good reason to expect that rapidly growing children with most of their lives ahead of them are more likely to develop cancer than adults. However, the very act of placing the daminozide analysis in this framework makes a value-laden statement: that the special sensitivity of young people should be considered in the public-policy process. By contrast, one of the authors of this article presented an analysis of daminozide's risk that made no special distinction regarding children. This analysis did include a broad range of possible hazards based on risks found in other chemicals in the same family of hydrazines. Both analyses were factual and "objective," but they clearly were not value neutral.

Since risk assessment cannot be wholly insulated from value judgments and is rife with uncertainty, it lacks the special credibility that some would claim for it. Therefore, agencies such as the EPA, the Occupational Safety and Health Administration (OSHA), and the Food and Drug Administration (FDA) should not expect their risk analysts to come up with "bottom-line" answers to questions about whether use of a hazardous chemical or process should be banned or encouraged. Instead, these agencies should encourage their analysts to share the uncertainties involved in assessing a risk with policymakers and the public, who can then make more informed—albeit more complicated—decisions about regulating the risk.

## Linking Cause and Effect: A Difficult Task

Risk assessment is a relatively new discipline, if it can be called a discipline at all, and there is no consensus on which basic rules and procedures to apply in solving particular problems. Although government analysts have been assessing risks of various kinds for decades, the call for quantitative risk assessment—i.e., expressing the extent of risk in numerical form—is relatively recent. Until the 1970s, risk was perceived as a simpler, more black-and-white problem: is or is not DDT carcinogenic and mutagenic? The answer is an unequivocal yes. However, data on the risks of many other toxic substances are more ambivalent, and the risk to people remains unclear. In such cases, policymakers are more likely to take into account the economic consequences of restricting the chemical in question. More sophisticated detection technologies also allow scientists to measure the effect of these substances in smaller and smaller amounts, producing results that are not as clearcut as previous tests were.

Evaluating an environmental hazard to see how many people might be at risk is difficult because analysts must follow that hazard through whatever twists and turns it takes in the real world. Analysts have to determine how potential threats are released into, and move through, the environment. They have to figure out how much of the substance people might eat, breathe, or otherwise take up, and then estimate how much of it they would absorb. Finally, analysts must determine just how much of a hazard the absorbed level of the substance poses.

Figuring out how much of a toxic chemical reaches people—the degree of exposure—can be quite challenging. For example, analysts generally use computer models to assess how a toxic plume of different-sized particles of arsenic disperses through the air from a smelting plant, or how a chemical would be carried through the ground to wells supplying water. Yet it is difficult to incorporate all the important information into these models. In most cases groundwater flows more readily in horizontal directions (out from a waste site) than vertically. But measurements of the way water flows in different directions may not be available, making it hard to predict when wells at different depths and distances from the waste site might be affected. Even when data of this sort are available, the models themselves often oversimplify the system in question. One EPA model of smelter emissions into air assumed that the smelting plant was on a flat plain when in fact it was on a steep hill. As a result, EPA scientists misstated the wind patterns and initially overestimated the concentrations of arsenic that would affect the nearby town.

Even if analysts know how much of a substance is in the environment, they can't necessarily predict how much people will actually absorb. People breathe at different rates depending on their level of activity: workers laboring heavily at a construction site, garment workers sitting at sewing machines, and people sleeping in the surrounding community will all receive different doses of an airborne contaminant. Individuals also have widely different breathing rates and dietary habits, profoundly affecting the doses of specific substances they receive from air and food. Finally, people absorb substances in varying amounts depending on the thickness of their skins and the properties of their nasal mucous, and even on whether they tend to breathe through their noses or their mouths.

Determining how much of a hazard the absorbed substance poses is another complicated problem. For example, evidence of cancer or other toxic effects in humans attributable to a specific cause would seem to constitute the ideal basis for regulation. A high incidence of lead poisoning in people living near lead smelters would seem to indicate the need for better pollution controls on those plants. Unfortunately, such clear-cut instances are rare, because epidemiological studies are notoriously insensitive in detecting health effects from relatively low levels of exposure. As David Ozonoff, chief of the Environmental Health Section at the Boston University School of Public Health, has said, "A good working definition of a catastrophe is an effect so large that even an epidemiological study can detect it."

The problem is that the rates of specific illnesses from a given hazard often must be several times above average before one can conclude that they aren't simply random fluctuations. In one celebrated case now being tried, a group of parents from Woburn, Mass., is suing two chemical companies for dumping toxic wastes in the neighborhood. The parents claim that these wastes leaked into local drinking water and caused an unusually high incidence of leukemia among their children. This high incidence, which has already resulted in the deaths of five children, could indeed be due to the companies' toxic-waste dumping, or it could be a random fluctuation. It could also be the result of a completely different and unknown phenomenon.

Another complication is the fact that unless scientists perform special monitoring measures at the time of exposure, there is rarely good information about how heavily certain populations have been exposed to a chemical. Nor is information about "confounding" factors such as smoking, alcohol use, and the toxic's interactions with other environmental hazards readily available. These difficulties do not always render epidemiological information useless: solid positive results can provide a good indication of a specific level of risk from a given substance. For example, studies among workers exposed to arsenic, and among residents exposed to arsenic compounds in well water, have unequivocally revealed the carcinogenic properties of inorganic arsenic. However, a negative result is usually not proof that there is no risk, but rather an indication that the risk, if any, is less than the study is capable of detecting.

## The Trouble with Animal Studies

In the great majority of cases where the epidemiological evidence is incomplete or ambiguous, using animal studies to make projections may make more sense. However, such studies suffer from their own serious uncertainties. Experimental animals are generally exposed to high concentrations of chemicals to ensure that if there are any toxic effects, they will appear at statistically significant levels. A mathematical model called a "dose-response curve" has to be fitted to the resulting data to assess the probability that people will get cancer at the much lower levels they might realistically encounter. However, such high doses can complicate the interpretation of results in a number of different ways.

For example, molecular biologists have discovered the existence of certain enzymes in cells that convert chemicals to more toxic metabolites, beginning the march toward cancer. At very high doses of the toxic substance, these enzymes are fully occupied and cannot generate toxic by-products at a greater rate. Thus, increased doses do not necessarily lead to more cancers. In extrapolating downward to realistic levels of human exposure, risk assessors often don't take this saturation effect into account and, in effect, underestimate the risk to humans.

Ideally, the dose-response relationship would be derived from a detailed theory about how the chemical actually works to produce cancers in animals and in humans. Such a theory would be based on knowledge about how the chemical is absorbed, metabolized, and excreted from animal and human systems. Risk assessors do not generally attempt to draw

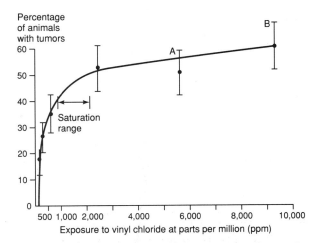

Extrapolating the results of animal studies to humans can be complicated. Animals are exposed to large amounts of chemicals so that toxic effects will appear at significant levels. However, at these doses, the animal enzymes that convert the chemicals to cancer-causing substances become saturated and cannot make any more toxic by-products. This effect occurred when rats were exposed to increasing doses of vinyl chloride. The curve showing how many animals developed tumors leveled off shortly after the saturation point.

Risk assessors usually [only have data corresponding to] the two highest dose points (A and B). . . . [If that had been the case here,] the curve would have been a straighter, more gradual line, . . . underestimating the human risk at low doses.

together much of the latest knowledge available, and as a result their models do not capture the details of cause-and-effect relationships. Instead, these analysts generally fit one of a number of statistical formulas to the data on tumor formation.

Unfortunately, the different mathematical models produce widely varying results. "Multistage" models, favored by molecular biologists, are based on the belief that cancer occurs as the end result of a series of genetic changes in specific cells. If a toxic chemical interferes with the copying of the genetic information in a cell's DNA, the errors, or mutations, will usually be passed on to the descendent "daughter" cells. Even if the error is confined to a single nucleotide base on the DNA code, the results can be severe if important genetic information is altered.

For example, in some cases a cancer-causing gene known as an oncogene becomes active because of a single mutation: one specific nucleotide is replaced with another. That one change instructs the gene to produce a specific amino acid instead of another. Chains of amino acids link together to form proteins, the basic molecules that help cells function. In this case, the single change in this amino acid produces a protein that begins the process of producing cancer, although researchers do not yet know how. The crucial implication of this model is that even the tiniest amount of a toxic substance that can affect DNA has some chance of inducing cancer. In principle, there is no exposure threshold below which the toxic substance does not pose a risk.

In contrast, "probit" models imply that individual organisms do in fact have a specific tolerance level, or threshold, below which exposure to a toxic chemical is safe. This type of model is traditionally favored by pharmacologists and toxicologists, who view biological processes as complex webs of processes, exquisitely balanced so that modest perturbations in the system will prompt corrective actions to restore normal functioning. As long as the biological insult the system suffers is not too great—i.e., is below a certain threshold—the system ought to be able to repair any damage that may be temporarily produced. The implication is that there is no benefit from regulating toxic substances when exposure falls below such a threshold.

Many scientists believe that this type of model is perfectly appropriate when applied to traditional types of acute toxic insult, such as the lung damage from methyl isocynate that occurred on such a devastating scale in Bhopal. However, some scientists find this model more questionable for carcinogens that seem to act directly or indirectly on DNA. Furthermore, given the difficulties of extrapolating human exposure from high-dose animal experiments, there is rarely, if ever, any way to specify actual thresholds for people.

Despite these reservations, risk analysts continue to use very different models depending upon their professional ideology. And the results continue to differ dramatically. In a recent experiment, Alice Whittemore, an epidemiologist at Stanford University School of Medicine, fit different models to experimental data for male rats exposed to the carcinogenic pesticide ethylene dibromide (EDB). Widely used to fumigate grain, EDB has recently been found in bread, cereal, cake mixes, and the like. Whittemore found that depending on the model chosen, the likelihood that an individual will get cancer from low-level exposure to EDB can differ by a factor of one million.

Given these enormous discrepancies and what is at stake for industry, workers, and the general public, there is considerable controversy as to which models are most appropriate for assessing risks. Analysts at EPA and OSHA tend to use the more conservative multistage model, favored by molecular biologists, or a similar model. For instance, OSHA used this model in a recent assessment of the risk of workers exposed to EDB. This assessment, which was based on extrapolation from an experiment with rodents, showed that the risk is very high indeed: after 45 years of intermittent exposure to 20 parts per million of EDB in the air—the maximum amount then permitted under OSHA standards—the chance that a worker would develop one common form of cancer was pegged at between 38 and 59 percent. However, very few workers were actually exposed at this level or for this duration; the calculation was done to show that there would be a "significant risk" to workers if they were exposed over a working lifetime to the permitted levels. OSHA, however, has not yet reduced its maximum level of EDB exposure in the workplace.

When results from the multistage model lead an agency to ban or severely restrict the use of a chemical, fireworks often result. For instance, the EPA's recent ban on the use of EDB as a pesticide prompted a vociferous protest from companies that manufacture the chemical. However, the ban has stood.

Other serious difficulties plague the process of interpreting animal studies and extrapolating them to humans. Because animals and humans metabolize substances differently, the level of the test chemical that reaches various parts of the animal and the human body can vary widely. Hence, animals and humans may suffer from different health effects. For example, bis-chlormethyl ether, used as a laboratory chemical, tends to produce nasal tumors in rats but lung tumors in people. The contrast may stem from differences in how deeply the chemical penetrates into the respiratory system before being absorbed.

Moreover, the metabolite or by-product of the test chemical, rather than the chemical itself, is often the toxic substance, and animal systems can differ from human ones in the type and concentration of metabolites they produce. For example, dogs and people primarily succumb to bladder cancer from some aromatic amines such as benzidine (used in the manufacture of dyes), while rodents get cancer of the liver. This is apparently because the different mammalian systems form metabolites that react at different sites in the body. As the National Research Council, which often advises the federal government on scientific issues, points out in a recent congressionally commissioned report,* correcting for these differences is not easy because researchers often lack enough information about human and animal systems.

## Disclosing Bias in Risk Assessment

It would be easy, but mistaken, to look at this litany of problems as a wholesale indictment of the entire concept of risk assessment. William Ruckelshaus himself came close to drawing such a conclusion in 1985 after leaving the EPA. He wrote that risk analysis "is a kind of pretense; to avoid paralysis of protective action that would result from waiting for 'definitive' data, we assume that we have greater knowledge than scientists actually possess and make decisions based on those assumptions."

Although now more aware of its limitations, Ruckelshaus still believes in risk assessment, and Lee Thomas, the current EPA administrator, is pursuing his predecessor's vision of higher-quality risk assessments. And properly so, since some form of risk assessment is essential in dealing with environmental hazards. There is real value in encouraging a conceptual and professional separation between risk analysts and risk managers. Such a separation will never produce the kind of ironclad and unimpeachable scientific analyses some would like. However, it will reinforce the preeminent duty of analysts to eliminate inappropriate biases from their thinking and work.

The sort of thing to avoid is the EPA's handling of formaldehyde under Gorsuch. In 1981 the agency in effect took an advocacy position to support a decision that regulation was not necessary. Its assessment suggested that there was no significant risk because its analysts lacked epidemiological evidence that the substance is carcinogenic. The analysts relied instead on arguments about thresholds to conclude that low-level exposure is safe. Most risk assessors know how slip-

pery epidemiological data are, and they are familiar with the vigorous debate regarding the viability of thresholds for carcinogenicity. The EPA ignored such ambiguities. Norton Nelson, a prominent environmental scientist, said that the EPA took "an extreme position" in deciding that the data on health risks of formaldehyde were not very significant.

Risk analysts should be encouraged to cultivate an approach—an ideology, if you will—that provides the public with exactly the opposite of extreme, advocacy positions. Since risk assessors often cannot answer the question "exactly how much risk does some hazard pose?" they must tackle the question they can answer: how much do we know about a particular hazard, and what are the important uncertainties in that picture?

The objective of this approach would be to help policymakers and the public make informed choices based upon the available information. Risk assessors of this stamp will have to be particularly open and sensitive in choosing which methods of analysis are appropriate in each situation. They will also need to draw on insights from various disciplines. Take, for example, the original studies assessing the risk of workers exposed to airborne lead. Since workers exposed to less lead didn't have correspondingly reduced blood lead levels, some scientists originally suggested that much of the lead found in workers' blood must have been introduced by some mechanism other than breathing. However, pharmacologists know that lead is stored in bone when blood is saturated, and can be released into the blood if blood lead levels fall. The original scientists didn't take that important fact into account and came up with the wrong assumption that workers were somehow eating the lead.

Risk assessors should be intimately familiar with such effects and be able to apply the insights and techniques of different specialties where appropriate. Unfortunately, risk assessment practices today tend to prevent such interdisciplinary familiarity. Many government risk assessors are not trained in toxicology, biochemistry, or other disciplines. Thus, they often do not have the background to understand and include in their assessments the detailed processes that produce disease. Instead, risk analysts usually review the research of other specialists and use statistical methods to draw quantitative conclusions.

---

*Editors' note: Risk Assessment in the Federal Government: Managing the Process (Washington, D.C., National Academy Press, 1983).

Specialists from different disciplines such as pathology, toxicology, and chemistry often form teams to perform risk assessments, which is a step in the right direction. But there's still a tendency to maintain strong disciplinary boundaries on such teams—to separate, for example, the analysis of the dose people are exposed to from an analysis of how that dose correlates with actual health effects.

Take exposure to formaldehyde. The substance can affect respiratory tissue in a number of ways. The formaldehyde can itself react with DNA and begin the process that ultimately can lead to cancer. It can also inhibit the enzymes responsible for repairing DNA. Finally, at high doses, it can kill cells and thereby stimulate cell replication to replace those that are lost. Enhanced cell replication reduces the time available for repair of DNA lesions, increasing the chance that permanent genetic changes will occur in the exposed cells. Given these multiple effects, there is some reason to suspect that short-term exposure to high doses of formaldehyde might cause more damage than longer term exposure to the same total amount. Hence, OSHA and EPA should express their results in terms of the amount of time people spend at specific exposure levels. Unfortunately, these agencies now express exposure levels as an average over time.

Some of this tunnel vision can be blamed on the narrow and often uncoordinated focus of the risk assessors. But the agencies themselves must take the blame for not acting quickly enough on important new information; pressure from agencies such as the White House Office of Management and Budget, and the potential for lawsuits from industry and citizen groups are two reasons for such lethargy. Some officials are reluctant to try innovative approaches to risk assessment because they may be more likely to be shot down in the tortuous process leading to a regulatory decision.

Risk assessors also must do a better job of deciding what the scope of their analyses ought to be. For example, they have to decide whether to include the possibility that a toxic chemical may interact dangerously with other chemicals in the environment, even though there are no hard data available on that interaction and its result. Risk assessors also have to decide who to consider when analyzing a chemical's toxicity. Take the case of ozone, a pollutant formed when hydrocarbon fumes evaporate from automobile exhausts, gasoline, and paint solvents, and react with sunlight. Should analysts focus on the risk to the majority of relatively healthy people, or to the relatively few elderly and asthmatic people who might be particularly susceptible to ozone smog?

Analysts must recognize that such choices are laden with value judgments and make an effort to avoid or at least state those judgments. Yet all too often, as John Holdren of the University of California at Berkeley has written, risk assessors tend to omit issues that they have decided are too uninteresting, difficult to quantify, speculative, or likely to be "misinterpreted." To prevent this kind of preselection, we should ensure that analysts publicly disclose the choices they make.

## A Cure for the "Bottom-Line" Illness

Finally, risk assessors must do a better job of identifying and assessing uncertainties and communicating them to policymakers and the public. Given the many difficulties with the science that risk assessment draws on, analysts should take care that it is never said of them, as it has been of the State Department: "They're never right, but they're always sure." The goal here should be, in the words of Nicholas Ashford, associate professor of technology and policy at M.I.T., to "bound the set of not clearly incorrect answers," rather than to focus solely on the most likely answer statistically.

Risk analysts should never present a "best estimate" of risk without some accompanying statement of statistical uncertainties and other ambiguities. Policymakers often suffer from "bottom-line illness"; all they want is the number at the end of the study. Risk assessors should carefully avoid that disease. Instead of one bottom-line estimate, they should present a range of likely estimates, including their different consequences.

We could formulate many other such prescriptions for analysts, but the basic principle is that they should communicate their findings so that policymakers and the public can fully understand the issues and uncertainties. Interested observers should be able to comprehend the important assumptions, data gaps, and choices almost as if they themselves had gone through the process. This is all the more important at a time when the U.S. has limited resources for making its environment "safe" from hazards. As

Ruckelshaus said recently, we have to "abandon the impossible goal of perfect security and accept the responsibility for making difficult and painful choices among competing goods."

In a more paternalistic society, an elite group of leaders might weigh competing costs and benefits and set safety limits for the rest of us according to the elite's view of what is best. But we live in a democracy and have a right to participate in the dialogue concerning which risks should be controlled and to what degree.

Moreover, we have to consider the fact that the cleanup itself is not generally risk free: workers can become injured and nearby residents can be exposed to air pollution from bulldozers cleaning up a site.

Risk assessment can help inform this important social and political dialogue. It can also help raise the quality of that dialogue, by revealing where science is—and is not—likely to be of service in resolving doubts about the nature and magnitude of environmental risks.

# Questions for Thought and Discussion

1. To illustrate their contention that risk assessment cannot be wholly insulated from value judgments, the authors cite one study that focused on the special risks of daminozide to infants and young children, and another that did not consider these risks at all. How would you structure a study that takes these risks into account, yet is as value-neutral as possible?

2. Give an example of a health risk that should be absolutely prohibited by regulation and an example of one that should merely be curtailed by regulation. Characterize in general terms the kind of health risks that should be prohibited and the kind that should be curtailed.

3. Based on the information in this paper, what do you believe are the two or three major flaws in the use of epidemiological studies and the two or three major flaws in the use of animal studies as a basis for risk assessment? Indicate how, based on the authors' suggestions or your own opinions, you would go about correcting each flaw.

4. It has been suggested that the process of health risk assessment would be improved by the development of guidelines that specify which exposure pathways, population subgroups, and interactions are to be considered. What do you think of this suggestion? What other methods might be used to improve the process and communicate the results?

# A Quantitative Estimate of Leukemia Mortality Associated with Occupational Exposure to Benzene

MARY C. WHITE and PETER F. INFANTE

*Occupational Safety and Health Administration, Washington, D.C.*

KENNETH C. CHU

*National Toxicology Program, Bethesda, Maryland*

## 1. Introduction

The risk of developing leukemia from occupational exposure to benzene at the current, 8-hr time-weighted average (TWA), permissible exposure level of 10 parts per million (ppm)[1] has been vigorously debated since the mid-1970s, when epidemiologic evidence conclusively demonstrated benzene's leukemogenic potential in humans. In 1978, the Occupational Safety and Health Administration (OSHA) promulgated a revised standard of 1 ppm TWA for occupational exposure to benzene.[2] OSHA took the

*Editors' note:* The current affiliation of Mary C. White is the Centers for Disease Control, Atlanta, Georgia, and of Kenneth C. Chu is the National Cancer Institute, Bethesda, Maryland.

"A Quantitative Estimate of Leukemia Mortality Associated with Occupational Exposure to Benzene" by Mary C. White, Peter F. Infante, and Kenneth C. Chu is reprinted, with permission, from *Risk Analysis*, vol. 2, no. 3 (September 1982), pp. 195–204. © 1982 Society for Risk Analysis.

position that "once the carcinogenicity of a substance has been established qualitatively, any exposure must be considered to be attended by risk when considering any given population." OSHA concluded that it was not possible to demonstrate a threshold level for benzene-induced carcinogenicity, or to establish a safe level for benzene exposure, and therefore decided that the permissible exposure level to benzene should be reduced to the lowest feasible level.

The revised standard was challenged in the courts, and in July 1980, the U.S. Supreme Court upheld a lower court's decision to vacate the revised standard.[3] A plurality of justices, in a split 5 to 4 decision, stated that OSHA had failed to make the threshold finding that exposure to benzene at the current standard is unsafe, in the sense that significant risks are present and can be eliminated or lessened by a change in practices. OSHA has interpreted the U.S. Supreme Court decision as requiring the consideration of significance of risk in the development of regulations for occupational carcinogens, including risk assessments when they can be appropriately performed. However, OSHA has stated that "when data are not available to perform a formal quantitative risk assessment, qualitative evidence, expert testimony and other evidence may be appropriately utilized to base a determination of significance of risk."[4]

Following the U.S. Supreme Court decision, the available epidemiologic evidence was evaluated for use in a quantitative risk assessment, since appropriate experimental data were not available. Although benzene also has been shown to induce nonmalignant blood disorders, chromosomal aberrations, and perhaps lymphomas,[5] attention was focused on the quantification of the risk of death from leukemia.

Risk assessments have been developed by the Environmental Protection Agency for environmental exposure to benzene[6] and by the Consumer Product Safety Commission for consumer exposure to benzene.[7] These risk assessments were prepared with regard to regulatory considerations at the time and have not been published. Our purpose in publishing this risk assessment for occupational exposure to benzene is to stimulate interest and discussion on the methodology and assumptions which can be used when the underlying scientific evidence is less than ideal for the development of a quantitative risk assessment, in a scientific rather than a regulatory context. We invite critical evaluations of the relative merits of our approach to the epidemiologic evidence as compared to other approaches that could be or have been taken.

## 2. Epidemiologic Evidence

From epidemiologic studies, relative risk values and the corresponding exposure estimates may be used to develop a mathematical model for predicting carcinogenic risk. Most often in studies of occupationally exposed persons, the relative risk of a given cause of death can be approximated by the standardized mortality ratio (SMR). However, because the number of workers who die from a particular cause of death in a population usually is small, the confidence interval surrounding the SMR may be quite large. In addition, adequate exposure measurements for the time period of interest may not be available.

The error component of the mathematical model used to predict carcinogenic risk will be a function of many factors, including the measurement errors associated with the SMR and exposure values. However, measurement error in the exposure values is an especially important problem, as exposure is the independent variable of any model. A large amount of error in the independent variable could lead to seriously biased results in the risk estimates.

Thus, in order for an epidemiologic study to contribute useful information to the development of a quantitative risk assessment, it must meet two absolutely minimum requirements. First, the study must provide an estimate of the excess risk in a population which was exposed to the substance in question, based on the experience of an appropriate control population. Second, reliable industrial hygiene measurements must be available to permit a reasonable characterization of previous exposure conditions.

The scientific literature is filled with case reports, case series, and epidemiologic studies which qualitatively link benzene exposure with leukemia in humans. This literature has been reviewed and summarized many times.[2,5,8] However, most of this scientific evidence has not been presented in a manner that can be used to quantify the magnitude of the risk of benzene-induced leukemia at a given exposure level. Only a handful of studies are available which attempt to measure the relative risk of leukemia under specified benzene exposure conditions.

Aksoy has published a series of reports which examined the incidence of leukemia among shoeworkers and other persons exposed to benzene in Istanbul.[9–11] Aksoy reported that 31 shoeworkers had been diagnosed as having leukemia in one Istanbul hospital between 1967 and 1975, out of a population of approximately 28,500 shoeworkers.[10] Aksoy calculated a crude incidence rate of leukemia in this group of 13.5/100,000, which was more than double the leukemia incidence rate of 6/100,000 for the general population. This comparison does not take into consideration differences in ages between shoeworkers and the general population, the fact that persons diagnosed as having leukemia at other hospitals in Istanbul would not have been counted, or the absence of an accurate leukemia incidence rate for Turkey.

Moreover, very little is known about past benzene exposures among the 28,500 shoeworkers in Istanbul. An early report by Aksoy[9] indicated that the duration of exposure to benzene ranged from 3 months to 17 years for a sample of 217 apparently healthy shoeworkers, and that the concentration of benzene in the working environment ranged from 15 to 30 ppm during nonworking hours to a peak of 210 ppm when benzene adhesives were being used. Another report[10] indicated that benzene levels as high as 640 ppm had been measured in the shoeworking environment. The representativeness of these few measurements to the entire shoeworking industry, on which the incidence rates are based, is impossible to judge. Therefore, risk values and exposure measurements based on Aksoy's work were considered too speculative to be included in a quantitative risk assessment.

In Italy, Vigliani and his co-workers also sought to assess the risk of leukemia among persons exposed to benzene. An early report[12] stated that 12 persons with leukemia attributed to benzene exposure had been diagnosed in the provinces of Milan and Pavia between 1960 and 1963; 11 had been diagnosed in 1962 or 1963. Assuming that about 5,000 people were exposed to benzene in these 2 provinces, Vigliani and Saita estimated that the incidence of acute leukemias in 1962 and 1963 was about 20 times higher than expected. The investigators acknowledged that an analysis of the incidence of acute leukemia based on a control group would have been preferable.

Even though there is considerable uncertainty in the relative risk of leukemia among these benzene-exposed workers, there is even greater uncertainty in the levels of previous benzene exposures. The only exposure information available is that benzene levels in some shoe factories in Pavia ranged from 25 to 600 ppm, with most between 200 and 500 ppm.[13] It is unknown how representative these figures may be of the exposure conditions of the 5,000 workers in Milan and Pavia who were estimated to be exposed to benzene. Thus, as with Aksoy's work, the information available from Vigliani's reports concerning relative risks of leukemia and past benzene exposure levels were considered too imprecise to be used as other than a qualitative indication of benzene's leukemogenic potential.

Only two epidemiologic studies were identified as meeting the minimum requirements for inclusion in a quantitative risk assessment for benzene.

The first was a mortality study of rubber workers in Ohio, conducted by Infante and his colleagues at the National Institute for Occupational Safety and Health (NIOSH). Initial results of this study were published in 1977,[14,15] and a more detailed presentation of the NIOSH study was published recently by Rinsky et al.[16]

In this study, 748 white male workers were identified who had been exposed to benzene in the production of Pliofilm, a rubber film material, in two essentially identical production facilities. To be included in the study, workers had to have worked for at least 1 day between January 1, 1940, and December 31, 1949, in a department with benzene exposure. The number of workers who died between January 1, 1950, and June 30, 1975, was compared to the expected based on U.S. white male mortality rates, using a modified life-table approach.

With the follow-up more than 90% complete,[15] the investigators found that 7 workers had died from leukemia, compared to an expected figure of 1.25. This excess in the number of workers who died from leukemia was statistically significant (SMR = 560, $P < 0.001$).*

Six of these seven workers had been diagnosed with acute myelogenous or monocytic leukemia. In order to take into account the distribution of specific cell types of leukemia, an expected number of deaths was generated using mortality rates for acute and monocytic leukemias from the National Cancer Institute.[15] The number of workers who died from acute or monocytic leukemia was found to be more than 8 times the expected figure of 0.70, yielding an SMR of 857.

In the most recent report of this study,[16] Rinsky et al. reported that most of the 748 workers had been exposed to benzene for a relatively short period of time; 58% of this group had been employed for less than 1 year. When the data were analyzed by length of employment, a significant excess in leukemia was observed among workers employed 5 or more years,

---

*Editors' note: SMR stands for standardized mortality ratio and $P$ pertains to the level of significance.

but not among workers employed for less than 5 years. Among the latter group, 2 workers had died from leukemia compared to 1.02 expected, an excess which was not statistically significant. However, among workers employed for more than 5 years, 5 had died from leukemia compared to 0.23 expected, yielding an SMR of 2100.

The second epidemiologic study selected for this quantitative risk assessment was a mortality study by Ott *et al.* of Dow Chemical Company employees exposed to benzene.[17] A total of 541 white males were identified who had been exposed to benzene in 3 chemical production areas of a Michigan plant, excluding workers who also had been exposed to arsenicals, vinyl chloride, or asbestos. All workers were employed at the plant between 1940 and 1970, and no minimum length of employment was required for inclusion in this study.

From 1940 to 1973, 91 workers had died. Three of these workers had died from leukemia; all were classified as myelocytic and two as acute myelocytic leukemia. This observed number was significantly ($P < 0.048$) greater than the expected number of cases of leukemia of other than lymphocytic or monocytic cell types (essentially acute and chronic myelocytic leukemia), which was calculated to be 0.8 based on incidence rates from the Third National Cancer Survey. If we assume that the expected incidence of myelocytic leukemia was very similar to the expected mortality from myelocytic leukemia, then the relative risk of death from myelogenous leukemia among these workers can be approximated by an SMR of 375.

## 3. Exposure Conditions

In recreating the exposure conditions of the workers included in both the NIOSH and Dow studies, an attempt was made to describe the range of average exposures, both by level and by duration.

With regard to the NIOSH study, the initial conclusion of the investigators was that "employees' benzene exposure was generally blow the recommended limit in effect at the time of each survey," (see table I). During the 1977 OSHA rulemaking for benzene, some parties disputed this conclusion and suggested that benzene exposures may have been higher than the recommended limits. The recent report of the NIOSH study[16] presented a very

**Table I. Estimates of 8-hr, TWA Benzene Concentrations for NIOSH Cohort Based on Recommended Environmental Levels[a]**

| Year | Benzene level |
|------|---------------|
| 1937–1940 | 150[b] |
| 1941–1946 | 100[c] |
| 1947 | 50 |
| 1948–1956 | 35 |
| 1957–1962 | 25 |
| 1963–1968 | 25[d] |
| 1969–1975 | 10 |

[a]References provided in Infante *et al.*[14] and Rinsky *et al.*[16]
[b]No recommended level was available before 1941.
[c]Recommended level was expressed as a maximum allowable concentration.
[d]Recommended level was expressed as a ceiling concentration.

detailed description of the processes and exposure levels at the studied facilities. The available information generally supports the NIOSH investigators' initial conclusion concerning exposure conditions. Although a few higher exposures to benzene were recorded (up to 680 ppm), often these measurements were taken in areas where workers were present infrequently for brief periods of time or not present at all. Moreover, records at the plant indicated that the company was aware of benzene's toxicity and that respirators were worn during some operations. For this risk assessment, we assumed that exposures were at the recommended standard for that time. We recognized that some workers may have been exposed occasionally to higher levels. However, we believed that the average exposure experience for the entire study population is adequately represented by the levels in table I, and that the use of these levels actually may have overestimated average benzene exposures for many workers.

The length of employment varied widely among the workers included in the NIOSH study. The minimum length of employment for inclusion in the study was 1 day, and the maximum length of time that anyone could have worked in one of the plants was 37 years. Well over half (58%) of the study population were employed for less than 1 year. In addition, the length of employment for the 7 workers who died from leukemia ranged from 1 month to 20 years, with an average of 8.5 years.

It was decided to base the risk assessment on the experience of workers who had been employed for 5 or more years, because most of the elevated leukemia risk was observed in this group. In addition, we did not know what proportion of the person-years for the total cohort had been contributed by persons with only very brief employment. The range of duration of employment was determined to be 5–30 years; the upper limit was based on the fact that workers who had been employed for more than 30 years contributed less than 0.01 to the number of expected leukemia deaths.

The range of average benzene exposures was estimated as follows. The earliest date any employee was exposed to benzene was 1937, since this was the date the oldest facility began Pliofilm production. Workers could have begun employment as late as 1949 and still have been included in the study. Follow-up ended in 1975, so exposure to benzene after this date could not have contributed to the observed risks. Thus, workers employed for 5 years could have been employed for any 5-year period between 1937 and 1954, and workers employed for 30 years could have been employed for any 30-year period between 1937 and 1975. Because benzene-exposure measurements were not available for the earliest years of Pliofilm production, we arbitrarily assumed that benzene exposures before 1941 were 50% higher, or 150 ppm TWA. The average benzene exposure over the period of 1937 to 1954, using 150 ppm for 1937–1940 and the values from table I, was 83 ppm. Over the period 1937–1975, the average benzene exposure was 50 ppm. Thus, the exposure to benzene which was associated with a 21-fold excess risk of leukemia was estimated to range from 83 ppm × 5 yr, or 415 ppm-yr, to 50 ppm × 30 yr, or 1500 ppm-yr.

Within the three production areas included in the Dow study,[17] the results of exposure monitoring from 1944 to 1974 indicated that benzene exposures were considerably lower than at the facilities studied by NIOSH. Although peak levels of benzene as high as 937 ppm were measured, estimated time-weighted average exposures ranged from 0.1 to 35.5 ppm for different job categories. The length of employment among workers in this study varied greatly; 23% were employed for less than 1 year and 17% were employed for 20 or more years. Ott and his colleagues examined the job histories of these workers and calculated cumulative ppm-month career doses for each worker. Their method was to multiply the mean TWA value for each job category by the

**Table II. Number of Observed and Expected Deaths from All Causes for Dow Cohort by Estimated Career Benzene Dosages**[a]

| Career dose category | Observed | Expected |
|---|---|---|
| 0–499 ppm-mo | 41 | 65.1 |
| 500–999 ppm-mo | 18 | 16.2 |
| 1000+ ppm-mo | 32 | 32.8 |
| Total | 91 | 114.1 |

[a]From Ott et al.[17] [Reprinted with permission of the Helen Dwight Reid Educational Foundation. Published by Heldref Publications, Washington, D.C. Copyright © 1978.]

number of months spent in that category. Information was not provided concerning the distribution of workers by ppm-months. However, the observed and expected number of deaths were analyzed by cumulative dose (see table II). From this analysis, we determined that workers who had exposures of less than 500 ppm-mo, or 42 ppm-yr, contributed 57% of the number of expected deaths from all causes. Workers who had exposures of more than 1,000 ppm-mo, or 83 ppm-yr, contributed 29% of the number of expected deaths. From this information, we estimated that approximately one-half of the study group were exposed for less than 42 ppm-yr, and nearly one-third were exposed for more than 83 ppm-yr. The investigators did not state what the largest number of ppm-months was for any worker in the study group. We estimated an upper limit of 1,500 ppm-mo, or 125 ppm-yr, for our range of exposure. Therefore, the range of benzene exposures estimated for the Dow study was 42–125 ppm-yr. It should be noted that the estimated cumulative doses for the three workers who died from leukemia were 545 ppm-mo (45 ppm-yr), 19 ppm-mo (1.6 ppm-yr), and 305 ppm-mo (25 ppm-yr). Given that one-half of the study population and 2 of the 3 leukemia cases may have had cumulative benzene exposures of less than 42 ppm-yr, we believe that our range of 42–125 ppm-yr is unlikely to be an underestimate of the benzene exposure that was associated with a 3.75-fold excess risk of leukemia in this study.

## 4. Selection of a Model

Several mathematical models have been developed to describe the relationship between the level of exposure to a carcinogen and the probability of developing cancer associated with that level.[18] We selected the one-hit model for a quantitative risk assessment for benzene, primarily because it is a simple model. The current biological evidence concerning the chemical induction of leukemia was considered insufficient to defend the use of a more complex model. In addition, the information available from epidemiologic studies regarding the relative risks of leukemia at different levels of benzene exposure was inadequate to statistically test the appropriateness of more complex models.

The one-hit model is based on an assumption that a single dose of a carcinogen can affect some biological phenomenon in the organism which subsequently will lead to the development of cancer. This model is a nonthreshold model; i.e., it assumes that every level of dose is accompanied by some amount of excess cancer risk.

Expressed mathematically, the one-hit model states that the excess cancer risk ($P_d$) is related to the dose ($d$) by the equation

$$(1) \qquad P_d = 1 - \exp(-B \times d).$$

If a "background" cancer risk ($P_0$) exists in the absence of exposure to $d$, then $P_d$ is not equal to the total probability of developing leukemia ($P_t$), but rather is related to $P_t$ and $P_0$ by the equation

$$(2) \qquad P_d = (P_t - P_0) / (1 - P_0)$$

as explained by Mantel and Bryan.[19]

Therefore, the excess risk ($P_d$) was redefined as

$$(3) \qquad P_d = [1 - \exp(-B \times d)](1 - P_0)$$

and the total leukemia risk as

$$(4) \qquad P_t = P_0 + [1 - \exp(-B \times d)](1 - P_0).$$

## 5. Converting Information from Epidemiologic Studies to the Model

The relative risk of leukemia mortality from both the NIOSH[16] and Dow[17] studies was approximated by the SMR. The SMR represents the ratio of observed deaths due to leukemia in the benzene-exposed group, divided by the number of expected leukemia deaths based on the death rates of a comparable age and sex-specific group

$$SMR = \frac{\text{number of deaths observed}}{\text{number of deaths expected}} \times 100.$$

For the risk assessment, the SMR is assumed to represent

$$SMR = (P_t/P_0)100$$

and thus,

$$(5) \qquad P_t = \frac{SMR}{100}(P_0).$$

To determine the excess probability ($P_d$) of developing leukemia from a given level of benzene ($d$) over a defined period of time, using Eq. (3), two values needed to be calculated.

The first was the working lifetime probability ($P_0$) of death due to leukemia (all cell types combined) and myelogeneous leukemia, independent of exposure to benzene. The $P_0$ values were calculated, using standard life-table methods, as the sum over 5-year age intervals (from 20 to 84) of the probability of death due to leukemia during the specific interval times the probability of survival to the beginning of that interval. This life-table method assumes the independence of these events.[20] Death rates for all causes and for all types of leukemia were taken from the 1975 mortality rates for U.S. white males.[21] Death rates for cell-specific leukemia were average annual rates for U.S. white males from 1973 to 1977, and were abstracted from mortality rates collected through the National Cancer Institute SEER program.[22] The calculated $P_0$ values were 0.00707\ for all types of leukemia and 0.00495 for myelogenous leukemia. (Appendicies providing detailed information concerning the calculation of these $P_0$ values are available upon request from the authors.)

The second value was $B$, which is associated with the rate at which the excess probability of leukemia increased with each increment in dose. It can be shown from Eqs. (4) and (5) that the solution for $B$ is:

$$(6) \qquad B = -\ln[(1 - (SMR/100)P_0)/(1 - P_0)]/d.$$

The SMR values and dose estimates which were used in these calculations are presented in table III.

As explained earlier, upper and lower estimates of the dose associated with the SMR were generated for both the NIOSH[16] and Dow[17] studies. Treating each study separately, the upper and lower dose

**Table III. Values for SMR, Dose, and _B_ Used to Calculate Excess Leukemia Risk**

| Study | SMR | Benzene dose estimate | _B_ value corresponding to dose estimate |
|-------|-----|----------------------|------------------------------------------|
| Rinsky et al.[16] | 2100 | 415 ppm-yr (lower) | 0.000370 |
| | | 1500 ppm-yr (upper) | 0.000102 |
| Ott et al.[17] | 375 | 42 ppm-yr (lower) | 0.000328 |
| | | 125 ppm-yr (upper) | 0.000110 |

estimates were used to calculate two _B_ values from Eq. (6). These _B_ values also are presented in table III. Separate risk estimates were calculated using each _B_ value and Eq. (3). The two resultant risk estimates represent a range of risk which reflects the uncertainty in the exposure conditions for each study.

## 6. Estimated Excess Leukemia Risk Under Different Exposure Conditions

Risk estimates were made for a working lifetime exposure to benzene, assumed to be 45 years, at both the current 10 ppm standard (450 ppm-yr) and at the vacated 1 ppm standard (45 ppm-yr). Recognizing that few employees would be exposed to benzene for as long as 45 years, estimates also were made for lengths of employment equal to 1, 5, 15, and 30 years at 10 ppm and at 1 ppm benzene.

Table IV presents the estimates of additional lifetime leukemia risk (per 1,000 workers) due to occupational exposure to 10 ppm or 1 ppm benzene for different lengths of time, based on the NIOSH study. Table V presents the corresponding risk estimates for myelogeneous leukemia based on the Dow

**Table IV. Estimates of Additional Lifetime Leukemia Risk per 1000 Workers[a] from Occupational Exposure to Benzene[b]**

| Years exposed | Exposure level | |
|---------------|----------------|--------|
| | 10 ppm | 1 ppm |
| 45 | 44–152 | 5–16 |
| 30 | 30–104 | 3–11 |
| 15 | 15–54 | 1.5–5 |
| 5 | 5–18 | 0.5–2 |
| 1 | 1–4 | 0.1–0.4 |

[a]Estimates were rounded to the nearest whole number (per 1,000) for risks greater than 1 per 1,000 workers.
[b]Based on data from Rinsky et al.[16]

**Table V. Estimates of Additional Lifetime Leukemia[a] Risk per 1000 Workers[b] from Occupational Exposure to Benzene[c]**

| Years exposed | Exposure level | |
|---------------|----------------|--------|
| | 10 ppm | 1 ppm |
| 45 | 48–136 | 5–15 |
| 30 | 32–93 | 3–10 |
| 15 | 16–48 | 2–15 |
| 5 | 5–16 | 0.5–2 |
| 1 | 1–3 | 0.1–0.3 |

[a]Estimates are based on leukemia cell types other than lymphocytic or monocytic.
[b]Estimates were rounded to nearest whole number (per 1,000) for risks greater than 1 per 1,000 workers.
[c]Based on data from Ott et al.[17]

study. We believe that the NIOSH study provides stronger evidence of an association between benzene and leukemia than does the Dow study, as well as better information on benzene exposure levels. Therefore, we have greater confidence in the risk estimates based on the NIOSH study and our discussion will focus on these risk estimates. The risk estimates based on the Dow study are consistent with those from the NIOSH study. In fact, the risk estimates in tables IV and V are much more similar than one may have expected, given the uncertainty associated with quantitative risk assessments.

As can be seen in table IV, 45 years of occupational exposure to benzene at 10 ppm were estimated to result in an excess lifetime leukemia risk of between 44 and 152 per 1,000 exposed workers. At 1 ppm benzene for 45 years, the excess lifetime leukemia risk was estimated to be between 5 and 16 per 1,000 exposed workers. A certain magnitude reduction in either length or level of exposure produced a reduction in risk of equal magnitude. The linear relationship between dose and excess risk resulted from the fact that the one-hit model is essentially linear at low doses.[23]

## 7. Discussion

Because any risk assessment must apply study results obtained under one set of conditions to another and sometimes very different set of conditions, the estimated risk contains uncertainty. Although the best available scientific data were used in this risk assessment for benzene, certain assumptions had to be made for several unknown variables. The use of these assumptions might have resulted in a large amount of error in the risk estimate. So that the risk estimates are viewed in the proper perspective, the major assumptions on which these estimates were based are discussed below.

First, it was assumed that the 748 white males in the NIOSH cohort and the 541 white males in the Dow cohort do not differ in their susceptibility to benzene-induced leukemogenesis from the more than 629,000 men and women, white and nonwhite, who are estimated to be occupationally exposed to benzene.[2] Even though this risk assessment was based on information derived from human populations, and therefore no species-to-species extrapolation was necessary, extrapolation from one population to another still may be problematic. Populations may vary from one another in their sensitivity and susceptibility to carcinogenic agents due to variations in genetic factors, age distributions, and exposures to potentially interacting environmental and social factors. However, no means are currently available to determine such variation in susceptibility.

With regard to age distributions, the SMR of a study population has been shown to be a useful approximation for the relative risk when the excess in mortality is consistent across all age groups.[24] It is not known whether the relative risk of benzene-induced leukemia is different at different ages. However, if the relative risk of benzene-induced leukemia is age-dependent, then the SMRs obtained from the NIOSH or Dow cohorts could overestimate or underestimate the relative risk of benzene-induced leukemia among another group with a different age composition.

Second, it was assumed that the risk ratios obtained from these studies represented working lifetime risks of benzene-induced leukemia. However, the risk ratios obtained from the NIOSH and Dow cohorts were based on exposure and follow-up periods which occurred over a limited fraction of the cohort member's lifetimes. For instance, in the Rinsky et al. report,[16] only 181 deaths were reported, meaning that 76% of the cohort was still alive or presumed to be alive at the end of the follow-up period. Similarly, only 91 deaths were reported in the Ott et al. study,[17] leaving 83% of the cohort still alive at the time of follow-up. The assumption of a constant SMR could either underestimate or overestimate the true risk, if the relative risk of leukemia among the cohort members actually increased or decreased during the period after the date of follow-up.

In addition, it was also assumed that the relative risk of leukemia would remain constant after exposure to benzene had ceased. Day and Brown[25] have demonstrated that, under the multistage theory, the effect of cessation of exposure will depend largely on whether the carcinogen affects an early or late stage of the cancer's development. If benzene were a late-stage carcinogen, then one would expect the excess leukemia risk to eventually decline in these cohorts as the length of time since termination of exposure increased.

However, Peto[26] has argued that the multistage theory probably is not applicable to nonepithelial cell cancers such as leukemia. Moreover, 3 of the 12 case reports described by Rinsky et al.[16] were of men who had died from leukemia 9 or more years after they left the Pliofilm operations. To date, there is no evidence to suggest that benzene may be a late-stage carcinogen or that the excess leukemia risk will decline after exposure has ceased.

Third, many of the industrial processes where benzene exposure occurs, such as petroleum refining, rubber manufacturing, and chemical processing, contain other known or suspected carcinogens. The presence of other carcinogens in the workplace might enhance (synergistically) the carcinogenic effect of benzene. However, it was assumed that no such synergism had occurred in the workplaces studied or would occur in other occupational settings.

Fourth, one of the most important assumptions made in this risk assessment was that the dose-response relationship between benzene exposure and leukemia development follows the one-hit model. Although this model is generally considered to be conservative at low exposure levels, it may actually underestimate the risk at lower levels if the risk of leukemia no longer increased with increasing dose after a certain level of exposure, i.e., the dose-response curve plateaus, and if the studied workers had been exposed to benzene at a level above the point at which the curve flattens out.

Fifth, it was assumed that for benzene exposure levels which are generally not associated with acute toxicity (TWA less than 100 ppm), a cumulative dose-effect relationship exists. That is, the level of risk associated with a certain dose was assumed to be constant, regardless of whether exposure occurred over a short period of time or at lower levels over a longer period of time. At present, there is no available method to test the validity of this frequently made assumption.

The cumulative dose model also assumes that the total amount of benzene to which a worker was exposed contributed to his risk of death from benzene-induced leukemia. Theoretically, a worker could be exposed to benzene after the point at which the irreversible but subclinical development of benzene-induced leukemia had begun. Thus, any exposure to benzene after leukemia had been induced would not add to the risk of mortality from leukemia and could be viewed as a "wasted" dose in terms of generating a dose-response curve. Therefore, a dose-response relationship which is based on total dose could overestimate the amount of dose that is responsible for a given level of excess risk.

Sixth, although these risk estimates take into consideration some of the uncertainty in the benzene levels from the NIOSH and Dow studies, these estimates do not reflect the uncertainty associated with the risk ratios from these studies. For instance, the upper 95% confidence interval for the SMR of 2100 from the Rinsky et al. report[16] is 5073. The use of this value would have resulted in much higher risk estimates. Thus, even though these risk estimates are expressed as ranges, the upper range of the risk estimates actually would have been greater if the uncertainty of the SMRs had been included in the analysis.

## 8. Conclusion

The risk estimates generated for total leukemia from the NIOSH study[16] and for myelogeneous leukemia from the Dow study[17] were almost identical. We estimated that a working lifetime exposure of 45 years to benzene at 10 ppm (the current OSHA standard) would result in a leukemia risk ranging from approximately 44 to 152 excess leukemia deaths per 1,000 exposed workers. A working lifetime exposure to benzene at 1 ppm would result in a leukemia risk ranging from 5 to 16 excess leukemia deaths per 1,000 exposed workers. At different lengths of exposure, the risk associated with 10 ppm was consistently 10 times greater than the risk at 1 ppm. This order of magnitude difference in risk was expected because the risk estimates were based on the one-hit model, and the one-hit model has been shown to be essentially linear at low doses.

This report demonstrates how some types of epidemiologic evidence can be used to estimate the level of carcinogenic risk associated with a certain level of exposure. However, it is far easier to quantify a level of carcinogenic risk than it is to quantify the level of confidence that should be placed in the risk estimate. When viewed in the context of scientific evidence which qualitatively indicates that benzene is a relatively potent human carcinogen, these risk estimates suggest that workers exposed to benzene at the current OSHA permissible exposure level of 10 ppm are at a substantially elevated risk of death from leukemia.

The opinions, findings, and views expressed in this paper are those of the authors and should not be attributed to the Occupational Safety and Health Administration or the National Toxicology Program.

## References

1. OSHA Safety and Health Standards for General Industry, 29 CFR 1910.1000, Table Z-2.

2. Occupational Safety and Health Administration, Occupational exposure to benzene; permanent standard, *Federal Register* 43, 5918–5970 (February 10, 1978).

3. *Industrial Union Department v. American Petroleum Institute*, 448 U.S. 607 (1980).

4. Occupational Safety and Health Administration, Identification, classification and regulation of potential occupational carcinogens; conforming deletions, *Federal Register* 46, 4889–4893 (January 19, 1981).

5. B. D. Goldstein, Hematoxicity in humans, in "Benzene toxicity: A Critical evaluation," *Journal of Toxicology and Environmental Health*, S. Laskin and B. D. Goldstein (eds), Suppl. 2, 69–105 (1977).

6. R. Albert, E. L. Anderson, C. Hiremath, R. McGaughy, S. Miller, R. Pertel, W. Richardson, D. Singh, T. W. Thorslund, and A. J. Zahner, "Carcinogen Assessment Group's Final Report on Population Risk to Ambient Benzene Exposures," U.S. Environmental Protection Agency, Washington, D.C. (1979).

7. P. D. White, M. S. Cohn, and W. K. Porter, *Benzene Risk Assessment* (U.S. Consumer Product Safety Commission, Washington, D.C., 1981).

8. National Research Council, Assembly of Life Sciences, Committee on Toxicology, *Health Effects of Benzene: A Review* (National Academy of Sciences, Washington, D.C., June 1976).

9. M. Aksoy, K. Dincol, T. Akgun, S. Erdem, and G. Dincol, Haematological effects of chronic benzene poisoning in 217 workers, *British Journal of Industrial Medicine* 28, 296–302 (1971).

10. M. Aksoy, S. Erdem, and G. Dincol, Types of leukemia in chronic benzene poisoning. A study in thirty-four patients, *Acta Haematologica* **55**, 65–72 (1976).

11. M. Aksoy, Different types of malignancies due to occupational exposure to benzene: A review of recent observations in Turkey, *Environmental Research* **23**, 181–190 (1980).

12. E. C. Vigliani and G. Saita, Benzene and leukemia, *New England Journal of Medicine* **271**, 872–876 (1964).

13. E. C. Vigliani, Leukemia associated with benzene exposure, *Annals New York Academy of Sciences* **271**, 143–151 (1976).

14. P. F. Infante, R. A. Rinsky, J. K. Wagoner, and R. J. Young, Leukemia in benzene workers, *Lancet* **ii**, 76–78 (1977).

15. P. F. Infante, R. A. Rinsky, J. K. Wagoner, and R. J. Young, Letter to the editor, *Lancet* **ii**, 868–869 (1977).

16. R. A. Rinsky, R. J. Young, and A. B. Smith, Leukemia in benzene workers, *American Journal of Industrial Medicine* **2**, 217–245 (1981).

17. M. G. Ott, J. C. Townsend, W. A. Fishbeck, and R. A. Langner, Mortality among individuals occupationally exposed to benzene, *Archives of Environmental Health* **33**, 3–10 (1978).

18. A. Whittemore, and J. B. Keller, Quantitative theories of carcinogenesis, *SIAM Review* **20**, 1–30 (1978).

19. N. Mantel and W. R. Bryan, "Safety" testing of carcinogenic agents. *Journal of the National Cancer Institute* **27**, 455–470 (1962).

20. M. Gail. Measuring the benefit of reduced exposure to environmental carcinogens, *Journal of Chronic Diseases,* **28**, 135–147 (1975).

21. U.S. National Center for Health Statistics, *Vital Statistics of the United States, 1975, Volume II—Mortality* (U.S. Department of Health, Education, and Welfare, Hyattsville, Maryland, 1979).

22. J. L. Young, C. L. Percy, A. J. Asire, J.W. Berg, M. M. Cusano, L. A. Gloeckler, J. W. Horm, W. I. Lourie, E. S. Pollack, and E. M. Shambaugh, "Cancer Incidence and Mortality in the United States, 1973–1977," National Cancer Institute Monograph #57, NIH Publication No. 81-2330, Bethesda, Maryland (June 1981).

23. D. G. Hoel, D. W. Gaylor, R. L. Kirschstein, U. Saffiotti, and M. A. Schneiderman, Estimation of risk of irreversible, delayed toxicity, *Journal of Toxicology and Environmental Health* **1**, 133–151 (1975).

24. M. J. Symons and J. D. Taulbee, Practical considerations for approximating relative risk by the standardized mortality ratio, *Journal of Occupational Medicine* **23**, 413–416 (1981).

25. N. E. Day and C. C. Brown, Multistage models and primary prevention of cancer, *Journal of the National Cancer Institute* **64**, 977–989 (1980).

26. R. Peto, "Epidemiology, multistage models, and short-term mutagenicity tests," in *Origins of Human Cancer,* Vol. 4, H. H. Hiatt, J. D. Watson, and J. A. Winsten, eds. (Cold Spring Harbor Laboratory, New York, 1977, pp. 1403–1428).

# Letter to the Editor:
# Benzene and the One-Hit Model

JERRY L. R. CHANDLER
*McLean, Virginia*

Future adverse health consequences subsequent to occupational exposure to benzene were estimated from epidemiologic data by White *et al.* A fundamental scientific *and* legal issue is the question of the

"Letter to the Editor: Benzene and the One-Hit Model" by Jerry L. R. Chandler is reprinted, with permission, from *Risk Analysis,* vol. 4, no. 1 (March 1984), pp. 7–8. © 1984 Society for Risk Analysis.

relative health risks associated with chronic "low" level exposure to benzene. If fact-based science and value-based social policy are to be articulated with veracity, then the best estimate of the true human risk should be sought. The best estimate of the true human risk demands the systematic evaluation of *all* relevant facts which may quantitatively influence the risk estimate. In addition, the assumptions and the implications of the predictive health model must be evaluated for consistency and coherency with the known facts.

White *et al.* "selected the one-hit model for benzene, primarily because it is a simple model." The one-hit model is not merely a statistical distribution but is mathematically derived from a set of specific assumptions which relate the "dose" to an irreversible biological process. Mathematically, the one-hit model describes a continuously increasing saturation function which starts at zero and asymptotically approaches one with increasing dose. Chemically, the logic of the one-hit model requires that a single unique chemical event constitutes the causal biological event. The causal biological event must be initiated in a single cell in order to be unique. In addition, the chemical event must be irreversible to be in accordance with the mathematical assumptions for a Poisson process. A critical determinant of the model's risk predictions is the assumption that the number of irreversible biological events is directly proportional to the dose, hence the *critical public health conclusion that a simple proportional relationship exists* between the dose and the response at "low" doses.

Is the one-hit model an accurate predictor of the future incidence of leukemia in exposed populations? White *et al.* state "the current biological evidence concerning the chemical induction of leukemia was considered insufficient to defend the use of a more complex model." Unfortunately, the nature of the evidence considered to be insufficient was not stated.

The notion that "simplicity" is adequate justification for predicting future incidence of human disease disguises at least four substantial scientific issues:

1. The fundamental issue of whether or not the dose-response relationships for benzene are continuous functions or are discontinuous functions is sidestepped by the simplicity argument. From a public health perspective, the critical issue of the existence of risk outside the observable range is thereby avoided.

2. Justification of selection of the linear model by noting its simplicity also leads to the notion that the response is proportional to dose. Of the infinite set of possible relationships between dose and response, the assumption that a simple proportional relationship exists needs to be justified. Since it is possible that the molecular mechanism of benzene-induced toxicity involves a set of complex nonlinear biochemical reactions, the simplicity argument may eventually be shown to be inconsistent with the molecular mechanism.

3. The argument of simplicity is insufficient scientific rationale to justify the strong conclusion that disease incidence is linear in a heterogeneous human population. Hit theory is based on the premise that rare stochastic events occur as first order processes in a *homogeneous* population. Discussion of the weight of evidence supporting the equisensitivity of all workers toward benzene intoxication is needed if this assumption of the model is to be credible.

4. Metabolic considerations may be critical to understanding the functional relationship between the dose and response. Since the parent benzene molecule is a relatively unreactive molecule at biologic pH and temperature, a spontaneous irreversible reaction with biomolecules at a biochemically significant rate is deemed unlikely. This suggests that the presence of benzene itself in a cell represents only the distribution of a foreign substance in the body and not the critical irreversible event assumed by the model. Benzene is metabolized to phenols and quinones by saturable enzymatic processes. Phenols, quinones, and unstable oxygenated benzene derivatives which lead to the formation of phenols and quinones are substantially more reactive than benzene itself. Kinetically, these reactions are not simple unimolecular reactions but follow higher order, saturable kinetics. Oxygenated metabolic derivatives are sufficiently reactive to covalently bind to tissue constituents, including genetic elements. The importance of these reactions in the molecular mechanisms which determine the dose response relationships for benzene toxicities could (and should) be evaluated in estimating the risk of benzene.

Simplicity in model selection is desirable when the selected model can account for the complete set of factual observations. However, our current understanding of human disease processes, including benzene toxicities, suggests that they are multifactorial in nature and more complex than a single, unique, irreversible event. To be credible, prediction of future disease incidence needs to be consistent and coherent with the intertwined biochemical, cellular, physiological, and genetic data. For these reasons, I believe that the best estimate of the true human risk for benzene will require a broader and deeper evaluation of the existing set of relevant facts.

# Letter to the Editor: Assessment of Leukemia Mortality Associated with Occupational Exposure to Benzene

PETER F. INFANTE
*Occupational Safety and Health Administration*
*Washington, D.C*

MARY C. WHITE
*Brookline, Massachusetts*

KENNETH C. CHU
*Gaithersburg, Maryland*

Dr. Chandler has questioned the appropriateness of using the one-hit, or linear model, and suggests that molecular mechanisms which determine the dose response relationship for benzene toxicity should be evaluated in estimating the risk of benzene.

We believe that an assessment of benzene toxicity could be improved by knowledge of biological factors such as the dose of the ultimate metabolite(s) in bone marrow (and other organs), the toxicity of these compounds, and the rates and sites of their transformation as well as the cumulative nature and pharmacokinetic distribution and action along with modifying effects of intrinsic and extrinsic factors. However, after decades of study, the basic mechanism by which benzene affects bone marrow precursor cells is still unclear. Furthermore, even if these factors could be identified, it would be difficult to determine the weight that should be given to each factor in the development of a mathematical model. Nevertheless, selection of this model seemed appropriate since linear dose response relationships exist for several end points measuring the toxic effects of benzene on the bone marrow. (These are discussed below).

For a number of carcinogens, the linear nonthreshold model has been shown to describe the dose response relationship (i.e., radiation and leukemia, arsenic and lung cancer, and asbestos and lung cancer). This model also has been used for carcinogenic agents by the World Health Organization's Arsenic Working Group, for ionizing radiation on an international basis, by a Task Group on Air Pollution and Cancer in Stockholm in 1977, as well as by regulatory bodies in the U.S. such as the Carcinogen Assessment Group (CAG) of the EPA.[1]

With specific regard to benzene and leukemia, OSHA requested several reviewers to comment on the model used by White *et al.* for the risk assessment.[2]

In response, Dr. Norman Breslow, Dr. Philip Cole, Dr. Charles Brown, and Dr. David Gaylor all thought the linear model was acceptable and that other more complex models were not justified. However, Dr. William Rowe felt other models should be evaluated in addition to the linear model, while Dr. Karrh felt the linear model was unverifiable. In our opinion, the issue of model selection can best be summarized by Dr. Charles Brown who stated[2]:

> The 'correct' model is unknown, and will remain so until we know the mechanistic relationship between benzene exposure and leukemia;

*Editors' note:* The current affiliation of Mary C. White is the Centers for Disease Control, Atlanta, Georgia, and of Kenneth C. Chu is the National Cancer Institute, Bethesda, Maryland.

"Letter to the Editor: Assessment of Leukemia Mortality Associated with Occupational Exposure to Benzene" by Peter F. Infante, Mary C. White, and Kenneth C. Chu is reprinted, in abridged form, by permission of the authors and the publisher, from *Risk Analysis*, vol. 4, no. 1 (March 1984), pp. 12–13. © 1984 Society for Risk Analysis.

however, I do not believe that the data warrant more sophistication than the simple linear model (to which the one-hit model is a very close approximation at low response rates); in addition, since the range over which the dose extrapolation is performed appears to be relatively small (one order of magnitude?), the dose-response model should have a small effect on the risk assessment results.

With regard to benzene and leukemia Dr. Norman Breslow recently stated that dose additivity and low dose linearity should be adopted as biologically reasonable and scientifically prudent assumptions in the absence of specific evidence to the contrary.[2] . . .

In order to further evaluate the issue of the level of benzene required to cause toxicity to the bone marrow, one can look to experimental studies where dose and dose level can be controlled. In this regard, the NTP bioassay on benzene demonstrates dose response relationships for multiple cancers as well as for leukopenia in both mice and rats.[3] The low dose in both species is equivalent to an 8-hour atmosphere exposure of 20 ppm. Additional experimental studies have now demonstrated that inhalation of 10 ppm benzene causes significant bone marrow depression (6 hrs/day, 178 days)[4] and disturbances of immune system function (6 hrs/day, 6 days).[5] Since most leukemias and related disorders in man seem to involve stem cell abnormalities and immune system deficiencies these findings may be highly relevant to benzene and leukemia.

Several recent experimental studies have demonstrated adverse effects on chromosomes and bone marrow in relation to low level benzene exposure. Collectively, these studies demonstrate chromosomal damage in bone marrow cells as a result of only one 6-hour exposure to 10 ppm[6] or one 4-hour exposure to 28 ppm,[7] or to an equivalent of two 8-hour exposures to 6 ppm.[8] . . .

## References

1. World Health Organization, *Environmental Health Criteria 18,* Arsenic, Geneva (1981).

2. Formal Comments Submitted to OSHA on Benzene Risk Assessment, Docket H-059B, Ex. No. 137, (1983).

3. NTP Technical Report on the Toxicology and Carcinogenesis Studies of Benzene, NTP 289, U.S. DHHS, NIH Pub. No. 84-2545 (in press).

4. K. A. Baarston, C. A. Snyder, and R. E. Albert, "Repeated Exposure of C57BL Mice to 10 ppm Benzene Markedly Depressed Erythropoietic Colony Formation," *Toxicology Letters* 20:337–342 (1984).

5. M. G. Rosen, C. A. Snyder, and R. E. Albert, "Depressions in B and T Lymphocyte Mitogen-Induced Blastogenesis in Mice Exposed to Low Concentrations of Benzene," *Toxicology Letters* 20:(1984).

6. A. D. Kligerman, G. L. Erexon, and J. L. Wilmer, "Development of Rodent Peripheral Blood Lymphocyte Culture Systems to Detect Cytogenetic Damage in Vivo," Paper presented at International Symposium on Sister Chromatid Exchange, Brookhaven Natl. Lab., (Dec. (1983).

7. R. R. Tice, J. L. Ivett, M. J. Sawey, D. L. Costa, R. T. Drew, and C. P. Cronkite, "Cytogenetic Manifestations of Benzene Induced Damage in Murine Bone Marrow," Paper presented at International Conference on Benzene, Collegium Ramazzini, New York (Nov. 1983).

8. M. M. Gad-El-Karim, B. L. Harper, and M. S. Legator, "Modifications in the Myeloclastogenic Effect of Benzene in Mice with Toluene, Phenobarbitol, 3-Methylcholanthrene, Aroclor 1254 and SKF-525A," *Mutation Res.* (in press).

# Questions for Thought and Discussion

1. Why are you in favor of or opposed to the notion of various government agencies doing independent risk assessments of the same chemical, as EPA, CPSC, and OSHA all did in the case of benzene?

2. This paper provides an informative description of studies involving the estimation of occupational exposure to benzene. How would you go about estimating environmental and consumer exposure?

3. Based on the accounts in this paper, how do you think that epidemiological evidence tends to be used for risk assessment? Compare this approach with the way in which you understand experimental evidence from laboratory testing of animals to be used. Describe some of the major advantages and pitfalls of each approach.

4. On a pair of axes for $P_t$ versus $d$, sketch two-dimensional plots of the one-hit model in equation (4) in this paper for $P_0 = 0$ and for $P_0 > 0$. Explain the meaning of the $P_t$ intercept and the significance of the parameter $B$. State three significant shortcomings of this model.

5. Interpret and discuss the implications of the following comment, which appears in the closing paragraph of the paper: "It is far easier to quantify a level of carcinogenic risk than it is to quantify the level of confidence that should be placed in the risk estimate."

# PART 5

# TECHNOLOGICAL RISK ASSESSMENT

# Introduction ——————————————

Technological systems function safely most of the time, but they are not fail-safe. The massive escape of toxic gas from the chemical plant in Bhopal, the enormous release of radioactivity from the nuclear power plant in Chernobyl, and the huge spill of crude oil from the oil tanker *Exxon Valdez* in Prince William Sound demonstrate how calamitous the failures can be. Technological risk assessment provides information about the frequency of system failures, the ways in which they might occur, and the potential severity of the consequences.

Given the complexity of some modern, large-scale systems, the task of assessing the risks can be formidable, especially when the technology is new. In such cases there is no operating experience and the uncertainties in the risk assessments may be enormous. The situation is complicated further by the virtual inevitability—despite the best efforts of risk assessors to evaluate the risks and the associated societal benefits—that objections will be raised by opponents of controversial technologies. In some cases these debates end up in the courts.

The three papers included in part 5 touch on these matters, dealing in turn with (1) the problem of making social choices about risky technologies that are otherwise beneficial, (2) the techniques for assessing the risks associated with such technologies, and (3) an illustration of such a technology—the importation of liquefied natural gas (LNG) into the United States in jumbo tankers.

In "Social Benefit Versus Technological Risk," Chauncey Starr argues that society seeks the "maximum benefit at minimum cost" in making technological choices. This is not to say that the relationships between benefits and risks are fully understood in a quantitative sense before a technology is adopted, but rather that they are revealed after considerable trial and error. The balances that are eventually struck, we are told, serve to show which trade-offs are socially acceptable. Starr shows, for instance, that acceptable risks of voluntary activities are about a thousand times greater than acceptable risks of involuntary activities.

Probabilistic risk assessment (PRA) techniques, their applications, and their shortcomings are the subject of "The Application of Probabilistic Risk Assessment Techniques to Energy Technologies," by Norman C. Rasmussen. The author's description of the possible consequences of the release of an airborne or waterborne pollutant highlights the many factors that must be considered when doing PRAs. They include, but are not limited to, weather conditions, population densities, features of the surrounding terrain, and knowledge of the adverse effects of the pollutant on humans and property. He also emphasizes that risk perception has a strong influence on the public acceptance of energy technologies.

"Assessing the Risk of an LNG Terminal," by Ralph L. Keeney, Ram B. Kulkarni, and Keshavan Nair, offers a vivid illustration of technological risk assessment. Liquefied natural gas (LNG) is transported in huge, oceangoing vessels at a volume 600 times smaller than the gaseous form, at a temperature of $-162°$ C. An assessment of the risks of operating a terminal for unloading these ships at a Texas port finds that the most serious risk is that a release of LNG onto water could result in subsequent ignition of the vapor cloud either at the site of the spill or downwind, causing numerous fatalities. Yet the calculations show that the proposed technology would be safer than the current means of generating electricity.

# Social Benefit Versus Technological Risk

## CHAUNCEY STARR
*University of California at Los Angeles*

The evaluation of technical approaches to solving societal problems customarily involves consideration of the relationship between potential technical performance and the required investment of societal resources. Although such performance-versus-cost relationships are clearly useful for choosing between alternative solutions, they do not by themselves determine how much technology a society can justifiably purchase. This latter determination requires, additionally, knowledge of the relationship between social benefit and justified social cost. The two relationships may then be used jointly to determine the optimum investment of societal resources in a technological approach to a social need.

Technological analyses for disclosing the relationship between expected performance and monetary costs are a traditional part of all engineering planning and design. The inclusion in such studies of *all* societal costs (indirect as well as direct) is less customary, and obviously makes the analysis more difficult and less definitive. Analyses of social value

*Editors' note:* The current affiliation of Chauncey Starr is the Electric Power Research Institute, Palo Alto, California.

as a function of technical performance are not only uncommon but are rarely quantitative. Yet we know that implicit in every nonarbitrary national decision on the use of technology is a trade-off of societal benefits and societal costs.

In this article I offer an approach for establishing a quantitative measure of benefit relative to cost for an important element in our spectrum of social values—specifically, for accidental deaths arising from technological developments in public use. The analysis is based on two assumptions. The first is that historical national accident records are adequate for revealing consistent patterns of fatalities in the public use of technology. (That this may not always be so is evidenced by the paucity of data relating to the effects of environmental pollution.) The second assumption is that such historically revealed social preferences and costs are sufficiently enduring to permit their use for predictive purposes.

In the absence of economic or sociological theory which might give better results, this empirical approach provides some interesting insights into accepted social values relative to personal risk. Because this methodology is based on historical data, it does not serve to distinguish what is "best" for society from what is "traditionally acceptable."

## Maximum Benefit at Minimum Cost

The broad societal benefits of advances in technology exceed the associated costs sufficiently to make technological growth inexorable. Shef's socioeconomic study (1) has indicated that technological growth has been generally exponential in this century, doubling every 20 years in nations having advanced technology. Such technological growth has apparently stimulated a parallel growth in socioeconomic benefits and a slower associated growth in social costs.

The conventional socioeconomic benefits—health, education, income—are presumably indicative of an improvement in the "quality of life." The cost of this socioeconomic progress shows up in all the negative indicators of our society—urban and environmental problems, technological unemployment, poor physical and mental health, and so on. If we understood quantitatively the causal relationships between specific technological developments and societal values, both positive and negative, we might deliberately guide and regulate technological developments so as to achieve maximum social benefit at minimum social cost. Unfortunately, we have not as yet developed such a predictive system analysis. As a result, our society historically has arrived at acceptable balances of technological benefit and social cost empirically—by trial, error, and subsequent corrective steps.

In advanced societies today, this historical empirical approach creates an increasingly critical situation, for two basic reasons. The first is the well-known difficulty in changing a technical subsystem of our society once it has been woven into the economic, political, and cultural structures. For example, many of our environmental-pollution problems have known engineering solutions, but the problems of economic readjustment, political jurisdiction, and social behavior loom very large. It will take many decades to put into effect the technical solutions we know today. To give a specific illustration, the pollution of our water resources could be completely avoided by means of engineering systems now available, but public interest in making the economic and political adjustments needed for applying these techniques is very limited. It has been facetiously suggested that, as a means of motivating the public, every community and industry should be required to place its water intake downstream from its outfall.

In order to minimize these difficulties, it would be desirable to try out new developments in the smallest social groups that would permit adequate assessment. This is a common practice in market-testing a new product or in field-testing a new drug. In both these cases, however, the experiment is completely under the control of a single company or agency, and the test information can be fed back to the controlling group in a time that is short relative to the anticipated commercial lifetime of the product. This makes it possible to achieve essentially optimum use of the product in an acceptably short time. Unfortunately, this is rarely the case with new technologies. Engineering developments involving new technology are likely to appear in many places simultaneously and to become deeply integrated into the systems of our society before their impact is evident or measurable.

This brings us to the second reason for the increasing severity of the problem of obtaining maximum benefits at minimum costs. It has often been stated that the time required from the conception of a technical idea to its first application in society has been drastically shortened by modern engineering organization and management. In fact, the history of technology does not support this conclusion. The bulk of the evidence indicates that the time from conception to first application (or demonstration) has been roughly unchanged by modern management, and depends chiefly on the complexity of the development.

However, what *has* been reduced substantially in the past century is the time from first use to widespread integration into our social system. The techniques for *societal diffusion* of a new technology and its subsequent exploitation are now highly developed. Our ability to organize resources of money, men, and materials to focus on new technological programs has reduced the diffusion-exploitation time by roughly an order of magnitude in the past century.

Thus, we now face a general situation in which widespread use of a new technological development may occur before its social impact can be properly assessed, and before any empirical adjustment of the benefit-versus-cost relation is obviously indicated.

It has been clear for some time that predictive technological assessments are a pressing societal need. However, even if such assessments become available, obtaining maximum social benefit at minimum cost also requires the establishment of a relative value system for the basic parameters in our objective of improved "quality of life." The empirical approach implicitly involved an intuitive societal balancing of such values. A predictive analytical approach will require an explicit scale of relative social values.

For example, if technological assessment of a new development predicts an increased per capita annual income of $x$ percent but also predicts an associated accident probability of $y$ fatalities annually per million population, then how are these to be compared in their effect on the "quality of life"? Because the penalties or risks to the public arising from a new development can be reduced by applying constraints, there will usually be a functional relationship (or trade-off) between utility and risk, the $x$ and $y$ of our example.

There are many historical illustrations of such trade-off relationships that were empirically determined. For example, automobile and airplane safety have been continuously weighed by society against economic costs and operating performance. In these and other cases, the real trade-off process is actually one of dynamic adjustment, with the behavior of many portions of our social systems out of phase, due to the many separate "time constants" involved. Readily available historical data on accidents and health, for a variety of public activities, provide an enticing stepping-stone to quantitative evaluation of this particular type of social cost. The social benefits arising from some of these activities can be roughly determined. On the assumption that in such historical situations a socially acceptable and essentially optimum trade-off of values has been achieved, we could say that any generalizations developed might then be used for predictive purposes. This approach could give a rough answer to the seemingly simple question "How safe is safe enough?"

The pertinence of this question to all of us, and particularly to governmental regulatory agencies, is obvious. Hopefully, a functional answer might provide a basis for establishing performance "design objectives" for the safety of the public.

## Voluntary and Involuntary Activities

Societal activities fall into two general categories—those in which the individual participates on a "vol-untary" basis and those in which the participation is "involuntary," imposed by the society in which the individual lives. The process of empirical optimization of benefits and costs is fundamentally similar in the two cases—namely, a reversible exploration of available options—but the time required for empirical adjustments (the time constants of the system) and the criteria for optimization are quite different in the two situations.

In the case of "voluntary" activities, the individual uses his own value system to evaluate his experiences. Although his eventual trade-off may not be consciously or analytically determined, or based upon objective knowledge, it nevertheless is likely to represent, for that individual, a crude optimization appropriate to his value system. For example, an urban dweller may move to the suburbs because of a lower crime rate and better schools, at the cost of more time spent traveling on highways and a higher probability of accidents. If, subsequently, the traffic density increases, he may decide that the penalties are too great and move back to the city. Such an individual optimization process can be comparatively rapid (because the feedback of experience to the individual is rapid), so the statistical pattern for a large social group may be an important "real-time" indicator of societal trade-offs and values.

"Involuntary" activities differ in that the criteria and options are determined not by the individuals affected but by a controlling body. Such control may be in the hands of a government agency, a political entity, a leadership group, an assembly of authorities or "opinion-makers," or a combination of such bodies. Because of the complexity of large societies, only the control group is likely to be fully aware of all the criteria and options involved in their decision process. Further, the time required for feedback of the experience that results from the controlling decisions is likely to be very long. The feedback of cumulative individual experiences into societal communication channels (usually political or economic) is a slow process, as is the process of altering the planning of the control group. We have many examples of such "involuntary" activities, war being perhaps the most extreme case of the operational separation of the decision-making group from those most affected. Thus, the real-time pattern of societal trade-offs on "involuntary" activities must be considered in terms of the particular dynamics of approach to an acceptable balance of social values and costs. The historical trends in such activities may therefore be more significant indicators of social acceptability than the existent trade-offs are.

In examining the historical benefit-risk relationships for "involuntary" activities, it is important to recognize the perturbing role of public psychological acceptance of risk arising from the influence of authorities or dogma. Because in this situation the decision-making is separated from the affected individual, society has generally clothed many of its controlling groups in an almost impenetrable mantle of authority and of imputed wisdom. The public generally assumes that the decision-making process is based on a rational analysis of social benefit and social risk. While it often is, we have all seen after-the-fact examples of irrationality. It is important to omit such "witch-doctor" situations in selecting examples of optimized "involuntary" activities, because in fact these situations typify only the initial stages of exploration of options.

## Quantitative Correlations

With this description of the problem, and the associated caveats, we are in a position to discuss the quantitative correlations. For the sake of simplicity in this initial study, I have taken as a measure of the physical risk to the individual the fatalities (deaths) associated with each activity. Although it might be useful to include all injuries (which are 100 to 1000 times as numerous as deaths), the difficulty in obtaining data and the unequal significance of varying disabilities would introduce inconvenient complexity for this study. So the risk measure used here is the statistical probability of fatalities per hour of exposure of the individual to the activity considered.

The hour-of-exposure unit was chosen because it was deemed more closely related to the individual's intuitive process in choosing an activity than a year of exposure would be, and gave substantially similar results. Another possible alternative, the risk per activity, involved a comparison of too many dissimilar units of measure; thus, in comparing the risk for various modes of transportation, one could use risk per hour, per mile, or per trip. As this study was directed toward exploring a methodology for determining social acceptance of risk, rather than the safest mode of transportation for a particular trip, the simplest common unit—that of risk per exposure hour—was chosen.

The social benefit derived from each activity was converted into a dollar equivalent, as a measure of integrated value to the individual. This is perhaps the most uncertain aspect of the correlations because it reduced the "quality-of-life" benefits of an activity to an overly simplistic measure. Nevertheless, the correlations seemed useful, and no better measure was available. In the case of the "voluntary" activities, the amount of money spent on the activity by the average involved individual was assumed proportional to its benefit to him. In the case of the "involuntary" activities, the contribution of the activity to the individual's annual income (or the equivalent) was assumed proportional to its benefit. This assumption of roughly constant relationship between benefits and monies, for each class of activities, is clearly an approximation. However, because we are dealing in orders of magnitude, the distortions likely to be introduced by this approximation are relatively small.

In the case of transportation modes, the benefits were equated with the sum of the monetary cost to the passenger and the value of the time saved by that particular mode relative to a slower, competitive mode. Thus, airplanes were compared with automobiles, and automobiles were compared with public transportation or walking. Benefits of public transportation were equated with their cost. In all cases, the benefits were assessed on an annual dollar basis because this seemed to be most relevant to the individual's intuitive process. For example, most luxury sports cars require an investment and upkeep only partially dependent upon usage. The associated risks, of course, exist only during the hours of exposure.

Probably the use of electricity provides the best example of the analysis of an "involuntary" activity. In this case the fatalities include those arising from electrocution, electrically caused fires, the operation of power plants, and the mining of the required fossil fuel. The benefits were estimated from a United Nations study of the relationship between energy consumption and national income; the energy fraction associated with electric power was used. The

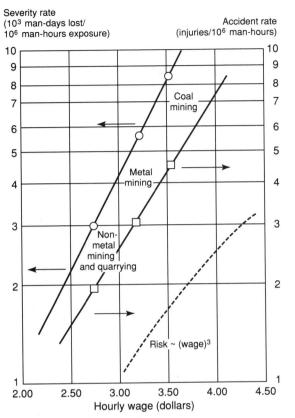

Figure 1.  **Mining accident rates plotted relative to incentive.**

contributions of the home use of electric power to our "quality of life"—more subtle than the contributions of electricity in industry—are omitted. The availability of refrigeration has certainly improved our national health and the quality of dining. The electric light has certainly provided great flexibility in patterns of living, and television is a positive element. Perhaps, however, the gross-income measure used in the study is sufficient for present purposes.

Information on acceptance of "voluntary" risk by individuals as a function of income benefits is not easily available, although we know that such a relationship must exist. Of particular interest, therefore, is the special case of miners exposed to high occupational risks. In figure 1, the accident rate and the severity rate of mining injuries are plotted against the hourly wage (2, 3). The acceptance of individual risk is an exponential function of the wage, and can be roughly approximated by a third-power relationship in this range. If this relationship has validity, it

may mean that several "quality of life" parameters (perhaps health, living essentials, and recreation) are each partly influenced by any increase in available personal resources, and that thus the increased acceptance of risk is exponentially motivated. The extent to which this relationship is "voluntary" for the miners is not obvious, but the subject is interesting nevertheless.

## Risk Comparisons

The results for the societal activities studied, both "voluntary" and "involuntary," are assembled in figure 2. (For details of the risk-benefit analysis, see the appendix.) Also shown in figure 2 is the third-power relationship between risk and benefit characteristic of figure 1. For comparison, the average risk of death from accident and from disease is shown. Because the average number of fatalities from accidents is only about one-tenth the number from disease, their inclusion is not significant.

Several major features of the benefit-risk relations are apparent, the most obvious being the difference by several orders of magnitude in society's willingness to accept "voluntary" and "involuntary" risk. As one would expect, we are loathe to let others do unto us what we happily do to ourselves.

The rate of death from disease appears to play, psychologically, a yardstick role in determining the acceptability of risk on a voluntary basis. The risk of death in most sporting activities is surprisingly close to the risk of death from disease—almost as though, in sports, the individual's subconscious computer adjusted his courage and made him take risks associated with a fatality level equaling but not exceeding the statistical mortality due to involuntary exposure to disease. Perhaps this defines the demarcation between boldness and foolhardiness.

In figure 2 the statistic for the Vietnam war is shown because it raises an interesting point. It is only slightly above the average for risk of death from disease. Assuming that some long-range societal benefit was anticipated from this war, we find that the related risk, as seen by society as a whole, is not substantially different from the average nonmilitary risk from disease. However, for individuals in the military-service age group (age 20 to 30), the risk of death in Vietnam is about ten times the normal

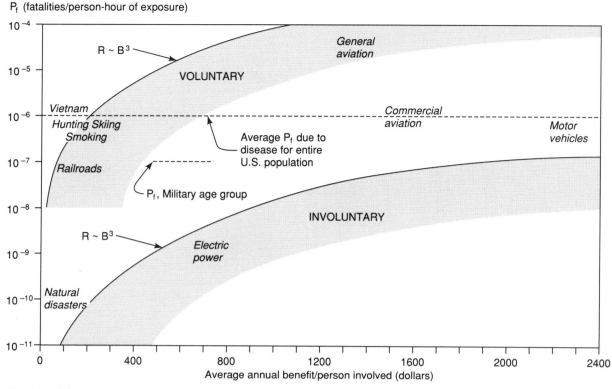

**Figure 2.   Risk (R) plotted relative to benefit (B) for various kinds of voluntary and involuntary exposure.**

mortality rate (death from accidents or disease). Hence the population as a whole and those directly exposed see this matter from different perspectives. The disease risk pertinent to the average age of the involved group probably would provide the basis for a more meaningful comparison than the risk pertinent to the national average age does. Use of the figure for the single group would complicate these simple comparisons, but that figure might be more significant as a yardstick.

The risks associated with general aviation, commercial aviation, and travel by motor vehicle deserve special comment. The latter originated as a "voluntary" sport, but in the past half-century the motor vehicle has become an essential utility. General aviation is still a highly voluntary activity. Commercial aviation is partly voluntary and partly essential and, additionally, is subject to government administration as a transportation utility.

Travel by motor vehicle has now reached a benefit-risk balance, as shown in figure 3. It is interesting

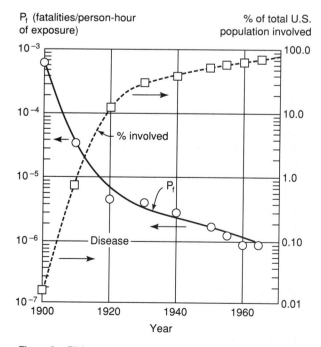

**Figure 3.   Risk and participation trends for motor vehicles.**

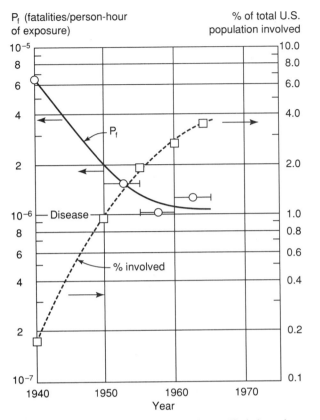

**Figure 4.  Risk and participation trends for certified air carriers.**

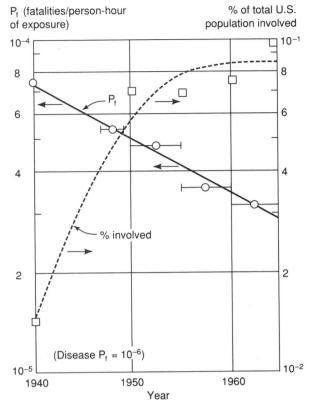

**Figure 5.  Risk and participation trends for general aviation.**

to note that the present risk level is only slightly below the basic level of risk from disease. In view of the high percentage of the population involved, this probably represents a true societal judgment on the acceptability of risk in relation to benefit. It also appears from figure 3 that future reductions in the risk level will be slow in coming, even if the historical trend of improvement can be maintained (4).

Commercial aviation has barely approached a risk level comparable to that set by disease. The trend is similar to that for motor vehicles, as shown in figure 4. However, the percentage of the population participating is now only 1/20 that for motor vehicles. Increased public participation in commercial aviation will undoubtedly increase the pressure to reduce the risk, because, for the general population, the benefits are much less than those associated with motor vehicles. Commercial aviation has not yet reached the point of optimum benefit-risk trade-off (5).

For general aviation the trends are similar, as

shown in figure 5. Here the risk levels are so high (20 times the risk from disease) that this activity must properly be considered to be in the category of adventuresome sport. However, the rate of risk is decreasing so rapidly that eventually the risk for general aviation may be little higher than that for commercial aviation. Since the percentage of the population involved is very small, it appears that the present average risk levels are acceptable to only a limited group (6).

The similarity of the trends in figures 3–5 may be the basis for another hypothesis, as follows: the acceptable risk is inversely related to the number of people participating in an activity.

The product of the risk and the percentage of the population involved in each of the activities of figures 3–5 is plotted in figure 6. This graph represents the historical trend of total fatalities per hour of exposure of the population involved (7). The leveling off of motor-vehicle risk at about 100 fatalities per hour

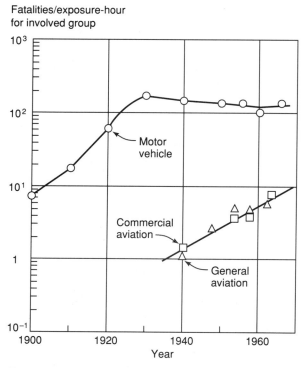

**Figure 6.  Group risk plotted relative to year.**

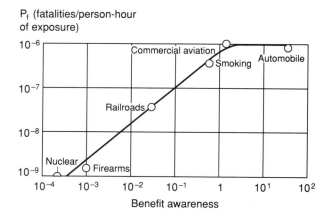

**Figure 7.  Accepted risk plotted relative to benefit awareness (see text).**

of exposure of the participating population may be significant. Because most of the U.S. population is involved, this rate of fatalities may have sufficient public visibility to set a level of social acceptability. It is interesting, and disconcerting, to note that the trend of fatalities in aviation, both commercial and general, is uniformly upward.

## Public Awareness

Finally, I attempted to relate these risk data to a crude measure of public awareness of the associated social benefits (see figure 7). The "benefit awareness" was arbitrarily defined as the product of the relative level of advertising, the square of the percentage of population involved in the activity, and the relative usefulness (or importance) of the activity to the individual (8). Perhaps these assumptions are too crude, but figure 7 does support the reasonable position that advertising the benefits of an activity increases public acceptance of a greater level of risk. This, of course, could subtly produce a fictitious benefit-risk ratio—as may be the case for smoking.

## Atomic Power Plant Safety

I recognize the uncertainty inherent in the quantitative approach discussed here, but the trends and magnitudes may nevertheless be of sufficient validity to warrant their use in determining national "design objectives" for technological activities. How would this be done?

Let us consider as an example the introduction of nuclear power plants as a principal source of electric power. This is an especially good example because the technology has been primarily nurtured, guided, and regulated by the government, with industry undertaking the engineering development and the diffusion into public use. The government specifically maintains responsibility for public safety. Further, the engineering of nuclear plants permits continuous reduction of the probability of accidents, at a substantial increase in cost. Thus, the trade-off of utility and potential risk can be made quantitative.

Moreover, in the case of the nuclear power plant the historical empirical approach to achieving an optimum benefit-risk trade-off is not pragmatically feasible. All such plants are now so safe that it may be 30 years or longer before meaningful risk experience will be accumulated. By that time, many plants of varied design will be in existence, and the empirical accident data may not be applicable to those being built. So a very real need exists now to establish "design objectives" on a predictive-performance basis.

Let us first arbitrarily assume that nuclear power plants should be as safe as coal-burning plants, so as not to increase public risk. Figure 2 indicates that the total risk to society from electric power is about $2 \times 10^{-9}$ fatality per person per hour of exposure. Fossil fuel plants contribute about ⅕ of this risk, or about 4 deaths per million population per year. In a modern society, a million people may require a million kilowatts of power, and this is about the size of most new power stations. So, we now have a target risk limit of 4 deaths per year per million-kilowatt power station (9).

Technical studies of the consequences of hypothetical extreme (and unlikely) nuclear power plant catastrophes, which would disperse radioactivity into populated areas, have indicated that about 10 lethal cancers per million population might result (10). On this basis, we calculate that such a power plant might statistically have one such accident every 3 years and still meet the risk limit set. However, such a catastrophe would completely destroy a major portion of the nuclear section of the plant and either require complete dismantling or years of costly reconstruction. Because power companies expect plants to last about 30 years, the economic consequences of a catastrophe every few years would be completely unacceptable. In fact, the operating companies would not accept one such failure, on a statistical basis, during the normal lifetime of the plant.

It is likely that, in order to meet the economic performance requirements of the power companies, a catastrophe rate of less than 1 in about 100 plant-years would be needed. This would be a public risk of 10 deaths per 100 plant-years, or 0.1 death per year per million population. So the economic investment criteria of the nuclear plant user—the power company—would probably set a risk level 1/200 the present socially accepted risk associated with electric power, or 1/40 the present risk associated with coal-burning plants.

An obvious design question is this: Can a nuclear power plant be engineered with a predicted performance of less than 1 catastrophic failure in 100 plant-years of operation? I believe the answer is yes, but that is a subject for a different occasion. The principal point is that the issue of public safety can be focused on a tangible, quantitative, engineering design objective.

This example reveals a public safety consideration which may apply to many other activities: The economic requirement for the protection of major capital investments may often be a more demanding safety constraint than social acceptability.

## Conclusion

The application of this approach to other areas of public responsibility is self-evident. It provides a useful methodology for answering the question "How safe is safe enough?" Further, although this study is only exploratory, it reveals several interesting points. (i) The indications are that the public is willing to accept "voluntary" risks roughly 1000 times greater than "involuntary" risks. (ii) The statistical risk of death from disease appears to be a psychological yardstick for establishing the level of acceptability of other risks. (iii) The acceptability of risk appears to be crudely proportional to the third power of the benefits (real or imagined). (iv) The social acceptance of risk is directly influenced by public awareness of the benefits of an activity, as determined by advertising, usefulness, and the number of people participating. (v) In a sample application of these criteria to atomic power plant safety, it appears that an engineering design objective determined by economic criteria would result in a design-target risk level very much lower than the present socially accepted risk for electric power plants.

Perhaps of greatest interest is the fact that this methodology for revealing existing social preferences and values may be a means of providing the insight on social benefit relative to cost that is so necessary for judicious national decisions on new technological developments.

## Appendix: Details of Risk-Benefit Analysis

*Motor-vehicle travel.* The calculation of motor-vehicle fatalities per exposure hour per year is based on the number of registered cars, an assumed 1½ persons per car, and an assumed 400 hours per year of average car use [data from *3* and *11*]. The figure for annual benefit for motor-vehicle travel is based on the sum of costs for gasoline, maintenance, insurance, and car payments and on the value of the time savings per person. It is assumed that use of an automobile allows a person to save 1 hour per working day and that a person's time is worth $5 per hour.

*Travel by air route carrier.* The estimate of passenger fatalities per passenger-hour of exposure for certified air route carriers is based on the annual number of passenger fatalities listed in the *FAA Statistical Handbook of Aviation* (see *12*) and the number of passenger-hours per year. The latter number is estimated from the average number of seats per plane, the seat load factor, the number of revenue miles flown per year, and the average plane speed (data from *3* ). The benefit for travel by certified air route carrier is based on the average annual air fare per passenger-mile and on the value of the time saved as a result of air travel. The cost per passenger is estimated from the average rate per passenger-mile (data from *3* ), the revenue miles flown per year (data from *12*), the annual number of passenger boardings for 1967 (132 × 10⁶, according to the United Air Lines News Bureau), and the assumption of 12 boardings per passenger.

*General aviation.* The number of fatalities per passenger-hour for general aviation is a function of the number of annual fatalities, the number of plane hours flown per year, and the average number of passengers per plane (estimated from the ratio of fatalities to fatal crashes) (data from *12* ). It is assumed that in 1967 the cash outlay for initial expenditures and maintenance costs for general aviation was $1.5 × 10⁹. The benefit is expressed in terms of annual cash outlay per person, and the estimate is based on the number of passenger-hours per year and the assumption that the average person flies 20 hours, or 4000 miles, annually. The value of the time saved is based on the assumption that a person's time is worth $10 per hour and that he saves 60 hours per year through traveling the 4000 miles by air instead of by automobile at 50 miles per hour.

*Railroad travel.* The estimate of railroad passenger fatalities per exposure hour per year is based on annual passenger fatalities and passenger-miles and an assumed average train speed of 50 miles per hour (data from *11* ). The passenger benefit for railroads is based on figures for revenue and passenger-miles for commuters and noncommuters given in *The Yearbook of Railroad Facts* (Association of American Railroads, 1968). It is assumed that the average commuter travels 20 miles per workday by rail and that the average noncommuter travels 1000 miles per year by rail.

*Skiing.* The estimate for skiing fatalities per exposure hour is based on information obtained from the National Ski Patrol for the 1967–68 southern California ski season: 1 fatality, 17 days of skiing, 16,500 skiers per day, and 5 hours of skiing per skier per day. The estimate of benefit for skiing is based on the average number of days of skiing per year per person and the average cost of a typical ski trip [data from "The Skier Market in Northeast North America," *U.S. Dep. Commerce Publ.* (1965)]. In addition, it is assumed that a skier spends an average of $25 per year on equipment.

*Hunting.* The estimate of the risk in hunting is based on an assumed value of 10 hours' exposure per hunting day, the annual number of hunting fatalities, the number of hunters, and the average number of hunting days per year

[data from *11* and from "National Survey of Fishing and Hunting," *U.S. Fish Wildlife Serv. Publ.* (1965)]. The average annual expenditure per hunter was $82.54 in 1965 (data from *3* ).

*Smoking.* The estimate of the risk from smoking is based on the ratio for the mortality of smokers relative to nonsmokers, the rates of fatalities from heart disease and cancer for the general population, and the assumption that the risk is continuous [data from the *Summary of the Report of the Surgeon General's Advisory Committee on Smoking and Health* (Government Printing Office, Washington, D.C., 1964)]. The annual intangible benefit to the cigarette smoker is calculated from the American Cancer Society's estimate that 30 percent of the population smokes cigarettes, from the number of cigarettes smoked per year (see *3* ), and from the assumed retail cost of $0.015 per cigarette.

*Vietnam.* The estimate of the risk associated with the Vietnam war is based on the assumption that 500,000 men are exposed there annually to the risk of death and that the fatality rate is 10,000 men per year. The benefit for Vietnam is calculated on the assumption that the entire U.S. population benefits intangibly from the annual Vietnam expenditure of $30 × 10⁹.

*Electric power.* The estimate of the risk associated with the use of electric power is based on the number of deaths from electric current; the number of deaths from fires caused by electricity; the number of deaths that occur in coal mining, weighted by the percentage of total coal production used to produce electricity; and the number of deaths attributable to air pollution from fossil fuel stations [data from *3* and *11* and from *Nuclear Safety* **5**, 325 (1964)]. It is assumed that the entire U.S. population is exposed for 8760 hours per year to the risk associated with electric power. The estimate for the benefit is based on the assumption that there is a direct correlation between per capita gross national product and commercial energy consumption for the nations of the world [data from Briggs, *Technology and Economic Development* (Knopf, New York, 1963)]. It is further assumed that 35 percent of the energy consumed in the U.S. is used to produce electricity.

*Natural disasters.* The risk associated with natural disasters was computed for U.S. floods (2.5 × 10⁻¹⁰ fatality per person-hour of exposure), tornadoes in the Midwest (2.46 × 10⁻¹⁰ fatality), major U.S. storms (0.8 × 10⁻¹⁰ fatality), and California earthquakes (1.9 × 10⁻¹⁰ fatality) (data from *11* ). The value for flood risk is based on the assumption that everyone in the U.S. is exposed to the danger 24 hours per day. No benefit figure was assigned in the case of natural disasters.

*Disease and accidents.* The average risk in the U.S. due to disease and accidents is computed from data given in *Vital Statistics of the U.S.* (Government Printing Office, Washington, D.C., 1967).

## References and Notes

1. A. L. Shef, "Socio-economic attributes of our technological society," paper presented before the IEEE (Institute of Electrical and Electronics Engineers) Wescon Conference, Los Angeles, August 1968.

2. *Minerals Yearbook* (Government Printing Office, Washington, D.C., 1966).

3. *U.S. Statistical Abstract* (Government Printing Office, Washington, D.C., 1967).

4. The procedure outlined in the appendix was used in calculating the risk associated with motor-vehicle travel. In order to calculate exposure hours for various years, it was assumed that the average annual driving time per car increased linearly from 50 hours in 1900 to 400 hours in 1960 and thereafter. The percentage of people involved is based on the U.S. population, the number of registered cars, and the assumed value of 1.5 people per car.

5. The procedure outlined in the appendix was used in calculating the risk associated with, and the number of people who fly in, certified air route carriers for 1967. For a given year, the number of people who fly is estimated from the total number of passenger boardings and the assumption that the average passenger makes six round trips per year (data from *3*).

6. The method of calculating risk for general aviation is outlined in the appendix. For a given year, the percentage of people involved is defined by the number of active aircraft (see *3*); the number of people per plane, as defined by the ratio of fatalities to fatal crashes; and the population of the U.S.

7. Group risk per exposure hour for the involved group is defined as the number of fatalities per person-hour of exposure multiplied by the number of people who participate in the activity. The group population and the risk for motor vehicles, certified air route carriers, and general aviation can be obtained from figures 3–5.

8. In calculating "benefit awareness" it is assumed that the public's awareness of an activity is a function of $A$, the amount of money spent on advertising; $P$, the number of people who take part in the activity; and $U$, the utility value of the activity to the person involved. $A$ is based on the amount of money spent by a particular industry in advertising its product, normalized with respect to the food and food products industry, which is the leading advertiser in the U.S.

9. In comparing nuclear and fossil fuel power stations, the risks associated with the plant effluents and mining of the fuel should be included in each case. The fatalities associated with coal mining are about ¼ the total attributable to fossil fuel plants. As the tonnage of uranium ore required for an equivalent nuclear plant is less than the coal tonnage by more than an order of magnitude, the nuclear plant problem primarily involves hazard from effluent.

10. This number is my estimate for maximum fatalities from an extreme catastrophe resulting from malfunction of a typical power reactor. For a methodology for making this calculation, see F. R. Farmer, "Siting criteria—a new approach," paper presented at the International Atomic Energy Agency Symposium in Vienna, April 1967. Application of Farmer's method to a fast breeder power plant in a modern building gives a prediction of fatalities less than this assumed limit by one or two orders of magnitude.

11. "Accident Facts," *Nat. Safety Counc. Publ.* (1967).

12. *FAA Statistical Handbook of Aviation* (Government Printing Office, Washington, D.C., 1965).

# Questions for Thought and Discussion

1. How would you develop a standard or set of standards for the degree to which technological benefits should exceed the associated costs in order for technological growth to be desirable? Which is a better measure of desirability: the difference between benefit and cost, or the ratio of benefit to cost?

2. This paper, which was written in 1969, states that "engineering developments involving new technology are likely to appear in many places simultaneously and to become deeply integrated into the systems of our society before their impact is evident or measurable." Do you believe that this statement still applies? Give two examples of current developments that support your conclusion.

3. Starr says that "society has generally clothed many of its controlling groups in an almost impenetrable mantle of authority and of imputed wisdom." To what extent do you think that society's increased awareness of environmental and technological risks has changed its view of the relevant controlling groups? Explain.

4. The conclusion is reached from figure 1 in this paper that the acceptance of individual risk among miners increases exponentially with their wages. How would you go about determining the degree to which the choice to work as a miner, given the wage levels, is a voluntary decision?

5. In figure 2, commercial aviation is shown to have a risk of about one fatality per million person-hours of exposure, and motor vehicle travel is shown to have a slightly lower risk. Does this mean that it is safer to drive from Chicago to New York than to fly? Explain.

# The Application of Probabilistic Risk Assessment Techniques to Energy Technologies

NORMAN C. RASMUSSEN
*Massachusetts Institute of Technology*
*Cambridge, Massachusetts*

## Introduction

It is well recognized that the procurement, production, distribution, and use of energy in its various forms all have the potential for causing adverse effects on people and the environment. Thus in choosing between various energy options one should consider these effects in addition to other obvious factors such as economics. It would be highly desirable to be able to quantify in some meaningful way the possible adverse effects of an activity with the goal of obtaining some figure of merit that could be used in directly comparing various options. Work toward achieving this goal for a variety of man's activities, especially in the area of energy, has been carried out at an ever increasing rate during the last two decades. This paper reviews some of these approaches in the energy field, but because of the very large amount of work that has been done, makes no attempt to be exhaustive. Rather, it discusses some of the more common approaches, how they have been carried out, and some of the major problems that still must be solved before the techniques are fully accepted. In fact a large part of the paper reviews the methods for dealing with rare accidents that have potentially large consequences, such as dam failures, large releases of radioactivity, and fires and explosions in liquid and gaseous fuels. An approach that is being widely used to analyze such risks is probabilistic risk analysis (PRA).

The term probabilistic risk analysis refers to a process of estimating the risk of an activity based on the probability of events whose occurrence can lead to undesired consequences. Although all uses of PRA fit within this very broad definition, there are many different methods employed in carrying out the detailed steps of the process. Thus it is common to find in the literature vastly different procedures for carrying out a PRA, sometimes even for the same technology and same type of accident. One of the main sources of these differences is differences in some of the basic definitions used. In particular, one finds in the literature fundamental differences in the definition and interpretation of the meaning of probability and of risk. In addition, there are a variety of ways currently used to estimate probability and also a variety of ways to express risk. Because these differences can lead to different results and to differences in the interpretation of the results, I review some of the key issues here.

## Some Definitions and Concepts of PRA

The terms "hazard" and "risk" often are interchangeable in common usage, but in PRA they have more precise definitions. The term hazard expresses the potential for producing an undesired consequence without regard to how likely such a consequence is. Thus one of the hazards of nuclear power is that a large amount of radioactivity could be released and, of course, this could produce a number of different undesired consequences. Clearly this hazard is associated with any nuclear power plant. The hazard is largest in the plant with the largest inventory of radioactivity and is the same for two plants with the same inventory. The term risk usually expresses not only the potential for an undesired consequence, but also how probable it is that such a consequence will occur. Thus two plants with the same radioactive inventory can pose vastly different risks, depending upon the effectiveness of their safety systems.

A mathematical definition of risk commonly found in the literature is:

$$\text{Risk}\left(\frac{\text{Consequence}}{\text{unit time}}\right) = \text{Frequency}\left(\frac{\text{event}}{\text{unit time}}\right)$$
$$\times \text{Magnitude}\left(\frac{\text{Consequence}}{\text{event}}\right)$$

Thus if we wished to calculate the annual risk of death from automobile accidents in the United States by this relation, we would have:

$$\left(15 \times 10^6 \frac{\text{accidents}}{\text{year}}\right)\left(\frac{1 \text{ death}}{300 \text{ accidents}}\right)$$
$$= 50{,}000 \frac{\text{deaths}}{\text{year}}$$

Clearly in this case the answer could have been arrived at just by looking up the number of automobile accident fatalities in the statistical records. How-

ever, if we wanted to know the risk from nuclear accidents, LNG explosions, etc., there is no statistical base so the formula becomes useful. Of course, we will need a method of estimating the frequency of accidents and the magnitude of these consequences, which is what PRA attempts to do by methods discussed below.

Although the above definition of risk is widely used, it has some serious shortcomings if used as a measure of risk for comparing various options. To illustrate a major problem, consider the following example. Suppose one activity has a frequency of occurrence of accidents of $10^{-6}$ per year with an average consequence of $10^6$ deaths per accident. A second activity has a frequency of occurrence of accidents of $10^{-1}$ per year, with an average of 10 deaths per accident. Since $(10^{-6} \times 10^6) = 1$ and $(0.1)(10) = 1$, the activities have the same risk by the above definition. However, most societies would prefer the latter to the former. A society or individual who feels this way is said to be risk averse to large consequence events. To reflect this type of risk aversion, the mathematical relationship would have to be changed to something other than a simple product. A number of different functional relations between frequency and magnitude can be found in the literature. However, none has yet gained widespread acceptance. It has become common practice in PRA to calculate the risk as defined above, but also to calculate a distribution function that expresses the frequency of accidents vs accident magnitude. The decision-making process then must decide how to account for risk aversion (see the section on the perception of risk for further discussion).

It may come as a surprise to those not familiar with the field of probability to learn that in the technical community there exist two quite different definitions of probability. The classical statisticians think of probability as a property of a process that can be determined from an infinite population of data. For example, the probability of "heads" ($P_H$) in a coin toss is defined as:

$$P_H = \frac{N_H}{N_T} \text{ as } N_T \rightarrow \infty$$

where $N_H$ = number of heads and $N_T$ = number of tosses. The statisticians have devised techniques for making estimates of $P_H$ and of the range of uncertainty in this estimate using less than an infinite population of data. Those who accept this definition are often called frequentists and believe that $P_H$ is a precise value and that information needed to make estimates of it can come only from observation of the process.

The other school of thought, accepted by those called subjectivists, holds that $P_H$ has a value at any time that represents the total available knowledge about the process at that particular time. Thus if the coin were examined carefully before the first toss and found to have a head and a tail and be well balanced, and an examination of the tossing process showed it to be fair, the subjectivist would say his best estimate of the value of $P_H$ is 0.5. Further, any logical observer given exactly the same information would reach the same conclusion. Note that the value of $P_H = 0.5$ is based on knowledge of the coin and the flipping process, but not on any observations.

Now suppose the coin is flipped ten times and we observe seven heads and three tails. At this point the frequentists' best estimate of $P_H$ is 0.7. However, the subjectivist will add this new information to his previous knowledge in a logical and consistent way. A well-known relationship called Bayes theorem (discussed in any text of probability and statistics) defines a logical and consistent way of doing this. Using Bayes theorem, the subjectivist finds his prior estimate of $P_H$ is modified only slightly upward when this new information is considered. If it is in fact a fair coin toss, then after a large number of tosses both groups will eventually obtain a value of $P_H$ that approaches 0.5.

The use of PRA in large accidents of low probability must employ the logic of the subjectivist (or Bayesian) approach since rarely will enough actual data exist to use the frequentists' definition. Although some controversy still exists between these two schools of thought, the Bayesian approach is gaining wider and wider acceptance. The rest of this paper discusses the use of this Bayesian approach for estimating risk. The reader should be aware that there still remains a group of applied statisticians who do not fully accept this approach.

## Determination of Accident Probability

As noted above, PRA requires knowledge of both the probability (or more precisely the frequency) of possible accidents and the magnitude of the consequences. First let us examine some of the commonly used methods for determining the accident frequency. The discussion below is limited to a few of the most commonly used approaches.

### Actuarial Method

In cases where there is substantial recorded experience the probability can be determined directly from this data. However, since one usually wants an estimate of the probability for some future time period, caution must be exercised to be sure the historical data is truly applicable. Care must be taken to account for such things as changes in the accident rate with time and any significant changes in the technology. Consider, for example, the expected rate of collision of LNG tankers serious enough to rupture the cargo tanks. Because of the limited number of LNG tankers, direct statistical data are not available. However, good estimates are available for ship collision rates and reasonable estimates for the collision probability of oil tankers that result in cargo tank rupture. There is a temptation to use this latter rate for LNG tankers (and some have succumbed to it). However, the hull construction of an LNG ship is much different from that of an oil tanker because of the significant insulation requirements, which led to a triple hull construction with insulation between two of the hulls. Thus it takes a much more severe collision to rupture an LNG tanker and so the oil tanker experience is likely to give a very poor estimate. Nevertheless in cases where there is appropriate data the actuarial method is the easiest to apply and generally produces reasonably accurate results. In addition, since the results are obtained simply and directly, they are easily understood and hence less controversial than the methods that follow.

### The Fault Tree Method

Fault tree analysis (FTA) is a technique used to predict the expected probability of failure of a system in the absence of actual experience of failure. This may occur because there is very little operating experience, or because the system failure rate is so low that no failures have been observed. The technique is applicable when the system is made up of

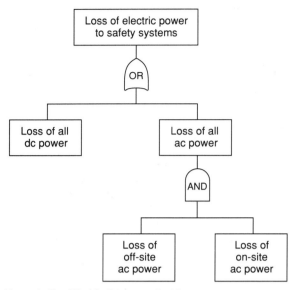

**Figure 1. Simplified fault tree on electric power.**

many parts and the failure rate of the parts is known. Thus, for example, the method would be suitable for estimating the probability of failure on demand of all electric power to the safety systems of a nuclear plant. The method is poorly suited for estimating the failure rate of a steel pressure vessel or a dam. To explain how the method works, let us consider the failure of the electrical supply to the safety systems in a nuclear power plant.

The fault tree analysis always starts with the definition of the undesired event whose probability is to be determined. The probability of failure can either be a failure frequency, which is the probability of failure per unit time, or a probability of failure on demand (such as the probability that a car will fail to start when the starter switch is turned). In our example the event will be "loss of power to safety systems on demand" and is shown in the top box of figure 1. In a typical nuclear plant the ac power can be supplied from any one of three sources: the utility grid (off-site power) or either one of two diesel generators at the plant (on-site power). In addition, the switches that control the power are operated by a dc circuit. Thus a loss of either dc or all ac sources leads to a loss of power. This relationship is shown in figure 1 by the second two boxes, which are coupled to the top event by an OR gate. This indicates that either one "OR" the other will produce the top event.

If the probability of failure of the ac and dc is small and independent, then the probability of losing power is just the sum of the probability of losing the ac and the probability of losing the dc. To further illustrate the process we develop the loss of ac power into loss of off-site and loss of on-site ac power. These events are coupled to loss of ac power by an "AND" gate to indicate both must occur simultaneously for loss of ac power to occur. If the loss of on-site and off-site power are independent, then the probability of loss of ac power is the product of the probabilities of the two lower events. Following this general approach, the tree is developed to lower and lower levels until the lowest events, which are called primary faults. These include such events as the switch fails to open, the solenoid sticks shut, the operator closes the switch accidentally, etc. For the fault tree method to work these primary faults must be events whose probability can be determined from experience. Since most of the components of a nuclear plant are standard commercial equipment that is widely used in many similar applications, this data base exists. In some cases, good, directly applicable data can be obtained; in other cases the data may be limited. In all cases the best available estimate should be used with an assigned uncertainty to reflect our knowledge of its accuracy.

Both equipment failures and human failures of omission and commission enter into a properly drawn fault tree. The widespread belief that these methods can not deal with human errors is false. However, the data on human failure rates tend to have significant uncertainties. In the Reactor Safety Study, commonly referred to as WASH-1400, we assigned a value of ± a factor of 10 to this uncertainty (1), which is about the same as the uncertainty of the most poorly known of equipment failure rates. Most of the equipment failure rates are somewhat better known. In WASH-1400 it was found by the fault tree method that human failures in operation and maintenance contributed more than half the risk.

Once all the primary event probabilities are assigned, the calculation of the probability of the top event becomes a problem of Boolean algebra. Numerous computer codes are now available for doing this calculation. It should be noted that for standby systems the top event will be a probability of failure to operate on demand, so all probabilities in the tree are true probabilities. However, for an operating system one determines a failure rate, and all probabilities in the tree must be failure rates. In figure 1 one must use the probability that a system such as the off-site power is available at some given time in the future. This probability is called the availability of the system.

Although the uncertainties in the primary event probabilities could, in principle, be propagated analytically through the fault tree if their distribution function is known, most often a Monte Carlo calculation is used to determine the uncertainty of the top event failure rate or probability. In this process one samples appropriately from each of the input distributions and calculates the top event probability. This process is repeated several thousand times to produce a distribution for the top event probability from which confidence limits can be derived. In the real fault tree for the problem described, each of the events must be further developed and the final tree would likely have hundreds of events and gates.

Rarely will it be correct to assume all events are independent, so any dependencies must be accounted for. The problem of several events with a common cause is a difficult one and greatly complicates the analysis. The development of techniques for dealing with the common cause problem is an active area of current research.

The reader will note that each box in a fault tree is a definition of failure. If one changed all these failure definitions to success, and interchanged all the "AND" and "OR" gates, the resultant success tree would be an equally logical and correct way to express the system relationships. In fact, such success trees are sometimes used. However, the fault tree is much more widely used because experience has shown that the analyst seems to be less likely to overlook an important failure mode if he continually asks himself "How can it fail?" rather than "How can it succeed?"

The fault tree technique has been evolving for the last two decades and is probably the most widely used method for the quantitative prediction of system failure. However, it becomes exceedingly difficult to apply in very complicated problems. For example, in its initial work, the group doing the WASH-1400 (1) study attempted to draw a fault tree whose top box was "accidental release of radioactivity." After some

effort with a group of experts on fault tree analysis, it was concluded that this definition was so general that it led to a hopelessly complicated fault tree. This led to the use of the event tree approach that enables one to break the problem into smaller parts, to which the fault tree can then be applied.

## The Event Tree Method

Event trees are similar in many respects to the decision trees that have been widely used in decision analysis. However, they contain no decision points. The logic used is almost the reverse of that in the fault tree. In the fault tree one starts with an undesirable event and reasons back to determine how it might have happened. In the event tree one starts with an initial event and asks to what states of the system it might lead.

To illustrate the approach we consider the case of a pipe break in the cooling system of a nuclear plant. This initial event allows the cooling water to be lost rapidly. If no other water is supplied to make up for this loss, the fuel will be uncooled and will melt, which would produce a very serious accident. There exist, however, a number of functions that can affect the actual outcome of this initial event. These functions include delivery of electric power, emergency cooling, removal of radioactivity from containment, and containment integrity. Figure 2 is a simplified event tree that illustrates the possible outcomes of the initial events, depending upon the success or failure of these functions. Thus if electric power is available we take the upward branch at the first branch joint, and the downward branch if electricity is not available. Similar choices are shown for each of the other plant functions. It should be noted, however, that the bottom line shows none of these choices. This is because if electric power fails, neither the emergency core cooling nor the radioactivity removal system can work. If both of these fail the core is assumed to melt and cause the containment to fail. The probability of any particular outcome is shown as the simple product of the failure probabilities along that path. Since the failure probabilities tend to be numbers of 0.01 or less, $(1 - P_i)$ is approximated as 1 so that the success probabilities are not included in the products shown. In actual practice it is often not correct to assume these probabilities are independent, so it becomes necessary

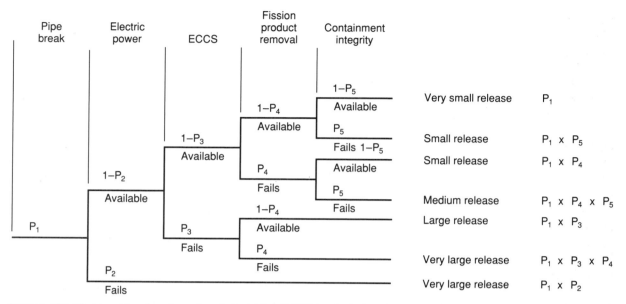

**Figure 2. Simplified event tree for a loss of coolant accident (LOCA) in a typical nuclear power plant.**

to use conditional probabilities in these products.

The values of the P's can now be determined by fault tree analysis and the event tree used to combine these values to get the probability of the various accident sequences For example, the fault tree in figure 1 could be developed to calculate $P_2$. The consequences of each accident sequence, in this case the amount of radioactivity released, must be determined from an engineering analysis of the condition of the plant for the defined failure states. The actual event trees for this initial event in the WASH-1400 analysis are much more complicated because the plant functions in the tree in figure 2 are usually accomplished by more than one system of hardware. Thus the number of headings across the top of the figure increases when the functions described are replaced by the hardware systems that actually perform these functions.

There are several problems encountered in the use of event trees. The first is to define a set of initiating events that, when fully developed, produces all the important accident sequences. A logic for doing this can usually be developed, but it varies depending upon the activity being analyzed. Although in the case of a nuclear power plant this was not too difficult, because essentially all the risk is associated with one event, i.e., serious overheating of the fuel, in

many other cases it may not be so simple. For example, the author has reviewed some recent attempts to apply the techniques to fusion machines where it seems to be much harder to develop a logic for defining a set of initiating events.

A second problem is defining the order of functions across the top of the event tree. This order turns out to be important when the performance of one system affects the performance of another. The WASH-1400 study group found it often took a number of iterations to produce a tree that handled this problem. It is essential that the analyst have a detailed understanding of all plant systems and how they operate and interact with each other. In most cases it requires the interaction of several analysts of different areas of expertise to achieve the final tree.

It must be recognized that there is never any proof in the mathematical sense that the analysis is complete. To ensure that no obvious factors have been missed requires analysts with considerable background and experience in the system being analyzed. One favorable characteristic of the method is that should it be discovered after the analysis has been finished that something has been omitted, it is usually a straightforward process to include it without having to redo the previous work.

## Calculation of Consequences

The calculation of consequence magnitude will of course depend upon the type of hazard. In many cases the hazard is associated with a pollutant that is released and spreads upon the influence of prevailing weather conditions. This would be the case for the release of radioactive gases and aerosols and vapor clouds from an LNG spill. The most commonly used method for such cases is the Gaussian Plume model of weather. This method requires knowledge of the wind direction, wind velocity, and the Pasquill weather stability class. A discussion of the method can be found in the literature (e.g., 2). It is generally believed that if carefully applied this model will yield fairly good results for distances of 5–10 miles. Beyond 10 miles its accuracy is questionable. However, if the local topography has unusual features (i.e., other than fairly flat ground around the release point) the results may be questionable even at close distances unless special consideration for handling these features is included. Fortunately, in postulated reactor accidents the health effects beyond about 10 miles depend mostly upon the total integrated population dose (person-rem), and errors in the Gaussian Plume model are not so important. To understand this, consider a case where the model predicted half the dose to twice the area, in contrast to what actually happened. If the areas affected are fairly large, as they usually are, then the population affected will be determined fairly well by using the average population density of the region. Thus in the case above one would still calculate the correct number of person-rem. When the population density is highly different for the two areas there will of course be an error, but in most cases the areas considered are large enough so that the error is not large. In the case of an LNG cloud the dilution process usually causes the downwind distance of a flammable mixture to be less than 10 miles even for very large spills.

In other cases entirely different approaches must be used. For example, the spreading of a major oil spill involves ocean currents, wave action, and wind. A dam failure requires a model to calculate the way the water spreads downstream of the dam. In all cases, however, the first part of the consequence calculation is a two-dimensional plot of the extent and severity of the particular hazard over the affected region. An excellent discussion of techniques for carrying out these types of calculations for different hazards can be found in (3).

The second step is then to compute the effect on people and property in the region. This requires knowing the population density at the time of release, the types of property in the affected area, and an exposure-effect relationship. Some of all of these inputs may require probability distributions.

Typically the calculation will require a fairly complicated computer code. In most cases the code is so complicated it is not possible to propagate the probability distributions of the input variables in closed form. Thus it has become common practice to use a Monte Carlo approach that samples appropriately from each distribution and repeats the calculation many times. The result is that the various consequences are determined in the form of probability distribution functions. Most often the results are presented in the form of cumulative probability distribution functions of accident frequency vs magnitude of consequence. Figure 3 shows such a plot (taken from the WASH-1400 study) for early fatalities (deaths within one year) from reactor accidents.

## Presentation of Results

The presentation of results poses several problems. One is that a given accident usually has a number of different consequences and it is often difficult to devise a method for combining them to give a single measure of the consequence suitable for measuring the risk, as is often desired in a risk-benefit analysis. A second problem is that there are several commonly used measures of risk, so the question arises as to which should be used.

The question of how to add property damage, measured in dollars, to health effects is a difficult one. There are numerous attempts in the literature to estimate the dollar value of life, injuries, and latent health effects. One finds answers that differ by as much as a factor of 100 or more. Thus the cost-benefit ratio is usually dominated by the choice of these health effect values. This is such a controversial issue that often PRA's only calculate the consequences individually and require the user of the information to supply personal judgments for these values.

Probability per year ⩾ X

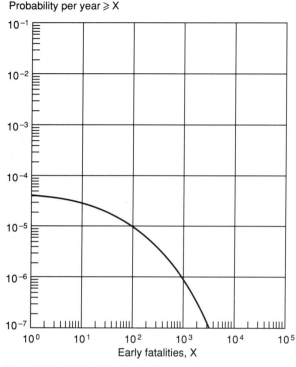

Early fatalities, X

**Figure 3. Probability distribution for early fatalities for 100 reactors at 68 US sites as given in WASH-1400 (1).**

The calculation of consequences usually generates the risk in the form of a probability distribution function such as the one shown in figure 3. In addition, risk can also be expressed as a societal risk or an individual risk. The societal risk is the average effect of the activity over some time period (usually one year) on society as a whole. Thus the societal risk of driving cars in the United States for the consequence of fatalities is about 50,000 deaths per year. The individual risk is the chance of a specific individual suffering a given consequence in a given time period. If for this illustration we assume all US citizens are equally exposed to accidental death as a result of automobiles, then the individual risk is 50,000 deaths per year for 220,000,000 individuals exposed to death, which equals $2.25 \times 10^{-4}$ per year or about 1 in 4400 per year for the individual risk. Each of these quantities has important uses. For example, in setting acceptable risk limits one usually considers limits on the societal risk and limits on the individual risk to the most highly exposed individuals. These two types of risk can be simply calculated

from the probability distributions by the appropriate averaging process. A recent report (4) of the Advisory Committee on Reactor Safeguards suggests acceptable criteria for reactor risks based on both of these types of risk, and gives an excellent discussion of the problems faced in setting risk criteria. It is common practice for a PRA to give not only the probability distribution but also the societal and individual risk.

## Perception of Risk

As noted in the introduction, the goal of a PRA is to obtain a quantitative measure of risk of a given activity that can be used in making comparisons between possible choices such as coal and nuclear power for the generation of electricity. So far we have described methodologies for calculating the probability distributions of various accident consequences. The quantities calculated by these procedures do not include any societal preferences for the type of risk. Since the goal of these analyses is to produce information that might aid decision makers, and since decision makers must be responsive to public preferences, clearly the information calculated by the PRA falls short of providing everything needed. The users therefore must develop some understanding of the public attitude towards risks, and there are many articles on this subject, e.g., Rowe (5), Lowrance (6), and Litai (7). These references contain a detailed review of much of this work. These and many other authors have tried to define those characteristics of risk that cause them to be perceived differently. Litai has reviewed the literature and identified some 26 different characteristics proposed by various authors. Since many of these are similar he suggests those listed in Table 1 as representing quite well the ideas included in the full list of 26. Following others, he suggests the use of a dichotomous scale for each characteristic. Thus any risk must be classified as either one or the other on each scale. For example, the risk from pollutants from burning coal to those living near the plant would be involuntary, ordinary, man-made, delayed, continuous, controllable, old, clear, and necessary. The risk of early death from a nuclear power plant would be involuntary, catastrophic, man-made, immediate, continuous, controllable, new, clear, and necessary. These two sources of electricity therefore differ in three characteristics: severity, manifestation, and familiarity.

**Table 1  Risk "Factors" Included by Litai (7)**

| Risk factor | Dichotomous scale |
|---|---|
| Volition | Voluntary: Involuntary |
| Severity | Ordinary: Catastrophic |
| Origin | Natural: Man-made |
| Effect manifestation | Immediate: Delayed |
| Exposure pattern | Continuous: Occasional |
| Controllability | Controllable: Uncontrollable |
| "Dread"/familiarity | Common/old hazard: "Dread"/new hazard |
| Benefit | Clear: Unclear |
| Necessity | Necessary: Luxury |

Clearly it would be desirable to have measures of how important these differences are to the public.

Several authors have developed numerical values called risk conversion factors (RCF) to express the importance of these differences to the public. These have been inferred by analyzing statistical data of various kinds that attempt to infer public preferences. Table 2 summarizes the results. The values found in Litai's work are the result of an extensive analysis of a major US life insurance company's data base and refer only to the consequence of death. If Litai's values are used, then one would penalize early deaths from nuclear energy by a factor of 30 for severity, a factor of 30 for manifestation, and a factor of 10 for familiarity. If we further assume that these are independent multiplicative factors, then nuclear energy must be penalized a factor of 9000 over coal for early nuclear fatalities, and a factor of 300 for latent nuclear fatalities. It further suggests that latent nuclear fatalities are 1/30 as important as early

fatalities, so that RCF suggests a way for combining these two results. Following this suggestion, figure 4 shows the controversial comparison from the WASH-1400 study of the risk of marked "early + delayed" is the original early fatalities curve plus the latent fatalities curve divided by a factor of 30. Of course the latent effects of the nonnuclear risks should be included in the same way but, with the possible exception of the chlorine releases, 1/30 of the latent effects is likely to be a very small correction. Since as perceived by most people the other risks are old and familiar, the second dotted line marked "early + delayed + new" is moved a factor of 10 higher. Although nuclear power is still below the other curves, it no longer appears to be as insignificant a risk as was shown in the original WASH-1400 comparison.

Clearly these results are far from being the final work on how to handle the difficult problems of comparing different types of risks. However, the above-cited studies suggest some interesting ways to look at risk comparisons that give important insights as to why some technologies have serious problems of public acceptance.

## Conclusions

During the last decade there has been a rapid expansion in the development of techniques for carrying out PRA's on a wide variety of activities, and also in the number of people trained in the use of these techniques. In the area of energy the technique is being applied to nuclear plants, LNG terminals and shipping, and to large petroleum processing facilities.

**Table 2  Comparison of RCF Values**

| RCF | Value (error factor) | | | | |
|---|---|---|---|---|---|
| | Litai (7)[a] | Rowe (5) | Starr (10) | Kinchin (11) | Otway & Cohen (12) |
| Natural/man-made | 20 | 10 (2) | | | |
| Ordinary/catastrophic | 30 | 50 | | | |
| Voluntary/involuntary | 100 | 100 (10) | ~1,000 | | 1–1,000 |
| Delayed/immediate | 30 (11) | 20%/yr[b] (2) | | 30 | |
| Controlled/uncontrolled | 5–10 | 100 (10) | | | |
| Old/new | 10 | | | | |
| Necessary/luxury | 1 (7) | | | | |
| Regular/occasional | 1 | | | | |

[a] Where no error factor is given a value of ~10 may be assumed.
[b] Must be compounded by number of years of exposure.

In the nuclear industry, especially since the Three Mile Island accident, there has been a rapidly expanding application of PRA by the US Nuclear Regulatory Commission to aid in their attempts to reduce nuclear power plant risks. The British have carried out a major PRA on the petrochemical complex on Canvey Island (8) and a PRA has been completed on the proposed LNG terminal at Point Conception in California (9). The use of this approach in understanding risk is becoming so widespread that it is doubtful that any new major energy activity that is perceived to have significant risks will be accepted before such an analysis.

Today the methods for estimating equipment failure probability are quite well developed. The dominant contributor to the uncertainty in the results is the uncertainty in the failure rates of components and in the human error rates of operators and maintenance personnel. These uncertainties should diminish as larger data bases are developed. The treatment of common cause failures is an area that requires further work. Although the calculation of consequences has rapidly improved during the last decade, these calculations still contribute significantly to the uncertainty and so are a fruitful area for further work. All in all, however, it is now possible to make estimates of the risks associated with many low probability, high consequence events from a variety of sources. Typically one can expect an uncertainty of plus or minus a factor of 10 in these calculations. Despite this large uncertainty the results can be useful in many ways. PRA will help develop a much better understanding of potential system weaknesses, and, if due consideration is given to the uncertainties, it can be used to make useful comparisons with other systems.

A major problem still remains in the interpretation of the results. The techniques for comparing risks of different types need to be more fully developed if widely accepted ways of handling this difficult problem are to be achieved. This problem is of course further complicated by those who deliberately use PRA results in incorrect and misleading ways. Unfortunately, the nature of statistics and probability is such that this is often easily done.

The ultimate goal in the use of these methods should be to provide a measure of risk of an activity that can be used in the regulatory process to provide assurance that an activity is acceptably safe. To achieve this goal we must develop confidence in the techniques and also a set of numerical criteria for an acceptable level of risk. The confidence in the technique will only be developed through the successful use of the method in many applications. If the expansion in the use of the method is too rapid it is likely that the work will be done by poorly trained and inexperienced personnel, which would lead to poor results. This should be avoided for it could significantly impede the adoption of PRA. The setting of numerical, acceptable safety criteria is just beginning and it will be a long and difficult job that must be undertaken by regulatory bodies as soon as possible. The recent work of Griesmeyer & Okrent (4) is a very important first step. Much more work of this type will be needed before the overall goal is achieved.

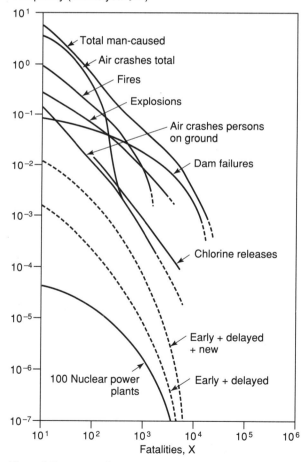

Figure 4. Frequency of man-caused events involving early fatalities based upon US experience as given in WASH-1400 (1). The two dashed curves result from the work of Litai (7).

## Literature Cited

1. US Nuclear Regulatory Commission. 1975. *Reactor Safety Study: WASH-1400,* Wash. DC: GPO

2. Slade, D. H., ed. 1968. *Meteorology and Atomic Energy 1968,* Wash. DC: US Atomic Energy Comm.

3. Netherlands Organization for Applied Scientific Research, Bureau for Industrial Safety TNO. 1980. *Methods for the Calculation of the Physical Effects of the Escape of Dangerous Material (Liquids and Gases), Parts I and II,* Voorburg, The Netherlands

4. US Nuclear Regulatory Commission. 1980. *An Approach to Quantitative Safety Goals for Nuclear Power Plants,* Rep. No. NUREG-0739. Wash. DC: GPO

5. Rowe, W. D. 1977. *An Anatomy of Risk.* New York: Wiley

6. Lowrance, W. W. 1976. *Of Acceptable Risk: Science and the Determination of Safety.* Los Altos, Calif: Kaufmann

7. Litai, D. 1980. *A risk comparison methodology for the assessment of acceptable risk.* PhD thesis, Mass. Inst. Technol., Cambridge, Mass.

8. Health and Safety Executive (United Kingdom). 1978. *Canvey: An Investigation of Potential Hazards from Operations in the Canvey Island, Thurrock Area.* London: HMSO

9. Science Applications, Inc. 1975. *LNG Terminal Risk Assessment Study for Los Angeles, California.* La Jolla, Calif.

10. Starr, C. 1969. Social benefit vs technological risk, *Science* 165:1232–38

11. Kinchin, G. H. 1978. Assessments of hazards in engineering work, *Proc. Inst. Civ. Eng.* 64:431–38

12. Otway, H. J., Cohen, J. J. 1975. *Revealed Preferences: Comments on the Starr Benefit-Risk Relationships,* Vienna: Int. Inst. Appl. Syst. Anal.

# Questions for Thought and Discussion

1. In discussing the use of the fault tree method to analyze standby systems and operating systems, Rasmussen says that the tree contains true probabilities in the former case and failure rates in the latter. What is the difference between a true probability and a failure rate? What is the relationship between availability and failure rate? In the case of operating systems, what concept corresponds to the standby-system concept of availability?

2. Construct a fault tree like the one in figure 1 for this unfortunate event: your car is damaged while parked on an incline in a public garage. Construct an event tree like the one in figure 2, assuming the initial event is that the parking brake does not hold while the car is parked on the incline.

3. Given that there were about 47,000 traffic fatalities and a population of 248 million in the United States in 1988, update the author's estimates of the societal and individual risks of driving cars.

Given EPA's estimate that phasing out asbestos products over the years 1990 to 2002 will avoid exposures that would otherwise cause 200 cancer deaths, what are the total societal and individual risks that will be avoided for that period, assuming that the population grows beyond the 1988 level at the rate of 1 percent a year and that everyone in the United States is exposed to asbestos products? On a mean annual basis, how do these risks compare to your updated driving risks?

Using the dichotomous scale for the factors in table 1, describe the characteristics of driving risks and asbestos risks. Based on these results, offer some reasons why the latter are being regulated so much more closely.

4. Interpret and discuss the significance of the following points in figure 4 of this paper:

• The points where the curve labeled "100 nuclear power plants" intersects the horizontal and vertical axes.

• The point where the curves labeled "Air crashes total" and "Air crashes persons on ground" intersect.

• The point where the curves labeled "Fires" and "Explosions" intersect.

# Assessing the Risk of an LNG Terminal

## RALPH L. KEENEY, RAM B. KULKARNI, and KESHAVAN NAIR
*Woodward–Clyde Consultants, San Francisco, California*

Natural gas is used to meet 25 per cent of the energy needs of the United States, and there are strong arguments that favor its continued use: it burns cleanly; an efficient distribution system exists; and consumers prefer it as the fuel for heating homes and other buildings. Natural gas is also essential in the production of fertilizers and other chemicals.

Though demand has been increasing, U.S. production of natural gas has been declining since 1971—a circumstance which leads us to plan to exploit the significant supplies of natural gas in many developing areas of the world where there is little or no demand for the gas (*see "Energy for the Third World" by William F. Martin and Frank Pinto, June/July, pp. 48–56*). A growing international trade in natural gas is likely in the years ahead. Indeed, by 1990 the world trade in natural gas could rise to between 5.3 and 8.12 trillion cubic feet, with the U.S. as the principal user. The

American Gas Association is of the opinion that U.S. imports of natural gas, which are currently 10 to 15 billion cubic feet per year, could reach 1.6 trillion by 1985 (about 10 per cent of the total gas supply), 2.4 trillion by 1990, and 3 trillion by 1995.

Natural gas is easily and cheaply transported by pipeline. But transportation where pipelines do not exist poses significant problems because of the large volume occupied by the gas at ambient temperatures. To solve this problem, a technology has been developed to convert natural gas to liquid by cooling it to a temperature of −259° F. (−162° C.), in which state it occupies 1/600th of the original volume. This liquefied natural gas (LNG) can then be shipped in specially constructed oceangoing tankers. A liquefaction facility is required at the source of the gas, and regasification is required before the fuel enters whatever distribution network is to bring the gas to its ultimate users.

Liquefied natural gas is colorless and odorless, and by itself it will not burn. It weighs about 28 pounds per cubic foot and therefore will float on water. LNG will vaporize rapidly if exposed to ambient temperatures; in the vapor state it is not poisonous but could cause asphixiation due to the absence of oxygen. When dispersed in the air and when the

*Editors' note:* The current affiliation of Ralph L. Keeney is the University of Southern California, Los Angeles, and of Keshavan Nair is Benjamin Nair, Inc., San Francisco, California.

concentration falls to between 5 and 15 per cent, the mixture is flammable.

## The Public Risks in LNG Commerce

The risks to the public in the handling of LNG arise because spilled LNG vaporizes rapidly. The vapor may either catch on fire at the location of the spill, resulting in a "pool fire"; or it may form a vapor cloud which can be carried downwind with the possibility of ignition and burnback toward the source.

In the unlikely event that there is a spillage of LNG on land from storage tank or piping failure, the LNG will vaporize quickly for a short period of time—two to three minutes—until the ground beneath it freezes. Thereafter, vaporization will continue to take place slowly. Vapor cloud formation is possible during the first few minutes and far less likely thereafter. LNG facilities are required to have fire suppressant equipment, and storage tanks must be provided with dikes which can, as a minimum, contain all the liquid stored (typically 90,000 cubic meters per tank); and a buffer zone is required between the dike and the boundary of the facility. Consequently, the public risk (that is, the risk to persons outside the facility) of LNG spills within the facility is considered minor.

The major concern for public safety is connected with an LNG spill on water or at a terminal as a result of a tanker-related accident. One postulated hazard is the possibility of a flameless vapor explosion when LNG comes in contact with water. This represents a very rapid vaporization of the LNG but does not involve combustion; it is a physical rather than chemical phenomenon. Tests indicate that the pressures generated by such an explosion—if in fact it could occur—are relatively small (100 pounds per square inch) even very close to the surface of the liquid and attenuate rapidly with distance.

A more serious hazard is presented by a scenario in which a vapor cloud from LNG spilled on water ignites either at the spill location—with a potential hazard to people and property in the vicinity of the spill—or after the vapor cloud is carried downwind. Lives and property within the cloud or close to its boundary would be affected.

An LNG detonation has never been observed in an unconfined space, and tests using a high explosive charge for ignition have failed to produce a detonation.

A potential risk to the public has been alleged from a phenomenon called "rollover," which results from the mixing of two or more LNG shipments with different composition and density in a storage system. This mixing may cause the pressure to build up in storage tanks and result in venting of natural gas. Modern facilities have procedures for preventing "rollover" by controlling the loading procedure, and this is no longer considered a significant problem.

## Weighing the Risks and Their Odds

Before any LNG import or export project can operate, more than 130 federal, state, and local permits must be obtained. The Federal Energy Regulatory Commission (formerly the Federal Power Commission), the Office of Pipeline Safety Operations, and Coast Guard are the major federal agencies involved. Among other matters, these agencies are concerned with evaluation of public risks, and that turns out to be a challenging problem in technology and policy.

Every system for producing and converting energy for human use presents hazards, and evaluating these in quantitative terms is critically important. It is a curious fact that we tend to take for granted the hazards in conventional energy systems, such as those based in coal and oil, and to focus our concern on the public risk involved in proposed new energy developments. To complicate matters, there is no clear definition of acceptable public risk for particular populations, particular activities, or society as a whole; clearly, one accepts a higher-than-average level of risk in certain occupations (coal mining, for example) and activities (competitive athletics and automobile transportation).

What is in fact the public risk associated with handling liquefied natural gas? The answer for any particular facility clearly depends on its design, size, location, and management. Perhaps the best way to put these issues in perspective is with an example—an analysis of the proposed La Salle Terminal, a marine terminal and LNG vaporization facility planned by the El Paso LNG Co. and its subsidiaries in Matagorda Bay, Texas.

Matagorda Bay, approximately 120 miles southwest of Houston on the Texas Gulf coast, is a sparsely populated area. The proposal is to receive, process, and distribute LNG from Algeria, delivered to the terminal by a fleet of LNG carriers. There would be approximately 143 carrier arrivals per year, each carrier delivering some 125,000 cubic meters of LNG; total production would be about one billion cubic feet of natural gas per day.

## Safety at the La Salle Terminal

In simplest outline, operations at the La Salle Terminal would be conducted in the following way: after an LNG carrier berths, its LNG cargo would be transferred through refrigerated (cryogenic) piping to one of three LNG storage tanks, each of 100,000 cubic meters capacity. The LNG would be withdrawn as required from these tanks, revaporized, and then piped to consumers through a high-pressure intrastate pipeline. No surface transportation of LNG is anticipated.

Special safety procedures and techniques are proposed in all phases of the design, construction, and operation of the La Salle Terminal and the LNG fleet serving it.

Storage tanks will be diked and will be designed to minimize spillage, and there will be additional spill impounding areas at the facility. This assures that any accidental release of LNG and its subsequent spreading will be contained at all times within the plant boundaries. The facility will be provided with automatic vapor dispersion and fire control systems adequate to minimize any hazards from thermal radiation or vapor dispersion at any plant boundary line under any credible weather conditions.

The LNG carriers will be of special double-hull and double-bottom construction and will use sophisticated anti-collision and navigational systems. Special U.S. Coast Guard operating procedures will be in effect in Matagorda Bay.

But despite all of these safety considerations designed to reduce to an extremely low level the likelihood of an LNG accident with consequences to the public, such an event is possible. It could be initiated by a spill of LNG resulting from a ship collision or a terminal-equipment malfunction. The circumstances required to cause such a spill suggest that the vapor being formed would be immediately ignited, resulting in a pool fire. If the released LNG were not immediately ignited the very cold liquid

(−260° F.) would rapidly evaporate, forming a cloud of gas, heavier than air, moving across the surface of the earth. If this cloud should come into contact with an ignition source, such as a gas pilot light, the flame of a cigarette lighter, or an electric spark, before the ratio of vaporized LNG to air becomes too low to allow ignition, the vapor cloud could ignite and burn, leading to property damage, injuries, and perhaps fatalities.

## The Risk Analysis Model

To analyze the risk of such accidents, we used a risk analysis process which included the development of accident scenarios and their associated probabilities, quantification of public risks, and evaluation of public risks. The components of the risk analysis model are indicated in [figure 1]. The complexities required that we make some simplifying assumptions, but the spirit of the analysis required that all assumptions be stated explicitly and conservatively so that our analysis would tend to overestimate the public risks.

Any risk analysis begins with an accident scenario, a sequence of events that must occur for public risk to exist. It must incorporate assumptions about the nature and location of the hypothetical LNG spill, the wind and weather conditions, the sources of ignition, and the effectiveness of spill and fire control systems. These are all built into an event tree, such as that shown in [figure 2]. A representative accident scenario could then be described as follows: an LNG carrier collision occurs in the harbor, releasing an LNG spill of a specified size. There is no immediate ignition, so a vapor cloud forms. The wind is from the east at 10 miles per hour; the eighth ignition source ignites the vapor cloud. The event tree shows that this chain of unlikely events must take place in a specific sequence.

In the risk analysis model, we calculate the annual probability of a particular scenario involving vapor cloud travel as the product of the following factors:

• The annual probability of the initiating accident,

• The probability of no immediate ignition for that accident,

• The probability of the wind direction,

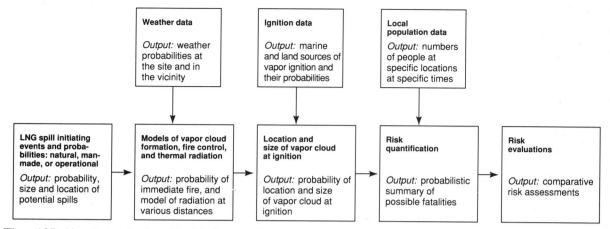

[Figure 1.] Problem: to examine the public risks from possible releases of LNG from the proposed La Salle Terminal. Solution: a formal risk analysis according to the model shown here. The analysis is accomplished in three stages: develop accident scenarios and their associated probabilities (the three columns at the left); quantify public risks associated with each probability; and, finally, evaluate the risks by comparing them with those involved in other human activities.

• The probability of the wind speed and the air stability, given that wind direction, and

• The probability that the nth ignition source ignites the vapor cloud.

Similarly, the probability of a particular accident scenario which results in a pool fire is equal to the annual probability of the initiating accident multiplied by the probability of immediate ignition.

## Analyzing the Risk at Matagorda

In the course of our analysis, we constructed accident scenarios and calculated probabilities for all combinations of these individual events, and finally we computed the public risks due to LNG terminal operations.

There are no generally accepted criteria for evaluating public risk. The approach used here was to compare risks generated from this project with existing risks to the public, with risks from alternate energy sources, and with levels of acceptable public risks suggested in the literature. We examined these risks using four criteria:

• *Societal risk*—the total expected fatalities per year.
• *Individual risk*—the probability of an exposed individual becoming a fatality per year.
• *Group risk*—the probability of an exposed individual in a specific group becoming a fatality per

year.
• *Risk of multiple fatalities*—the probability of exceeding specific numbers of fatalities per year.

We categorized the events that might cause LNG spills as follows:

• Natural hazards (for example, hurricanes and earthquakes) which affect the facilities.
• External man-made hazards (for example, aircraft crashes) which affect the facilities.
• Accidents involving the LNG carrier fleet.
• Accidents within the La Salle Terminal.

## Analyzing the Dangers of Natural Events

Our analysis of the various natural hazards, including earthquakes, severe winds, storm waves and tsunamis, and meteorites suggested that none of these represent significant public risk in comparison to the risks associated with other types of accidents. The likelihood that an earthquake would produce a ground acceleration at the La Salle Terminal site large enough to exceed the design specification of the storage tanks is approximately $10^{-11}$ per year. Even if the tank did rupture with such an earthquake, the analysis for the onshore facilities indicates that public risk is essentially nil. There is a higher probability of pipe breaks than of storage tank failure due to ground motions, but the analysis indicates that the results are inconsequential to public risk.

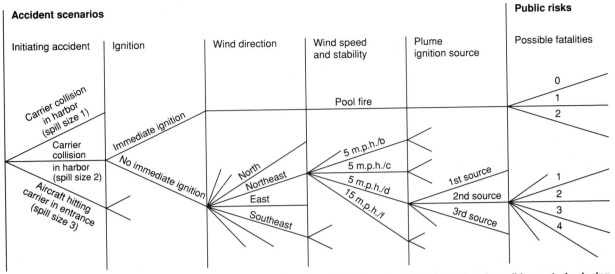

**[Figure 2.]** What could happen in an LNG accident at Matagorda Bay, Texas? This "event tree" shows chains of possible events, beginning with spills of three possible sizes resulting from a collision involving an LNG carrier. Spill size 1 represents a release of one tank of LNG—19,400 cubic meters; spill size 2 the release of two tanks, spill size 3 the release of 10,000 cubic meters. The problem is to establish the probabilities for each branch in this "event tree"—and therefore the probabilities of fatalities from the several hypothesized events.

The main source of severe winds in the vicinity of Matagorda Bay is hurricanes. The primary concern with wind is its effect on the storage tanks, since all LNG carriers will leave and remain outside of Matagorda Bay if any winds greater than 60 m.p.h. are forecast or observed. The tanks are designed to withstand an instantaneous gust of 217 m.p.h. and a one-minute wind of 166 m.p.h. Meteorological data indicate that the latter occurs once every 100 years and the former once every 200 years. Because of the safety factors in design standards, it is unlikely that a storage tank would begin to fail even when the design wind is exceeded. But even if the tank failed completely, under these conditions wind turbulence would disperse the vapor plume before it passed terminal boundaries if a pool fire had not been ignited on the site at the point of rupture.

There are no known faults capable of generating significant tsunamis in Matagorda Bay. Furthermore, operating policy will require that in storm conditions all LNG ships leave Matagorda Bay; and since warning of impending large waves would be available, the possibility that these could cause ship accidents and contribute to public risk is believed to be negligible.

The probabilities of a meteorite penetrating a ship's tanks in the entranceway to the harbor, in the harbor, or at the pier were calculated to be $3.23 \times 10^{-10}$, $5.49 \times 10^{-10}$, and $9.32 \times 10^{-9}$ per year, respectively—over two orders of magnitude smaller than the probability of ship collisions. We considered this probability essentially negligible and made no further analysis of this possibility. The likelihoods of meteorites penetrating terminal storage tanks and rupturing terminal pipelines were also essentially negligible, and since accidents in these cases would be contained within the terminal boundary, they were not investigated further.

## Man-Made Hazards: Aircraft, Sabotage, and Collisions

There are no major airports in the vicinity of Matagorda Bay, though small planes operate at the local Port O'Connor airstrip and there is a helicopter landing site in Port O'Connor from which approximately 25 flights leave daily for oil platforms in the Gulf of Mexico. It is assumed that there is no risk to the public from airplane crashes into the storage tanks or LNG pipelines, because the consequences of such events would be confined to the terminal area. For crashes of airplanes into ships, we assumed that

[Table 1.] How could LNG be spilled in Matagorda Bay, Texas? Accidents may be inevitable in handling LNG, but the authors' risk analysis of the El Paso LNG Co.'s proposed facilities in Matagorda Bay indicates that they will be infrequent. The initiating accidents listed below turn out to be the most serious of many postulated, and these were examined to determine the public risk associated with each under various conditions of weather and population.

| Location, cause, and size of spill | Annual probability of accident event | Spill size |
|---|---|---|
| **Entrance:** | | |
| Most credible spill due to collision | $3.46 \times 10^{-5}$ | 19,400 m³ instantaneously |
| Maximum credible spill due to collision | $9.28 \times 10^{-6}$ | 38,800 m³ in 9 minutes |
| Spill due to ramming | $7.59 \times 10^{-8}$ | 10,000 m³ in 12 minutes |
| Spill due to aircraft crash | $1.21 \times 10^{-8}$ | 10,000 m³ in 12 minutes |
| **Harbor:** | | |
| Most credible spill due to collision | $6.49 \times 10^{-7}$ | 19,400 m³ instantaneously |
| Maximum credible spill due to collision | $1.76 \times 10^{-7}$ | 38,800 m³ in 9 minutes |
| Spill due to ramming | $2.10 \times 10^{-7}$ | 10,000 m³ in 12 minutes |
| Spill due to aircraft crash | $9.23 \times 10^{-8}$ | 10,000 m³ in 12 minutes |
| **Pier:** | | |
| Most credible spill due to collision | $1.01 \times 10^{-7}$ | 19,400 m³ instantaneously |
| Maximum credible spill due to collision | $2.73 \times 10^{-8}$ | 38,800 m³ in 9 minutes |
| Spill due to ramming | $1.10 \times 10^{-5}$ | 10,000 m³ in 12 minutes |
| Spill due to aircraft crash | $5.93 \times 10^{-7}$ | 10,000 m³ in 12 minutes |

up to 10,000 cubic meters of LNG might be released in 12 minutes, and this scenario is included in the risk analysis. Accident possibilities from helicopter flights are not included in the analysis, because it is assumed that flight patterns can be arranged to avoid operations where LNG carriers are operating.

Qualitative examination of the potential risks due to sabotage indicates that the possibility of sabotage by determined terrorists cannot be completely eliminated by reasonable engineering or security systems. However, we believe that immediate ignition would very likely occur because of the violence required of a saboteur seeking to release the LNG, and this means that the consequences would be confined to the spill area. Furthermore, the decision to operate LNG terminals anywhere in the U.S. implies acceptance of some risk of terrorism.

The most serious accidents involve the collision of LNG carriers with other vessels, rammings of stationary or floating objects, and grounding of LNG carriers. Their probabilities were calculated for three areas of operations in the Matagorda Bay:

• The entranceway, the immediate approach route to and through the cut in Matagorda Peninsula;

• The harbor, the ship channel within Matagorda Bay; and

• The piers, the waters between the ship channel and the berth.

Either one or two cargo tanks may be involved in such a collision. The most credible spill event due to collision turns out to be the rupture of one cargo tank involving 19,400 cubic meters of LNG. The maximum credible spill event due to collision is considered to be the simultaneous rupture of two adjacent cargo tanks, involving up to 38,800 cubic meters of LNG. The most credible spill due to ramming is 10,000 cubic meters. The analysis of grounding accidents indicated a probability for cargo tank rupture so small as to be statistically insignificant. All these findings are summarized in [table 1]. Since land spills have been ruled out as insignificant to public risk, the attention in the rest of our study of public risks associated with the operation of the La Salle Terminal was devoted entirely to analyzing water spills.

Vaporization of the LNG would commence immediately following a cargo tank penetration. Any heat source of sufficient temperature and duration could cause ignition of this vapor. The primary ignition sources would be the friction and sparking generated by the immense forces involved in the penetration; these would probably generate temperatures from 1,600° to 2,700° F., far in excess of the ignition temperature of methane in air (1,000° F.). Secondary ignition sources on the carrier, such as boilers, galley fires, electrical cables, and light fixtures, will also be present and exposed to the spilled LNG as a result of whatever accident is taking place. We assume that immediate ignition will almost surely occur under such circumstances; the probability assumed is 0.99.

If immediate ignition does not occur, and in cases of accidental spills caused by negligence on board carriers (where immediate ignition may not occur) the characteristics of wind, on-shore ignition sources, and public activities become important in computing public risk. The movement and behavior of an LNG plume would depend on wind direction, wind speed, and air stability. For the most credible spill of one tank (19,400 cubic meters spilled instantaneously), with a 10 m.p.h. wind and typical air stability, the vapor cloud could travel up to 3.31 miles, if not ignited, in roughly 25 minutes. Thereafter the LNG would be dispersed so much that its average concentration would be below the lower flammable limit of 5 per cent. It is within this 25-minute period that the probability of ignition—and then of injury and death due to ignition—must be critically analyzed.

A vapor plume can be ignited from a variety of sources such as spark plugs, open flames, pilot lights, and electrical sparks due to short circuits. Because of the difficulty of tabulating all possible ignition sources, we conservatively assumed that each house or building and each recreational boat or commercial fishing ship has one ignition source. To assume fewer sources is conservative because it implies a larger likelihood that a vapor cloud would cover a larger area before ignition. It was assumed that the probability that any source ignited the vapor plume was 0.1.

In analyzing injuries and fatalities due to burns, we assumed for this study that a thermal radiation of 5,300 B.t.u. per square foot per hour would be the lower limit resulting in fatalities. (This standard is conservative; one would actually expect only blistering of skin exposed for five seconds to such thermal radiation.) Such a radiation level would be found 1,100 feet from the center of an LNG pool fire resulting from 10,000 cubic meters of LNG spilled in 12 minutes, 2,500 feet from a pool fire resulting from either the 19,400 or 38,800 cubic meter spills of LNG, and 525 feet from a vapor plume fire.

Having agreed on these assumptions, we needed information on the distribution of population in the vicinity of the La Salle Terminal to determine the expected annual number of public fatalities and the annual risk levels to individuals (that is, the probability that any individual might become a fatality per year) due to the LNG terminal operation. This turned out to present some unexpected complications because of large seasonal and weekly fluctuations in the population. After preliminary analysis, we decided that three combinations—nontourist season (November through March), tourist weekends, and tourist weekdays—would be necessary to adequately describe population distributions. The risk quantification considered each of these cases separately. In the end, our population study indicated that we could reasonably assume 3.62 people per permanent household, 4.02 people per occupied transient household, 4.5 people per recreational boat, and 2.5 people per commercial fishing boat.

## An Illustrative Calculation

To show how all this information was used, we illustrate the calculation of the average number of fatalities under one accident scenario: a collision between an LNG carrier and another ship in the harbor at the intersection of the ship channel and intercoastal waterway (see [figure 3]) on a weekday during the tourist season. This collision produces an instantaneous one-tank spill of 19,400 cubic meters of LNG. Furthermore, the spill does not immediately ignite, so that a vapor cloud forms. The wind is from the east at 10 m.p.h. with typical air stability. The eighth ignition source ignites the vapor cloud.

The probability of this accident scenario is the product of the following factors:

• The annual probability of a collision releasing one tank of LNG (most credible spill) in the harbor during a weekday in tourist season (the annual probability is $6.49 \times 10^{-7}$ [table 1]; the probability that this will occur on a tourist season weekday is $2.33 \times 10^{-7}$);

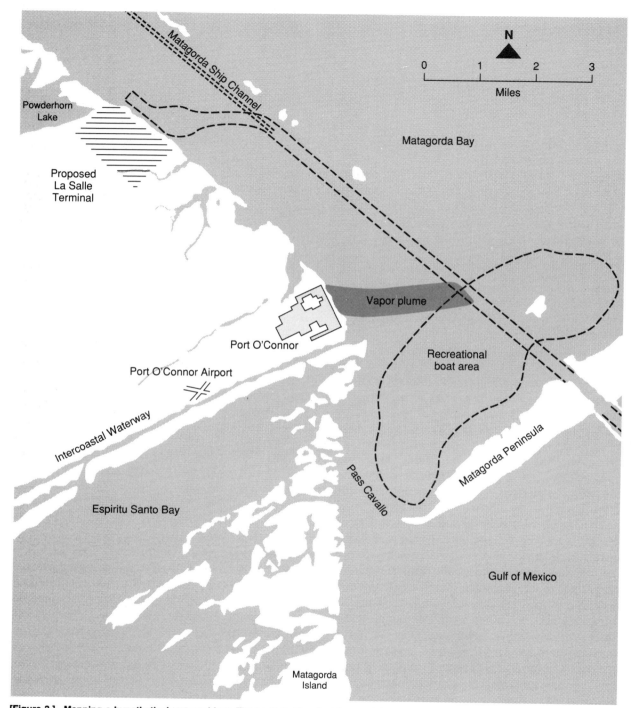

[Figure 3.] **Mapping a hypothetical LNG accident.** The La Salle Terminal for receiving and vaporizing liquefied natural gas from Algeria would be built by the El Paso LNG Co. and its subsidiaries northwest of Port O'Connor on Matagorda Bay, between Galveston and Corpus Christi, Texas. To reach it, LNG carriers would enter the Matagorda ship channel from the Gulf of Mexico. As part of the authors' risk analysis, they studied the probability of a collision scenario in which one tank of an LNG carrier was ruptured so as to instantaneously release 19,400 cubic meters of LNG. A 10 m.p.h. east wind would move the resulting vapor cloud toward the town of Port O'Connor, where fatalities from possible ignition of the vapor cloud might exceed 400. But the probability of this specific episode is shown to be $2.71 \times 10^{-12}$, and the expected fatalities per year among residents and visitors to Port O'Connor due to the proposed LNG facility is $1.7 \times 10^{-5}$.

[Table 2.] Should there be an LNG accident in Matagorda Bay, Texas, how will the winds carry the flammable vapor clouds which may be formed? These probabilities of various winds and air turbulences (increasing turbulence is represented by higher "stability classes") were an essential input into the LNG risk analysis described in the accompanying article. To assure a conservative result, winds of 20 miles an hour and greater were included in the 15-m.p.h. category since this results in longer time periods until the vapor cloud is dispersed.

| Wind speed (m.p.h.) | Stability class | Wind direction | | | | | | | |
|---|---|---|---|---|---|---|---|---|---|
| | | North | Northeast | East | Southeast | South | Southwest | West | Northwest |
| 5 | C | 0.0775 | 0.1042 | 0.0545 | 0.0230 | 0.0357 | 0.1986 | 0.1615 | 0.0916 |
| 5 | D | 0.0619 | 0.0763 | 0.0499 | 0.0273 | 0.0382 | 0.0906 | 0.0817 | 0.0833 |
| 5 | F | 0.0872 | 0.1341 | 0.2582 | 0.1374 | 0.2379 | 0.4742 | 0.4912 | 0.1277 |
| 10 | C | 0.0968 | 0.1200 | 0.1085 | 0.0814 | 0.0664 | 0.0614 | 0.0364 | 0.0763 |
| 10 | D | 0.1517 | 0.1809 | 0.1856 | 0.1460 | 0.1378 | 0.0673 | 0.0841 | 0.1527 |
| 10 | F | 0.0549 | 0.0600 | 0.0853 | 0.1050 | 0.1284 | 0.0635 | 0.0465 | 0.0392 |
| 15 | C | 0.0065 | 0.0197 | 0.0260 | 0.0448 | 0.0358 | 0.0034 | 0.0058 | 0.0093 |
| 15 | D | 0.4636 | 0.3048 | 0.2320 | 0.4351 | 0.3198 | 0.0411 | 0.0929 | 0.4198 |

• the probability (0.01) of no ignition in this collision;

• the probability (0.131) that the wind is from the east;

• the probability that, given an east wind, it is 10 m.p.h. with the assumed stability (0.1856—[table 2]); and

• the probability that the eighth ignition source ignites the vapor cloud (0.0478).

The calculation, using the probabilities indicated, is:

$$(2.33 \times 10^{-7}) (0.01) (0.131) (0.1856) (0.0478)$$

$$= 2.71 \times 10^{-12}.$$

To calculate the expected fatalities from this accident scenario on a tourist-season weekday, we calculate the maximum extent of the flammable vapor plume (if it is not ignited) and superimpose this area on a map of the Matagorda Bay area (see figure 3). Then we assemble a count of the population in this plume area, using the tourist season weekday distribution of occupied houses, boats in the harbor, and transient population. The expected number of recreational boats in the plume's path is two; no commercial boats are assumed in this area on tourist season weekdays. On such days an average of 1,000 daytime transient visitors are in Port O'Connor, essentially all of them on the beach to the east of the town. The plume covers 34 per cent of the beach, and so we assume that 340 daytime transient visitors are within

the vapor cloud. Thirty-seven permanently-occupied dwellings and 103 houses occupied by transients—a total of 140 households and as many ignition sources—are covered by the maximum possible plume. Assuming one person is away from each household, the average weekday daytime occupancy of these houses is taken as 2.62 and 3.02, respectively, a weighted average of 2.91.

If the vapor cloud is ignited by the eighth ignition source, we know that it is ignited by the sixth house encountered; the two boats count as potential ignition sources. All people within this cloud at the time of ignition are assumed to be fatalities, so the expected number of fatalities in the cloud is

$$2(4.5) + 340 + 6(2.91) = 366.46.$$

In addition, we assumed that all individuals within 525 feet of the vapor cloud fire would be fatalities if exposed to thermal radiation. Based on the average population density of Port O'Connor, this could be 230 individuals, of whom 20 per cent might be outdoors and hence fatalities. Hence, the total expected fatalities from the illustrative scenario on a summer weekday is 412.46. If the probability of such an accident is $2.71 \times 10^{-12}$, the contribution of this particular accident scenario during a tourist season weekday to the overall annual expected fatality rate is $1.118 \times 10^{-9}$.

Similar calculations were made for all other accident scenarios leading to [table 3], showing annual expected fatalities among different population groups in the Matagorda Bay area. The probability

**[Table 3.]** This table summarizes the results of the risk analysis of the proposed La Salle Terminal near Port O'Connor, Texas. The authors computed these probabilities of death in any single year for individuals in population groups which might be affected. There is one chance in $2.5 \times 10^{-11}$ that a permanent resident in the city would be a victim. Risks to visitors using Port O'Connor's beach and to boaters in the harbor are somewhat greater—but remain absolutely very small. Sensitivity analyses indicated these results were not significantly affected by the basic assumptions in the model.

| Group | Expected fatalities per year | Number of people sharing the risk | Risk per person per year |
|---|---|---|---|
| Permanent population in Port O'Connor | $2.0 \times 10^{-8}$ | 800 | $2.5 \times 10^{-11}$ |
| Permanent population in Indianola | $1.3 \times 10^{-7}$ | 80 | $1.7 \times 10^{-9}$ |
| Transient daytime visitors | $2.5 \times 10^{-6}$ | 2500 | $9.9 \times 10^{-10}$ |
| Individuals in boats | $1.35 \times 10^{-5}$ | 3000 | $4.5 \times 10^{-9}$ |
| All individuals exposed to risk | $1.7 \times 10^{-5}$ | 9000 | $1.9 \times 10^{-9}$ |

of exceeding a specific number of fatalities in a given year is shown in [figure 4].

## Risk Evaluation: How LNG Compares

To put these figures in perspective, it's necessary to compare them with similar figures for other forms of energy production.

The La Salle Terminal is designed to receive and vaporize approximately one billion cubic feet of natural gas per day. This is equivalent to the power produced by eighteen 1,000-megawatt electric power plants operating at a 70 per cent capacity factor. Based on 1970 data for the State of Wisconsin used by W. A. Buehring, the expected number of deaths to the public due to transporting fuel or direct deaths due to plant accidents for a 1,000-megawatt coal facility was 0.695 per year. The implication of 18 such plants is 12.51 expected fatalities per year; this compares with La Salle's expected level of 0.000,017. Buehring's corresponding number for 18 1,000-megawatt nuclear plants is 0.36 expected fatalities per year.

Other individual risk levels due to government and private activities have been computed. The risk to an average individual in the U.S. due to fire is 16,000 times greater than the risk to an individual exposed to

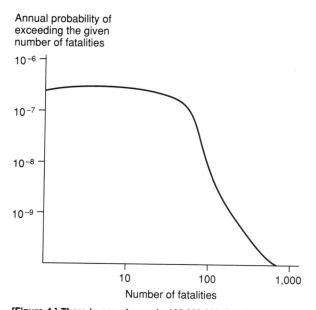

Annual probability of exceeding the given number of fatalities

**[Figure 4.]** There is one chance in 100,000,000 that there will be more than 100 fatalities in Port O'Connor, Texas in any one year due to an accidental spill of LNG near the proposed La Salle Terminal of El Paso LNG Co. This chart is the result of the authors' risk analysis, utilizing the likelihood of the original accident, the likelihood of adverse winds and air conditions, the likely number of people and their distribution in Port O'Connor, and the likelihood that vapor from the spilled LNG will be ignited before it disperses. The chances of a larger loss of life is less—one in 10 billion for 750 fatalities in any given year.

the operations of the proposed La Salle Terminal. The group with the highest annual risk from the proposed LNG facility is people in boats. The risk is $4.5 \times 10^{-9}$ per person—one chance in approximately 220 million. From the La Salle Terminal, the annual risk per person in Port O'Connor is $2.5 \times 10^{-11}$. The expected public risk due to gas distribution systems in the U.S. is $5.15 \times 10^{-7}$ per individual per year, which represents one chance in 1.9 million; this is 271 times as great as the possibility of death due to the operations of the proposed La Salle Terminal. Public fatalities due to electric shock in electrically wired residences are $1.11 \times 10^{-6}$ per individual per year.

To help interpret these risks, consider the following. Approximately 65 meteorites weighing more than one pound hit the United States each year; if one owns a one-floor house with 3,050 square feet, the probability that one of these meteorites will hit that house within a year is $1.9 \times 10^{-9}$. This is identical to the average individual risk to operation of the La Salle Terminal.

## Quantified Public Risk

The method of formal risk analysis described in this article has several important features: it permits integration of judgments from experts in various fields into a logical framework, assumptions can be stated explicitly, sensitivity analyses can be conducted to appraise the significance of the assumptions, and the public risk can be systematically estimated. In addition, because explicit risks are assigned, strategies to reduce risks can be identified.

The analysis showed clearly that the risks to the public from the operation of the La Salle Terminal were below those to which the population in the vicinity of the terminal is exposed at the present time. The study was used in preparing the safety analysis report submitted to the Federal Power Commission. The final Environmental Impact Statement issued by the Federal Power Commission has stated that the levels of public risk associated with the La Salle Terminal facility are acceptable.

# Questions for Thought and Discussion

1. Why do you suppose that, as observed by the authors, some people might take for granted the hazards associated with obtaining and using conventional fuels such as coal to generate electricity, while objecting to newer forms of fuels such as LNG?

2. The authors present a chart with a curve that shows the probability of exceeding specific numbers of fatalities per year for the proposed level of LNG imports. Suppose that another curve was produced under the assumption that current methods of energy production would continue to be used to satisfy the same expected level of demand. If this curve was above the LNG curve to the left of "100 fatalities" and then crossed it and was below it to the right of "100 fatalities," what conclusion would you draw about the relative safety of LNG versus the current form of energy? Which form of energy would you prefer, and why? How would the curves change if the workers in the power plants where the fuel is actually consumed were to be accounted for in the risk calculations?

3. Based on the information in this paper, answer the following:

- Do you have enough data to conclude that you would approve or disapprove of importing LNG into the Matagorda Bay area? If you do have enough data, what is your conclusion and why? If not, what other data would you want?
- Would your conclusion change if you were a resident of this area? Why?
- How would you try to convince the local residents to agree with your conclusion?
- What additional measures could be taken to reduce the risk further?

4. Suppose that the utility agreed to invest in hazard mitigation and emergency response to reduce the risks below the levels estimated in this paper. These estimates indicate that, due to the La Salle Terminal, people in boats have an annual risk of death of $4.5 \times 10^{-9}$ per person compared to an annual risk per permanent resident in nearby Port O'Connor of $2.5 \times 10^{-11}$. Do you believe that more should be spent to protect people in boats than to protect the general population of nearby residents at risk from LNG accidents? How much more? Explain. Alternatively, if you do not think that anything at all should be spent to protect anyone from these low-probability accidents, defend that position.

# PART 6

# RISK COMMUNICATION

# Introduction ————————————————

The experts trained and experienced in risk assessment are not always successful in communicating with the general public about how they assess health and safety risks and how their findings should be interpreted. And the risk managers who base their decisions on the experts' advice sometimes fail to convince the public that risks can be kept to acceptable levels. Lessons from the arenas of toxic waste disposal and nuclear power plant safety bear witness to these facts.

When risk communication breaks down, concerned citizens may end up feeling that they are not being taken seriously or that they are being treated unfairly, while the risk managers and their expert advisers may think that the public does not understand the problem or is unwilling to cooperate. The study of risk communication has recently received a great deal of attention because it has become clear that without good communication, risk assessment and risk management may be largely in vain. Although it is the newest among the research areas represented in this collection of readings, some theories, principles, and applications of risk communication have already begun to emerge. The three papers in part 6 demonstrate the progress that has been made. The first provides an overview of risk communication issues and research, the second suggests ways for risk managers to communicate better with communities that are being asked to host hazardous waste facilities, and the third explores the effectiveness of different ways of communicating information about the risks of radon exposure in the home.

The paper by Alonzo Plough and Sheldon Krimsky, "The Emergence of Risk Communication Studies: Social and Political Context," states that risk communication is hampered both by federal agencies that selectively draw on risk research results for their own ends and by citizen groups that, having learned of the political nature of risk decisions, brush aside the technical issues and focus on political actions to achieve their goals. But an even more fundamental obstacle, the authors observe, is the discord that exists between the "technical rationality" of the risk experts and the "cultural rationality" of citizens who must live with the consequences of risk decisions. For risk communication to be a two-way process, the experts must recognize the validity of the citizens' views.

In "Getting to Maybe: Some Communications Aspects of Siting Hazardous Waste Facilities," Peter M. Sandman focuses on conflicts among three parties involved in siting hazardous waste facilities in New Jersey: the community, the

Hazardous Waste Siting Commission, and the developer of the facility. By acknowledging the community's power to stymie the selection process by legal and political actions, the author argues, the commission and the developer would enable the community to say "maybe" to the proposal rather than "no," especially if the community realizes that it can always retreat to "no" should things go badly. The specific suggestions that he offers for furthering the discussion, Sandman warns, are likely to "fail abysmally" if the community's power is not acknowledged.

What kind of written information is best for communicating the risks of radon in the home and informing people how to make decisions about whether or not to take action to reduce exposure? F. Reed Johnson, Ann Fisher, V. Kerry Smith, and William H. Desvousges, in "Informed Choice or Regulated Risk? Lessons from a Study in Radon Risk Communication," report preliminary results from testing the effectiveness of five different booklets and a one-page fact sheet designed to inform people about these risks and to influence them to take appropriate mitigative actions. Their experiments show that the various ways of presenting the same information have decidedly different effects on learning, risk perception, and mitigation behavior.

# The Emergence of Risk Communication Studies: Social and Political Context

ALONZO PLOUGH and SHELDON KRIMSKY
*Tufts University, Medford, Massachusetts*

Why has the concept of risk communication suddenly become a widely discussed framework for public policy in the environmental and health areas? Prior to 1986 there were only a few essays in the scholarly and policy literature with "risk communication" in their titles. Since that year, however, scores of titles with the term have appeared[1] along with conferences, special sessions in scientific meetings, agency-sponsored workshops, and grants. From one perspective this is not so unusual. New problems often capture the attention of researchers and become the centerpiece of academic and policy research for a period of time. Scientific subfields both from within and across disciplines are constantly emerging. Risk communication might be just another fashionable rubric for the activity of a specialized group of researchers. If that were the case, then its birth as an area of study would be of interest primarily to historians and sociologists of science.

Alonzo Plough and Sheldon Krimsky, "The Emergence of Risk Communication Studies: Social and Political Context," *Science, Technology, & Human Values*, vol. 12, nos. 3 and 4 (Summer/Fall 1987), pp. 4–10, copyright 1987 by the Massachusetts Institute of Technology and the President and Fellows of Harvard College. Reprinted by permission of Sage Publications, Inc.

The emergence of risk communication as a research theme cannot be fully appreciated or accounted for without understanding its link to a set of issues that symbolize the discord between scientific experts and the public around the issue of risk. These tensions are played out in disputes between different research traditions on fundamental questions regarding the perception of risk and the essential nature of human rationality. Federal regulatory agencies in the health and environmental areas draw on this diverse and at times conflicting body of research very selectively, choosing analytical frameworks that are most compatible with their policy agendas. Citizen groups, less concerned with formal theories, are increasingly aware that getting a message across to government in disputes over health and environmental hazards is essentially a political activity. On the other hand, experts in risk analysis have come to appreciate the gap between their analytically derived conclusions and the conceptions of popular culture. In the final analysis, those who control the discourse on risk will most likely control the political battles as well.

In this essay we will analyze the emergence of risk communication as a significant new organizing theme for a set of diverse but conceptually related problems concerning the political management of public risk perceptions and individual behavioral responses to risks. Our discussion begins with a brief look at the historic conditions that gave rise to the current thematic focus on risk communication in the social and behavioral sciences; we then examine the role of heuristics in defining the legitimate areas of study; and finally we argue that the research activities centered on risk communication have precipitated new debates over technical and cultural meanings of rationality.

Risk communication is more than a research framework. It has become a concept that is strongly marketed by specific interest groups and used instrumentally to achieve particular ends. At the federal policy level, the Environmental Protection Agency (EPA) has been the strongest marketer of the concept of risk communication. Its chief administrator has stated: "On the national level we will build risk communication into regulatory policy whenever possible."[2] The EPA has elevated the concept of risk communication to a strategic level of importance in both its regulatory activities and its research agenda. Industries that are regulated by the EPA also see risk communication as a key policy and management issue.

In the public health area the terminology of risk communication is slightly different (federal health officials speak about *educating* the public about risks) but the concept is equally important. The current policy discourse on preventing health problems such as AIDS, teen pregnancy, and substance abuse focuses on communicating risk to a target population.[3] This emphasis has gone beyond the health education strategies of an earlier era in public health. It incorporates slick national media campaigns developed by public relations or advertising firms in conjunction with science-based strategies in the attempt to change "unhealthy" personal behaviors.

Risk communication has always been an important component in medical practice, particularly in doctor–patient relationships. More recently, with the rise of chronic disease as a major source of health problems and the uncertainty associated with the risk factors for these diseases and the course of treated illnesses, risk communication in medicine has attained a more central and visible role.[4]

Although the study of risk communication is considered new, the practice of it may be as old as human culture itself. There is also nothing new about the social problems that shape the public discourse on risk. Throughout human history individuals and groups have had to contend with a variety of risks for survival and personal well-being. Risk is inherent to the human condition. The vagaries of an ever-changing environment, incurable diseases, and man-made threats like war have presented risks to both individual and collective survival.

The origins of structured risk analysis have been traced to the Babylonians in 3200 B.C.[5] All of human history is very much a story of assessing and adapting to risks. Methods used by the ancients to predict risks and to communicate knowledge about avoiding hazardous situations were based on myths, metaphors, and ritual. Risk communication was embedded in folk discourse.

## The Professionalization of Risk

The transition from folk discourse about risk to an expert-centered communication was preconditioned by a series of key historical events beginning in the late 18th century and continuing through the present time. One necessary condition was the rise of the modern state with an implied responsibility for general social welfare functions. The modern state's legitimacy, in part, derives from its claim to protect the population from physical harm.

By the early 20th century the development of public health institutions provided a second condition that furthered the professionalization of risk. Medicine emerged as an influential profession which defined health risks and controlled intervention strategies. Public health departments began the first large-scale environmental risk monitoring in the sanitation and food safety areas under new governmental mandates.[6] Formal risk messages communicated the danger of hazards such as unsafe water and unpasteurized milk and indicated that experts should be trusted to apply new technologies to reduce the risks of infectious diseases.

A third tributary to the expert-driven field of risk assessment is decision analysis. During World War II the government's need for scientifically based decision methodologies gave rise to a new era of federal research support that spawned fields like operations research and systems analysis. In the late 1940s a variety of quantitative methodologies were introduced to promote understanding of chance processes and to create a rational framework for economic and strategic military decisions. Eventually, these models were applied to the practical problems of predicting and altering the course of risk factors in public health, medicine, and the environment. The methods first used in mathematics, economics, and statistics slowly diffused into the social and health sciences and gave rise to a number of hybrid approaches for calculating risks and quantifying decisionmaking. By the late 1960s the decision methodologies brought promise of a systems science that could provide a rational basis for complex policy decisions concerning technological risks. This was a compelling idea to federal regulators. The schools of rational risk analysis assumed a great burden, namely, to create a bridging logic between the *assessment of risk* and political decisions concerning the *types*, *levels* and *distribution* of risks acceptable to society. The results have not fulfilled the original expectations. It has been difficult to find common ground between the social world of risk perceptions guided by human experience and the scientists' rational ideal of decisionmaking based on probabilistic thinking.

The modern field of risk analysis is concerned primarily with predicting or quantifying the risks of "scientifically identified hazards" (i.e., toxicity of chemicals, probabilities of accidents, spread of a disease).[7] The communication of information about risks usually occurs within a context of fear and uncertainty. Nuclear radiation, toxic wastes, AIDS, asbestos, and other hazards invoke a range of responses in the scientific, regulatory, and lay communities. Questions of the acceptability of any identified risk are deeply connected to perceptions of fairness and justice. There is wide agreement within the professions that value and equity problems in making decisions about risk are serious. There is less consensus about the extent to which political and ethical choices are an unavoidable part of *risk assessment*.

With the passage of the Environmental Policy Act of 1969 and the Occupational Safety and Health Act of 1970, the creation of the Office of Technology Assessment in 1972 and the development of a host of other programs in the early 1970s, the institutionalization of risk evaluation has been realized, making equal weighting of technical and value considerations difficult to achieve in practice. Legislation requires a formal and legally defensible assessment of risk as exemplified by the requirement of environmental impact statements for most federal agencies. Consulting firms marketing their quantitative expertise blossomed in response to the national emphasis on risk and risk management. Concurrently, graduate schools of public policy and public health, responding to the new role of quantitative decisionmaking within the federal government, began to require courses in decision analysis. In schools of public policy the effective management of environmental and health risks is synonymous with quantitative assessment of problems. Within this orientation, the social and cultural context of risk is of marginal concern.

Quantitative models of risk including comparative risk assessment disregard the many value issues embedded in risk analysis. These problems re-emerge in the domain of risk management when regulatory agencies must decide what actions to take once risk assessments have been accomplished. William Ruckelshaus defines risk management as "the distribution of current resources to shape some desirable future state; risk management in its broadest sense means adjusting our environmental policies to obtain the array of social goods—environmental, health-related, social, economic, psychological—that forms our vision of how we want the world to be."[8] While these certainly are the factors that must be considered if equitable decisions on risk are to be made, there is no public consensus that government can conduct this broad social management of risk in a fair and equitable manner. Thus, the rise of environmental advocacy in the 1970s challenged the expert model of risk management.

Environmentalism became a powerful social movement. Regulatory policies evolved that gave greater attention to the need for public participation. Inevitably, conflicts arose between the rational quantitative approach to risk assessment and public perceptions of risk. The dilemmas of reconciling democratic ideals and citizen-centered values with the rationality of elite institutions and formal decision processes became more pronounced. While some experts and regulators lamented the rise of "irrational" discourse in environmental debates (the dreaded "not-in-my-backyard" response to siting issues), this same discourse was an expression of their own democratic ideals, which included the opportunity to oppose official decisions that ignored the experiential context of risk.

In the field of public health, the risk debate during the 1970s began to focus on the individual's responsibility for poor health outcomes. Health care costs increased from 7 to 10% of the gross national product, and the complex patchwork of federal health policy seemed unmanageable. To some policymakers, the greatest risk area in the late 1970s was containing costs of medical care. Prevention became more focused on poor health outcomes related to lifestyle variables such as smoking, diet, exercise, and sexual behavior.

Risk factor research and intervention programs increasingly focused on the risky individual and less on the social and cultural context of risk. Personal health risk assessment shares with environmental risk assessment the notion of the "irrational individual." In the former, the individual does not make rational choices about risky behaviors such as smoking and not wearing a seat belt and therefore the individual takes irresponsible risks. In the latter case the faulty logic is reversed: The individual maintains an "exaggerated" fear of hazards which experts consider to be relatively safe. The field of comparative risk assessment actually connects both of these constructions of the irrational individual.

A partial answer to the question of why risk communication has emerged as a framing issue for environmental issues can be found in the differences between professional risk analysts and popular culture. To explore this further we shall look at how risk communication has been conceptualized within the policy sciences.

## From Definitions to Heuristics

The term "risk communication" describes a wide range of activities. It has both a conventional definition and a symbolic definition. The former reflects the use of the term in risk management while the latter derives from the role of risk in political discourse. To frame the concept exclusively in conventional terms restricts the meaning of risk communication to surface behavior or what natural scientists like to call "the phenomenon of the event." The conventional account neglects cultural themes, moti-

vations, and symbolic meanings which may be of equal or greater importance to the technical understanding of how and why a risk message gets transmitted. When risk communication becomes embedded in the political arena, it is less about risks per se than about responsibility or accountability for certain events. We shall discuss the symbolic definition of risk communication after we discuss its conventional definition.

"Risk communication" can refer to any public or private communication that informs individuals about the existence, nature, form, severity, or acceptability of risks. In this broad use of the term, risk communication may be directive and purposeful or nondirective and fortuitous. It may describe the controlled release of information toward certain well-defined ends or it may represent the unintended consequences of informal messages about risks. This broad interpretation has been adopted by Kasperson and Palmlund; they state that risk communication "enters our lives in a multitude of forms, sometimes part of the imagery of advertising, sometimes a local corporation's formal statement, or its failure to say anything, sometimes a multi-volumed and impenetrable technical risk assessment."[9] Almost any communication from any source that speaks to the issue of risk satisfies the authors' definition.

In risk management a narrower use of the concept usually focuses on an intentional transfer of information designed to respond to public concerns or public needs related to real or perceived hazards. Thus, risk communication incorporates tacit or explicit goals for targeted groups about specific events or processes. The information is channeled from experts to a general audience.

The conventional definition of risk communication centers on the intentionality of the source of information and the quality of the information. Covello and his colleagues have defined risk communication as "any purposeful exchange of scientific information between interested parties regarding health or environmental risks."[10] This definition constraining risk communication to scientific information between interested parties raises several questions. For example, are the claims of lay people in news reports about environmental hazards excluded under this definition? Further, it is not clear what relevance the term "interested parties" has in this

**Table 1. Definition Latitude of Risk Communication**

|  | Broad | Narrow |
| --- | --- | --- |
| Intentionality | Risk communication goal unnecessary | Intentional and directed; outcome expectations about the risk message |
| Content | Any form of individual or social risk | Health and environmental risks |
| Audience directed | Targeted audience not necessary | Targeted audience |
| Source of information | Any source | Scientists and technical experts |
| Flow of message | From any source to any recipient through any channel | From experts to nonexperts through designated channels |

definition; the Surgeon General may receive a bored or disinterested response to his reports of the risks of cigarette smoking, but few would disqualify these messages as risk communication.

We can envision risk communication as having five components in its definition: intentionality; content; audience directed; source; and flow. Different definitions of risk communication are narrow or broad depending on the latitude of interpretation of these elements (table 1). The most restricted interpretation would define a risk communication as a plan executed by a regulatory body targeted to a special audience and embodying specific outcome goals for behavior or attitudinal change.

The conventional definition of risk communication restricts the purview of its study to how "experts" inform others about the truth. Under this notion, some researchers view the exchange of risk information as flowing from technical elites to the polity as a form of scientific noblesse oblige. The process involves the transfer of "scientific" facts and a set of conclusions drawn from those facts. In the broader and more comprehensive definition, technical elites are not the exclusive trustees of risk information. For those who study social process, this broader definition highlights the importance of nonelites as risk communicators.

To understand the symbolic meaning of risk communication, we have to study risk in its social context. A scientist speaking to a community about the health risks of a chemical dump may be carrying out a ritual that displays confidence and control. The technical information (the message) is secondary to the real goal of the communicator: "Have faith; we are in charge." Local residents citing a litany of symptomatology that they attribute to contaminated water are using risk communication as a channel for their anxieties over environmental overload. "Popular" or "barefoot" epidemiologists are lay people who spot disease clusters and use risk communication to organize for additional public health studies and eventual toxic cleanup. For a company, risk communication is frequently not about risks but about safety and confidence. When a contract research firm in Cambridge, Massachusetts, wrote a "Dear Neighbor" letter to thousands of residents on the issue of its research with chemical warfare agents, risk communication meant "we wouldn't dream of doing anything hazardous to the community, trust us."[11] In the area of public health, risk communication is the final result of a complex political process through which a problem gets a relatively simple policy construction and is brought to the public consciousness in a restricted form. For community organizers, risk communication is a strategy for solidifying a movement and rebuilding the social complexity of an adverse health or environmental outcome.[12] The symbolic definition of risk communication differs substantially from the conventional view. While the former includes cultural and experiential inputs, the latter generally is reductionist, focusing on quantifiable variables.

## Risk, Rationality, and Culture

Risk communication evolved out of a need of risk managers to gain public acceptance for policies grounded in risk assessment methodologies. The prevailing view of many experts and risk managers is that local communities and the general public react to limited, false, or inadequate information. These lay groups exercise a personal or democratic prerogative in response to "bad" information that is inconsistent with the more fully informed conclusions of risk assessment experts on whom policymakers depend for developing rational responses to complex problems. Frequently, the exercise of local democracy and personal choice is at odds with the rationality of technical experts. Quantitative risk analysis, rather than narrowing differences, may actually exacerbate antagonisms between the technosphere (the culture of experts) and the demosphere (popular culture). Casting the issues in a technical language reduces the possibility of a dialogue between the public and elites.

Recent studies in the risk perception literature reinforce the conception that rationality and democracy are antagonistic to one another. There are many areas where public perceptions about hazards are inconsistent with so-called objective information. For example, people are said to exhibit too little concern about some hazards (smoking, auto accidents, geological radon, and exposure to sunlight) and too much concern about others (nuclear power, toxic wastes, pesticides, and genetic engineering).

Recently, the EPA convened experts from its principal divisions to determine which events, technologies, and situations represented the greatest environmental risks.[13] The results were not surprising. The experts' inventory of environmental priorities did not correlate positively with the agency's regulatory priorities. The EPA was allocating a large share of its resources for reducing adverse environmental effects in areas its own experts did not consider to warrant the most attention.

The discrepancy between what experts deem most important and what the public demands of its government raises difficult policy questions. Two things deserving respect, namely, scientific rationality and democracy (the rights of local communities to express their will on issues of health and safety), are in conflict. How does one proceed? The options that policymakers usually consider include:

1. Circumventing the public by avoiding disclosure, by distraction, by preemption, or by citing social contract doctrine according to which agencies represent the public through their elected officials and can decide for the people.

2. Appealing to some exemplary and independent authoritative body that will apply the rational decision framework and secure public confidence.

3. Communicating the risks and educating the public into thinking about the problem the way the experts do. Public perception must be brought into conformity with scientific rationality.

Of the three options, only the last is directed specifically at reducing the opposition between the demosphere and the technosphere from the experts' perspective. The emphasis is on a restricted notion of informed democratic practice. Within this context, risk communication is the responsibility of elites and falls into the general rubric of "public understanding of science." The success of risk communication is measured by the degree to which popular attitudes reflect the technical rationality of risk and the extent that popular behavior conforms to technocratic values. A lack of convergence is attributed to a failure of risk communication.

The rapid growth in research on risk perception began to cast doubt about the public education model. Popular conceptions of risk resisted the conclusions of elites despite clear presentation of the "facts." Studies appeared that purported to explain the discrepancies between expert and lay perceptions of risk. Variables that are intended as proxies for cultural determinants were introduced to account for the differences. Two events with the same risk (probability of mortality) evoke different risk perceptions in experimental studies. One event is *perceived as* voluntary while the other is *perceived as* involuntary. Lay people do not compare events strictly in terms of actuarial risks.

Psychologists began codifying these and other factors that appeared to explain the discrepancies between technical risk assessment and public perceptions. This has resulted in a labeling of risk events according to the restricted logic of cognitive science. The conventional risk communication approach was modified to accommodate adjustment parameters (voluntary vs. involuntary; familiarity vs. unfamiliarity; natural vs. man-made). However, instead of building a culture-based theory of risk perception, psychometricians isolated the cultural factors and treated them as another variable in an experimentally derived technical framework. Every risk event possesses objective hazard estimates and certain qualities that begin to take on ontological status. Thus, a risk event that is voluntary would not be compared on pure rational grounds to one that is involuntary. This system preserves the dichotomy between expert judgment and lay perception of risk. It merely categorizes "irrationalities" and does not explore the cultural underpinnings of risk perception. Moreover, cultural noise affecting the popular response to risk is rationalized. This form of the analysis treats the cultural inputs into risk perception as deviant but comprehensible. Risk communication is still viewed as information transfer from experts to lay people.

A cultural approach that seriously considers popular behavior and symbolic dimensions distinguishes two forms of rationality applied to risk: technical and experiential. Both make contributions to the problem of constructing and analyzing a risk event, but neither is sufficient. Deviance is not the appropriate metaphor to understand differences between the demosphere and the technosphere. Once these distinct modes of rationality are understood, the problem of risk communication is transformed; the problem becomes one of mutual understanding and mutual learning. This cultural model is based on the notion that expert and popular approaches to a risk event can each be logical and coherent on their own terms but may exhibit differences in how the problem is articulated, in the factors relevant to the analysis, and in who the experts are.

## Technical Rationality

This form of thinking rests on explicitly defined sets of principles and scientific norms. These include hypothetico-deductive methods, a common language for measurement, and quantification and comparison across risk events. In its more advanced forms technical rationality encompasses a mature theory with predictive power. The emphasis is on objective (non-personal) inputs rather than subjective (experiential) information. Perceived responses to risk are important only in understanding the extent to which ordinary people's ideas deviate from the truth. Logical consistency is an imperative. Two events that have an identical risk profile are treated the same—they are interchangeable.

## Cultural Rationality

One of the common mistakes in attempting to codify the public attitudes about risk is to measure people's responses to hypothetical questions. Cultural rationality can only be understood when people's cognitive behavior is observed as they are threatened by a real risk event. It is only then that the full panoply of factors comes into play that create a complete picture of a public response. To understand cultural rationality, one must address anthropological and phenomenological issues as well as behavioral ones. Technical rationality, on the other hand, believes that risk can be studied independently of context. Mary Douglas provides some insight on this point: "The question of acceptable standards of risk is part of the question of acceptable standards of morality and decency, and there is no way of talking seriously about the first while evading the task of analyzing the cultural system in which the second take their form."[14]

Lay people bring many more factors into a risk event than do scientists. For technical experts, the event is denuded of elements that are irrelevant to the analytical model. Table 2 illustrates some of the differences. Many events that are deemed to have very low or insignificant risk by experts are viewed as serious problems by the laity. Burial of low level radioactive wastes and releasing a natural organism minus a few genes into the environment are among such cases. Where there has been discussion of rationality, it has focused on the scientific grounds of a decision. And yet there are clear instances of reasonable decisionmaking at the community level that are inconsistent with expert opinion. Once it is accepted that two inconsistent decisions can be rational and consistent *on their own grounds*, it is possible to reach beyond the deviant model of risk communication.

Cultural reason does not deny the role of technical reason; it simply extends it. The former branches out, while the latter branches in. Cultural rationality does not separate the context from the content of risk analysis. Technical rationality operates as if it can act independently of popular culture in constructing the risk analysis, whereas cultural rationality seeks technical knowledge but incorporates it within a broader decision framework.

If these forms of rationality are unalterably antagonistic, technical reason and popular response to risk may be truly incommensurable. But forms of rationality must be capable of responding to a process of mutual learning and adjustment. If the technosphere

**Table 2. Factors Relevant to the Technical and Cultural Rationality of Risk**

| Technical rationality | Cultural rationality |
|---|---|
| Trust in scientific methods, explanations; evidence | Trust in political culture and democratic process |
| Appeal to authority and expertise | Appeal to folk wisdom, peer groups, and traditions |
| Boundaries of analysis are narrow and reductionist | Boundaries of analysis are broad; include the use of analogy and historical precedent |
| Risks are depersonalized | Risks are personalized |
| Emphasis on statistical variation and probability | Emphasis on the impacts of risk on the family and community |
| Appeal to consistency and universality | Focus on particularity; less concerned about consistency of approach |
| Where there is controversy in science, resolution follows status | Popular responses to scientific differences do not follow the prestige principle |
| Those impacts that cannot be uttered are irrelevant | Unanticipated or unarticulated risks are relevant |

begins to appreciate and respect the logic of local culture toward risk events and if local culture has access to a demystified science, points of intersection will be possible.

## Conclusion

In this essay we have described the historical and social context of risk communication as a fundamental and enduring human problem and, recently, as a focus of academic study. The research activities in risk communication are closely linked to the requirements of governmental agencies with a mandate to protect the public from technological, environmental, and health risks. As a body of work, the first wave of risk communication studies and strategies generally have followed a useful but limited framework for defining the central problem of the field: divergence between expert, policy, and lay communities on matters of risk.

Two issues seem important from this review. First, experts and institutions have developed models of risk and communicated them in various forms throughout history. Conflicts between groups over risk assessments (should we go to war; will there be a famine) are fundamental to all cultures. The current discourse on risk and its reliance on sophisticated quantitative models of assessment and cognitive typologies of perception represent a change in the form of risk communication but not in the underlying value controversies over the social context of risk. What we have are new methods addressing persistent structural problems.

The second issue centers on the fit between these new methods and the structural problems of risk communication. Research on the risks described in this essay has an overwhelming tendency to avoid the experiential context of risks—that is, actual people considering real threats to their well-being or other persons' well-being. Laboratory experiments of cognitive psychologists represent risk perception within the bounded models of experts. These models reveal more about the cognitive context of the research tradition than about how persons construct and experience a risk event in a social context. Researchers who view the differences between popular culture and technical rationality as a form of deviance are not likely to generate better strategies for risk communication.

We argue that risk communication has emerged from a context of political conflict. The analytic models attempted to respond to real problems. However, the dominant model of risk communication creates a template that trivializes the complexity of cultural factors. We regard this as an imbalance in the research agenda which should be corrected.

As Mary Douglas reminds us, ideas about the world come directly out of human experience.[15] This, we would argue, holds true for both experts and lay persons. There is a social context of expertise and officialdom as well as of lay communities. Bias, irrational action, and narrow-interest group behavior intrude into both of these contexts.

## Notes

1. Vincent Covello, "Introductory Remarks," Workshop on the Role of Government in Health Risk Communication and Public Education, 21–23 January 1987.

2. Lee Thomas, "Risk Communication: Why We Must Talk About Risk," *Environment*, Volume 28, Number 2 (March 1986): 40.

3. See, for example, U.S. Public Health Service, *Promoting Health/Preventing Disease: Objectives for the Nation* (Washington, DC: U.S. Public Health Service, 1980).

4. Alonzo L. Plough, *Borrowed Time: Artificial Organs and the Politics of Extending Lives* (Philadelphia: Temple University Press, 1986).

5. Vincent Covello and Jeryl Mumpower, "Risk Analysis and Risk Management: An Historical Perspective," *Risk Analysis*, Volume 5, Number 2 (1985): 103–120.

6. Barbara Gutmann Rosenkrantz, *Public Health and the State: Changing Views in Massachusetts: 1842–1936* (Cambridge, MA: Harvard University Press, 1972).

7. James F. Short, "The Social Fabric at Risk: Toward the Social Transformation of Risk Analysis," *American Sociological Review*, Volume 49 (December 1984): 711–725.

8. William D. Ruckelshaus, "Risk, Science, and Democracy," *Issues in Science and Technology*, Volume 1, Number 3 (Spring 1985): 19–38.

9. Roger Kasperson and Ingar Palmlund, "Evaluating Risk Communication," unpublished paper, Center for Technology, Environment, and Development, Clark University, Worcester, MA, January 1987.

10. Vincent T. Covello, Detlof von Winterfeldt, and Paul Slovic, "Communicating Scientific Information about Health and Environmental Risks: Problems and Opportunities from a Social and Behavioral Perspective," in V. Covello, A. Moghissi, and V. R. R. Uppulori, eds., *Uncertainties in Risk Assessment and Risk Management* (New York: Plenum Press, forthcoming).

11. The letter contained the sentence: "As neighbors, I would like you to know what we are doing and to be accurately informed about the safety and security precautions we have built into our new toxic materials laboratory so that you can feel as safe as my colleagues and I do when we come to work . . . each day." For an overview of the case, see Sheldon Krimsky, "Research under Community Standards: Three Case Studies," *Science, Technology, & Human Values*, Volume 11, Issue 3 (Summer 1986): 14–33.

12. Kasperson notes that in local controversies "risk communication becomes a vehicle of protest by which community groups create resources with which to bargain with government in the risk management process." Roger E. Kasperson, "Six Propositions on Public Participation and Their Relevance for Risk Communication," *Risk Analysis*, Volume 6 (1986): 276.

13. U.S. Environmental Protection Agency, *Unfinished Business: A Comparative Assessment of Environmental Problems* (Washington, DC: EPA, February 1987).

14. Quoted in Short, *op. cit.*, p. 720.

15. Mary Douglas, *Risk Acceptability According to the Social Sciences* (New York: Russell Sage, 1985), pp. 100–101.

# Questions for Thought and Discussion

1. Offer five examples of risk communication from everyday experience (e.g., the warnings on cigarette packs or the placement of traffic safety signs). Which of these do you think is most ignored by the public and why? What would you do to improve communication with regard to this risk?

2. Suppose that your community faced the prospect of becoming the site of a hazardous waste facility (a "not-in-my-backyard" issue) and a corridor for hazardous waste trucking (a "not-on-my-street" issue). What risk information would you like to have communicated to you and in what form? What organizations or kinds of organizations would you look to for information?

3. The authors distinguish between the conventional and symbolic definitions of risk communication. Explain and illustrate this distinction in terms of the risks of one of the following global environmental problems: acid rain, depletion of the ozone layer, or the greenhouse effect. Provide one or two suggestions for improved risk communication for the problem that you select.

4. The possibility is raised in this paper that technical rationality and cultural rationality are "unalterably antagonistic," and an appraisal is given where progress might be made to reconcile the two views. Do you think that this is a neutral appraisal, or is most of the burden of reconciliation placed either on technocrats or on the public? Explain. What suggestions would you make to contribute to the progress of reconciling the two views?

# Getting to Maybe:
# Some Communications Aspects of
# Siting Hazardous Waste Facilities

PETER M. SANDMAN
*Rutgers University*
*New Brunswick, New Jersey*

## Introduction

The United States generates roughly fifty million metric tons of non-radioactive hazardous wastes annually.[1-2],3 While much can be done to reduce this figure, a healthy economy will require adequate facilities for transporting, treating, storing and disposing of hazardous wastes for the foreseeable future. Current facilities are far from adequate; new ones and safer ones must be sited and built. The alternatives are dire—economic and technological slowdown on the one hand, or "midnight dumping" and similar unsafe, illegal and haphazard disposal practices on the other.

The principal barrier to facility siting is community opposition: "not in *my* backyard." Experience amply justifies this opposition. Communities have learned, largely from the media, that hazardous waste facilities endanger public health, air and water quality, property values, peace of mind and quality of life. They have also learned, largely from the environmental movement, that they can mobilize politically to block the siting of a facility, eminent domain statutes notwithstanding.

Technical improvements have reduced, though not eliminated, the risk of "hosting" a hazardous waste facility. State governments have learned how to regulate facilities more effectively. Responsible hazardous waste generators have come to terms with the need to reduce waste flow and handle remaining wastes properly. Responsible environmentalists have come to terms with the need to accept some waste and some risk in its disposal. A government-industry-environmentalist consensus is emerging in behalf of state-of-the-art facility design, development and siting. However, this consensus is not enough. The community typically rejects the consensus, and may well enforce its dissent through its exercise of a *de facto* veto.[4]

The comments that follow are predicated on several assumptions: (1) A facility can be designed, managed and regulated so that risks are low enough to justify community acceptance (without this, the task of siting is unethical); (2) Community acceptance is more desirable and more feasible than siting over the community's objections (without this, the task of meeting with a community is unnecessary); and (3) The positions of the siting authority and the developer are sufficiently flexible—legally, politically and economically—to permit meaningful concessions to community demands (without this, the task of gaining community approval is unachievable).

## Acknowledge the community's substantial power to slow or stop the siting process.

Despite the preemption and eminent domain provisions of New Jersey's Major Hazardous Waste Facilities Siting Act,[5] many observers are convinced that a facility cannot be sited over a community's objections. The resources in the community's hands are many: legal delay, extralegal activities, political pressure, legislative exemption, gubernatorial override. The subtitle of one of the leading books on the siting problem testifies to the conviction of authors David Morell and Christopher Magorian that the community has something close to a veto. The book is entitled *Siting Hazardous Waste Facilities: Local Opposition and the Myth of Preemption*.[6] Moreover, in a January 25, 1985 interview with *The New York Times,* Department of Environmental Protection (DEP) Commissioner Robert E. Hughey agreed. "Siting," he said, "will be fought everywhere. I think everything else but this has an answer."[7] At the Seton Hall Symposium on siting, Douglas Pike of Envirocare International acknowledged the veto power of communities when he stated: "We have to operate as if there is no eminent domain."

Ironically, nearly everyone is impressed by the community's power of opposition—except the community, which sees itself as fighting a difficult, even desperate uphill battle to stop the siting juggernaut. From a communication perspective, this is the worst possible state of affairs. Suspecting that the "fix" is in, the community judges that it simply cannot afford to listen, to consider alternatives, to negotiate modifications. Intransigence looks like its best shot, perhaps its only shot. But suppose the Commission and the developer were to acknowledge *to the community* its considerable power: "Look, we probably can't site this thing unless you agree, and there are plenty of chances for you to stop it further on down the

pike. Why don't we put the possible battle on ice for now and explore whether there is any possible agreement. If the talks fail, you can always go back to the fight." It will not be easy, of course, to persuade the community that this is not a trick, that it is forfeiting nothing by negotiating now, that it can switch its stance from "no" to "maybe" while protecting the road back to "no." It will take some effort not to *over*state the community's power. Though more powerful than it thinks, the community is not omnipotent, and the risk of override is real. The goal is to let the community know, publicly, what other participants already know privately: that it will be extremely difficult to site a facility over community objections, and that the siting authority would greatly prefer not to try. Formal acknowledgments of community power, such as a developer's pledge to honor a community referendum on any agreement that might be negotiated, are sometimes possible. But even an informal acknowledgment will reduce intransigence and encourage open discussion.

Acknowledging the community's substantial power will have three other desirable impacts. First, it will reduce community resentment of what is seen as a power imbalance, an outrageous imposition of state control over local self-determination. This resentment and the deep-seated feeling of unfairness that accompanies it are major factors in community rejection of hazardous waste facilities. Residents look at New Jersey's siting law and note that in the final analysis, state action prevails over local preference. Angrily, they resolve to resist. Open acknowledgment of *de facto* power will lessen the anger at the imbalance of *de jure* power.[8]

Second, acknowledging community power will reduce fear about the health effects of a hazardous waste facility. One of the best documented findings in the risk perception literature is that we fear voluntary risks far less than involuntary ones. According to one study people will accept *one thousand times* as great a risk if it is chosen than if it is imposed by others.[9] Therefore, to the extent that the community feels itself in control of the siting decision, the risks of the facility become much more acceptable and much less fear-arousing.

Third, acknowledging community power will put the dialogue on a more frank footing than the classic "one-down/one-up" pattern that tends to dominate siting discussions. Under this pattern a community tries to prove itself the equal of the developer and the siting authority, while secretly feeling that it is not. The developer and the authority adopt a parental "the-decision-is-not-yours-but-we-value-your-input" attitude, while secretly fearing the community's *de facto* veto. Negotiations are much easier when the parties are acknowledged equals.

## Avoid implying that community opposition is irrational or selfish.

Nothing interferes so thoroughly with the settlement of a dispute as the suggestion from either side that the other is being irrational or selfish. Yet developers, siting authorities and their expert consultants often aim this charge at community opponents. The acronym "NIMBY"—Not In My Back Yard—has become a sarcastic code, implying that opponents approve of siting in principle but oppose it in their neighborhoods for insupportable reasons. Some community groups, by contrast, still use the phrase as an anthem of their battle to prevent the Love Canals of the future. For example, Nicholas Freudenberg's book on how to organize community opposition is entitled *Not In Our Backyards.*[10] But the sarcastic meaning prevails. Opponents now take offense when developers or siting authorities start talking about "the NIMBY syndrome"—and they are correct to be offended.

Some opponents disapprove of siting new facilities anywhere, but choose to fight only in their own communities where their stake is greatest and their power base strongest. Some argue that source reduction and recycling can eliminate the need for new facilities, or that facility siting should be conditioned on policies that will reduce the waste stream, or that expansion of existing facilities is a wiser alternative, or that we should wait for improvements in waste treatment technology. Some take the position that the type of facility proposed is unduly dangerous, or that the site chosen is environmentally inappropriate, or that the developer's record is unsatisfactory. Others assert that equity dictates a different location. Rural dwellers argue that they should not serve as host to a facility because they did not produce the waste in the first place. Urbanites argue, on the other hand, that they have suffered enough pollution already. These are *all* coherent positions that deserve respectful responses. Dismissing them as a manifestation of the NIMBY syndrome is not fair, accurate nor strategically wise.

Similarly, community distrust of risk estimates by experts is not irrational. The experts generally work for interests with a stake in reassuring answers. Even with total integrity, non-resident experts in pursuit of a site can be expected to reach less cautious conclusions than residents with no special interest in siting. Moreover, there is ample precedent in the last several decades of siting experience to justify fears of a lack of integrity, or of incompetence or callousness. At best, the field is new and risk estimates are inherently uncertain. It is rational to distrust the experts even without any expertise of one's own. People who are trying to sell a hazardous waste facility are no different from people who are trying to sell, say, insulation for a home. One does not have to understand what they are saying technically to suspect that they are not to be trusted.

Furthermore, many siting opponents have acquired impressive expertise of their own. They have sifted the evidence in pursuit of technical arguments to support their position. In some cases, the opponents have become impressively knowledgeable. When pro-siting experts dismiss *all* objections as ignorant because *some* are without foundation, they are fighting *ad hominem,* inaccurately and unfairly.

It is important to note that many siting questions have no technical answers: How much risk is too much? What should you do when the answers are uncertain? These are "trans-scientific" questions, sometimes couched in technical language but unanswerable by technical methods.

Sociologists divide people into the categories "risk-aversive" and "risk-tolerant." What separates them is a fundamental values difference. The risk-aversive believe that if you are not sure of what you are doing you should not do anything, that meddling usually makes things worse. The risk-tolerant believe that problems should be solved incrementally, that the new problems caused by their tinkering will be solved later by someone else's tinkering. Neither position is unreasonable, and neither can be supported or refuted by technical information.

It takes courage for community activists to pit their newly acquired knowledge and deeply felt values against the professional stature of the experts. Unsure of their technical ground, these activists defend it all the more tenaciously, sensitive to the merest hint of disrespect. They deserve respect instead and they will not listen until they feel they have it.

### Instead of asking for trust, help the community rely on its own resources.

Most of the people working to site a hazardous waste facility consider themselves moral and environmentally responsible people. Many are incredibly dedicated to meeting society's need for a decent facility. They also view themselves as professionals, as careful specialists who know what they are doing. In both of these roles they feel that they deserve at least trust, if not gratitude. They experience community distrust— sometimes even community hatred—with great pain. The pain often transforms into a kind of icy paternalism, an "I'm-going-to-help-you-even-if-you-*don't*-know-what's-good-for-you" attitude. I suspect that much of the rhetoric about community irrationality, selfishness and the "NIMBY syndrome" has its origins in hurt feelings. It is entirely reasonable for socially responsible experts to want to be trusted, to feel that they deserve to be trusted, and to resent the fact that they are not trusted.

It is sometimes said that the solution to the siting problem is to build trust. To be sure, the siting authority and the developer must make every effort not to trigger still more mistrust. For example, any hint of *ex parte* discussions between the siting authority and the developer must be avoided. But just as it is reasonable for siting experts to expect to be trusted, it is also reasonable for local citizens to withhold their trust, to insist on relying on their own judgment instead. The Commission must not only accept this, but also encourage and facilitate it.

Information policy is an excellent case in point. As noted earlier, one need not understand a technology in order to distrust experts with a vested interest. One, however, *must* understand the technology in order to decide whether the experts are right despite their vested interest. There is wisdom in the Siting Act's provision of research grants to the community at two stages in the siting process.[11] Methods should be found for the Commission to help the community inform itself even earlier in the process, when positions are still relatively fluid. The advantage of an independently informed community is not only that

citizens will understand the issues, but that they will be *satisfied* that they understand the issues, and thus feel less pressure to construct a rejectionist front. A community that believes it has the knowledge to decide what should be done and the power to do it can afford to be reasonable. A community that believes it lacks sufficient knowledge and power, even if it has them, must conclude that the undiscriminating veto is the wisest course.

Similarly, communities want to know that if a facility *is* built they will not need to rely on outside experts for monitoring and enforcement. Many mechanisms can provide this autonomy:

(1) training of local health authorities, and citizen activists, to monitor effluents;

(2) funding for periodic assessments by consultants accountable to the community;

(3) duplicate monitoring equipment in a public place, so citizens can check, for example, the incinerator temperature for themselves;

(4) establishment of a trust fund, with trustees acceptable to the community, to supervise compensation in the event of accident, so citizens need not rely on the state courts.

Do not underestimate the depth of community disillusionment. Modern society depends on letting experts decide. When experts fail to decide wisely we are jolted into belated and reluctant attention. We feel betrayed. We are angry because we must now pay attention. We feel guilty for having relinquished control in the first place. We do not know what to do but are convinced we cannot trust others to decide for us. Above all, we fear that others will impose their unwise decisions on us even now that we are paying attention.

When the community grimly demands its autonomy, it is too late to ask for trust. Experts must instead presume distrust while helping the community exercise its autonomy wisely.

### Adapt communications strategy to the known dynamics of risk perception.

When people consider a risk, the process is far more complex than simply assessing the probability and magnitude of some undesired event. Departures from statistical accuracy in risk perception are universal and predictable. Communications strategy can therefore take the departures into consideration. It is crucial to understand that the following patterns of risk perception are "irrational" only if one assumes that it is somehow rational to ignore equity, uncertainty, locus of control and the various other factors that affect, not "distort," our sense of which risks are acceptable and which are not. Rational or not, virtually everyone considers getting mugged a more outrageous risk than skidding into a tree on an icy highway. And virtually everyone is more frightened by a hazardous waste facility than by a gasoline storage tank. Our task is not to approve or disapprove of these truths, but to understand why they are true and how siting communication can adapt to them.

The points in the following section deal with why communities fear hazardous waste facilities more than technical experts judge that they "should," and how communication can be used to reduce the discrepancy. It might be possible to employ this counsel to the exclusion of all else in this article, hoping to pacify community fears without acknowledging, much less honoring, community power. Such an effort would, I think, fail abysmally. Communications strategy must be part of fair dealing with the community, not a substitute for it.

## Patterns of Risk Perception

*1. Unfamiliar risks are less acceptable than familiar risks.* The most underestimated risks are those, such as household accidents, that people have faced for long periods without experiencing the undesired event. The sense of risk diminishes as we continue to evade it successfully. Thus, the perceived riskiness of a hazardous waste facility is, in part, a reflection of its unfamiliarity. Stressing its similarity to more familiar industrial facilities can diminish the fear; so can films, tours and other approaches aimed at making the facility seem less alien. Even more important is to make the wastes to be treated seem less alien. Detailed information on the expected waste stream—what it is, where it comes from and what it was used to make—should reduce the fear level considerably.

*2. Involuntary risks are less acceptable than voluntary risks.* As mentioned earlier, some studies show acceptance of voluntary risks at one thousand times the level for involuntary risks.[12] Eminent domain, preemption and the community's general feeling of outside coercion thus exacerbate the level of fear. Acknowledging the community's power over the siting decision will lessen the fear and make siting a more acceptable outcome.

*3. Risks controlled by others are less acceptable than risks under one's own control.* People want to know that they have control over not only the initial decision but also the entire risky experience. To some extent this is not possible. Once a facility is built it is difficult to turn back. But credible assurances of local control over monitoring and regulation can be expected to reduce risk perception by increasing control. Similarly, trust funds, insurance policies, bonds and such contractual arrangements can put more control in local hands. Quite apart from any other advantages, these arrangements will tend to diminish the perception of risk.

*4. Undetectable risks are less acceptable than detectable risks.* A large part of the dread of carcinogenicity is its undetectability during its latency period. As a veteran war correspondent told me at Three Mile Island, "In a war you worry that you might get hit. The hellish thing here is worrying that you already got hit." While it is not possible to do much about the fear of cancer, it *is* possible to make manifest the proper, or improper, operation of the facility. For instance, a local monitoring team, or a satellite monitoring station in the City Hall lobby, can make malfunctions more detectable, and can thereby reduce the level of fear during normal operations. Not coincidentally, these innovations will also improve the operations of the facility.

*5. Risks perceived as unfair are less acceptable than risks perceived as fair.* A substantial share of the fear of hazardous waste facilities is attributable to the fact that only a few are to be sited. A policy requiring each municipality to manage its own hazardous waste would meet with much less resistance. A more practical way of achieving equity is to negotiate appropriate benefits to compensate a community for its risks and costs (this is, of course, after all appropriate health and safety measures have been agreed to). In a theoretical free market, the negotiated "price" of hosting a facility would ensure a fair transaction. The point to stress here is that compensation does not merely offset the risk faced by a community. It actually *reduces* the perceived risk and the level of fear.

*6. Risks that do not permit individual protective action are less acceptable than risks that do.* Even for a very low-probability risk, people prefer to know that there are things they can do, as individuals, to reduce the risk still further. The proposed protective action may not be cost-effective, and the individual may never carry it out, but its availability makes the risk more acceptable. Discussion of hazardous waste facility siting has appropriately focused on measures to protect the entire community. Some attention to individual protective measures may help reduce fear.

*7. Dramatic and memorable risks are less acceptable than uninteresting and forgettable ones.* This is generally known as the "availability heuristic": people judge an event as more likely or frequent if it is easy to imagine or recall.[13] The legacy of Love Canal, Kin-Buc [a large, abandoned landfill in New Jersey that is now a Superfund site], Chemical Control and the like has made hazardous waste dangers all too easy to imagine and recall. A corollary of the availability heuristic is that risks that receive extensive media treatment are likely to be overestimated, while those that the media fail to popularize are underestimated. The complex debate over media handling of hazardous waste goes beyond the scope of this article.

*8. Uncertain risks are less acceptable than certain risks.* Most people loathe uncertainty. While probabilistic statements are bad enough, zones of uncertainty surrounding the probabilities are worse. Disagreements among experts about the probabilities are worst of all.

Basing important personal decisions on uncertain information arouses anxiety. In response, people try either to inflate the risk to the point where it is clearly unacceptable or to deflate it to the point where it can be safely forgotten. Unfortunately, the only honest answer to the question "Is it safe?" will sound evasive. Nonetheless, the temptation, and the pressure, to offer a simple "yes" must be resisted. Where fear and distrust coexist, as they do in hazardous waste facility siting, reassuring statements are typically seen as facile and self-serving. Better to acknowledge that the risk is genuine and its extent uncertain.

*9. Cross-hazard comparisons are seldom acceptable.* It is reasonable and useful to compare the risks

of a modern facility to those of a haphazard chemical dump such as Love Canal. The community needs to understand the differences. It is also reasonable and useful to compare the risks of siting a facility with the risks of not siting a facility—midnight dumping and abandoned sites. This comparison lies at the heart of the siting decision. On the other hand, to compare the riskiness of a hazardous waste facility with that of a gas station or a cross-country flight is to ignore the distinctions of the past several pages. Such a comparison is likely to provoke more outrage than enlightenment.

*10. People are less interested in risk estimation than in risk reduction, and they are not interested in either one until their fear has been legitimized.* Adversaries who will never agree on their diagnosis of a problem can often agree readily on how to cope with it. In the case of facility siting, discussions of how to reduce the risk are ultimately more relevant, more productive and more satisfying than debates over its magnitude. Risk reduction, however, is not the only top priority for a fearful community. There is also a need to express the fear and to have it accepted as legitimate. No matter how responsive the Commission is to the issue of risk it will be seen as cold and callous unless it also responds to the *emotional* reality of community fear.

## Do not ignore issues other than health and safety risk.

The paramount issue in hazardous waste facility siting is undoubtedly the risk to health, safety and environmental quality. But this is not the only issue. It is often difficult to distinguish the other issues so they can be addressed directly—especially if legal and political skirmishes have thrust the risk issue to the fore.

Negotiated compensation is especially useful in dealing with these other issues. Moreover, negotiation helps to distinguish them from the risk issue. It is not uncommon, for example, for a community group to insist in adversary proceedings on marginal protective measures at substantial expense. In negotiations where other issues can more easily be raised, the group may reveal that it is also worried about the possible fears of prospective home purchasers and the resulting effect on property values. The developer may find it easy to bond against *this* risk. The homeowners have thus protected their property at a cost that the developer, who plans to establish an excellent safety record, expects will be low. It is extremely useful, in short, to probe for concerns other than risk, and to establish a context, such as mediated negotiation, where such concerns can be raised.

Aside from health risk, the impacts of greatest concern are: (1) the decline in property values; (2) the inability of the community to keep out other undesirable land uses once one has been sited; (3) the decline in quality of life because of noise, truck traffic, odor and the like; (4) the decline in the image of the community; (5) the overburdening of community services and community budgets; and (6) the aesthetically objectionable quality of the facility.

Apart from these possible impacts, a number of non-impact issues may create adverse community reaction to a proposed facility:

1. Resentment of outside control, including the threat of preemption and eminent domain.

2. The sense of not being taken seriously; resistance to one-way communication from planners and experts who seem to want to "educate" the community but not to hear it; perceptions of arrogance or contempt.

3. The conviction that the siting process is unfair, that "the fix is in."

4. The conviction that the choice of this particular community is unfair, that the community is being asked to pay a high price for the benefit of people who live elsewhere, and that it would be fairer to ask someone else to pay that price. This feeling is especially strong in communities that are poor, polluted or largely minority. These communities see their selection as part of a pattern of victimization.

5. Support for source reduction and recycling instead of new facilities.

Another issue that often surfaces is whether the facility will accept non-local waste. In a recent Duke University poll of North Carolina residents, only seven percent approved of allowing out-of-state waste to be disposed of in their county.[14] By contrast, thirty-eight percent would allow waste from other North Carolina counties and forty-nine percent would allow waste from within the county.[15] Tech-nically, it may well be impractical to require each community to cope with its own waste. Psychologically, however, this is far more appealing than central facilities, for at least three reasons: (1) It seems intrinsically fairer to have to dispose of one's own waste than to be forced to dispose of everyone else's; (2) A strictly local facility will not earn a community an image as the hazardous waste capital of the state or region; and (3) Local wastes already exist, either stored on-site or improperly dumped, and a new local facility thus represents no net increase in local risk. Enforceable guarantees to limit "imported" waste should alleviate in part at least one source of opposition to a facility.

## Make all planning provisional, so that consultation with the community is required.

A fatal flaw in most governmental public participation is that it is grafted onto a planning procedure that is essentially complete without public input. Citizens quickly sense that public hearings lack real provisionalism or tentativeness. They often feel that the important decisions have already been made, and that while minor modifications may be possible to placate opponents, the real functions of the hearing are to fulfill a legal mandate and to legitimize the *fait accompli*. Not surprisingly, citizen opponents meet what seems to be the charade of consultation with a charade of their own, aiming their remarks not at the planners but at the media and the coming court battle.

This scenario is likely even when the agency sees itself as genuinely open to citizen input. For legal and professional reasons, experts feel a powerful need to do their homework *before* scheduling much public participation. In effect, the resulting presentation says to the citizen: "After monumental effort, summarized in this 300-page document, we have reached the following conclusions. . . . Now what do you folks think?" At this point it is hard enough for the agency to take the input seriously, and harder still for the public to believe it will be taken seriously. Thus, Siting Commission Chairman Frank J. Dodd complained that the siting hearings "have turned into political rallies. The last thing that was discussed was siting criteria. It was how many people can you get into an auditorium to boo the speakers you don't like and cheer for the ones you support."[16]

The solution is obvious, though difficult to implement. Consultations with the community must begin early in the process and must continue throughout. Public participation should not be confined to formal contexts like public hearings, which encourage posturing. Rather, participation should include informal briefings and exchanges of opinion of various sorts, mediated where appropriate. The Commission must be visibly free to adjust in response to these consultations, and must appear visibly interested in doing so. Above all, the proposals presented for consultation must be provisional rather than final—and this too must be visible. A list of options or alternatives is far better than a "draft" decision. "Which shall we do?" is a much better question than "How about this?"

This sort of genuine public participation is the moral right of the citizenry. It is also likely to yield real improvements in the safety and quality of the facilities that are built. As a practical matter, moreover, public participation that is not mere window-dressing is probably a prerequisite to any community's decision to forego its veto and accept a facility. This is true in part because the changes instituted as a result of public participation make the facility objectively more acceptable to the community. Public participation has important subjective advantages as well. Research dating back to World War II has shown that people are most likely to accept undesirable innovations, such as rationing, when they have participated in the decision.[17]

Much in the Siting Act and in the behavior of the Commission represents important progress away from the traditional "decide-announce-defend" sequence, whereby an agency ends up justifying to the public a decision it has already made. Holding hearings on siting criteria instead of waiting for a site was progress.[18] The money available for community research is progress.[19] There is also progress evidenced in a recent statement by Commission Executive Director Richard J. Gimello that hearings have persuaded him that two incinerators would be wiser than the one originally proposed in the draft hazardous waste management plan.[20] However, there is a long history of "decide-announce-defend" to be overcome before we achieve what communication theorists call "two-way symmetric communication" and politicians call "a piece of the action."

## Involve the community in direct negotiations to meet its concerns.

The distinction between community input and community control is a scale, not a dichotomy. Planning expert Sherry Arnstein describes an eight-rung "ladder of public participation," as follows: manipulation; therapy; informing; consultation; placation; partnership; delegated power; citizen control.[21] She adds:

> Inviting citizen's opinions, like informing them, can be a legitimate step toward their full participation. But if consulting them is not combined with other modes of participation, this rung of the ladder is still a sham since it offers no assurance that citizen concerns and ideas will be taken into account.[22]

A really meaningful participation program, Arnstein argues, involves some framework for explicit power-sharing with the community.[23]

In hazardous waste facility siting, today's community has two kinds of power: (1) the legally guaranteed right to provide input at many stages of the siting process; and (2) the political ability to delay, harass and quite possibly stop that process. The first, as Arnstein points out, is not enough to reassure a community that feels little trust for those at whom the input is directed.[24] That leaves the other source of power, the *de facto* veto.

This sort of analysis has led many observers to propose siting legislation that accords greater power to the community. Indeed, one state, California, makes siting virtually contingent on community acceptance.[25] Others, such as Massachusetts and Connecticut, do not go so far as to provide a *de jure* community veto, but do require the community to negotiate with the developer, with binding arbitration in the event of deadlock.[26] Still other states permit local regulation of the facility, but grant to a state agency the authority to override community regulations that make siting impossible.[27] As Morell and Magorian note, "expanded public participation procedures in a preemptive siting process are a far cry from such a balance of state and local authority."[28]

While New Jersey's Siting Act does not require negotiations with the community, it certainly does not foreclose the option—an option far more useful to the community than mere input, and far more conducive to siting than the *de facto* veto. The most productive option is probably negotiation between the developer and the community, with or without a mediator. If they are able to come to terms, the Commission could incorporate these terms in its own deliberations while still retaining its independent responsibility to protect health and environmental quality. If they are *un*able to come to terms, the Commission could retain its preemptive capabilities and the community its political ones. For the community, then, the incentive to negotiate is the likelihood that it can secure better terms from the developer than it can get from the Commission in the event of deadlock. For the developer, the incentive is the considerable possibility that there will be no facility at all unless the community withdraws its objections.

What is negotiated? What the community has to offer is of course its acceptance of the facility. What the developer has to offer is some package of mitigation (measures that make an undesirable outcome less likely or less harmful), compensation (measures that recompense the community for undesirable outcomes that cannot be prevented) and incentives (measures that reward the community for accepting the facility). The terms are value judgments. For example, a developer is likely to see as an incentive what the community sees as mere compensation. The distinctions among the three nonetheless have great psychological importance. Communities tend to see mitigation as their right. Compensation for economic costs is seen as similarly appropriate, but compensation for health risks strikes many people as unethical. Incentive offers, especially where health is the principal issue, may strike the community as a bribe.

Of course some forms of mitigation, compensation, and incentives are built into the Siting Act; among the most notable provisions are the five percent gross receipts tax[29] and the provision for strict liability,[30] which permits compensation for damage without proof of negligence. Clearly a still more attractive package is needed to win community support. What can help the parties in negotiating the package? I suggest training in negotiation for community representatives. An impartial mediator might also be provided, perhaps from the Center for Dispute Resolution of the Public Advocate's Office. Finally, a clear statement from the Siting Commis-

sion on how it will deal with a settlement if one is achieved would be useful.

Much will depend, of course, on the delicacy and skill of the developer. Compensation, in particular, should be tied as closely as possible to the damage to be compensated. A straight cash offer may be hotly rejected, whereas a trust fund to protect water quality would be entirely acceptable. Similarly, cash for damage to health is much less acceptable than cash for damage to community image. Where possible, compensation and incentive proposals should come from the community or mediator to avoid any suggestion of bribery. Some risks, of course, are so terrible that they are, and should be, unacceptable regardless of the compensation. No negotiation is possible unless the community agrees that a hazardous waste facility does not pose an unacceptable risk.

A great advantage of negotiation is that it encourages an openness about goals and concerns that is inconceivable in an adjudicatory process. Citizens concerned about property values may find themselves in a hearing talking instead about safety—but in a negotiation they will talk about property values. Similarly, a developer in an adjudicatory proceeding tends to understate risk. In a negotiation the community will insist that if the risk is so low the developer should have no objection to bonding against it. Suddenly both the developer and community will have an incentive to estimate the risk accurately. This pressure to be open affects not only the compensation package but the actual facility design as well. If developers must contract to compensate those they injure, they will be more likely to take the possibility of injuries into account in their planning than if they are merely instructed to "consider" social costs.

## Establish an open information policy, but accept community needs for independent information.

Former EPA Administrator William D. Ruckelshaus was fond of quoting Thomas Jefferson: "If we think [the people are] not enlightened enough to exercise their control with a wholesome discretion, the remedy is not to take it from them, but to inform their discretion." Ruckelshaus usually added, "Easy for him to say."

Part of the problem of informing the public about

hazardous waste facility siting is that the skills required to explain technical information to the lay public are uncommon skills. They are especially uncommon, perhaps, among those who possess the requisite technical knowledge. There are techniques to be learned: a standard called "communicative accuracy" to help determine which details may be omitted and which may not; various sorts of "fog indexes" to measure readability and comprehensibility; and other ways of simplifying, clarifying and dramatizing without distorting. The range of media available for the task also extends well beyond such standbys as pamphlets and formal reports.

The desire to explain technical issues in popular language is at least as difficult to acquire as the ability to do so. Experts in all fields prefer to confine their expertise to fellow professionals; "if laypeople misunderstand me I will have done them a disservice, and if they understand me what will have become of my expertise." All fields ostracize their popularizers. When the information is uncertain, tainted with values, and potent ammunition in a public controversy, the case for professional reticence becomes powerful indeed.

Nonetheless, it is essential to the success of the siting effort that information policy be as open as humanly possible. Unless legally proscribed, *all* information that is available to the Commission should be available to the community. The Commission should also make available simplified summaries of key documents and experts to answer whatever questions may arise. It is particularly important that all risk information be available early in the siting process. Failure to disclose a relevant fact can poison the entire process once the information has wormed its way out—as it invariably does. The standard is quite simple: any information that would be embarrassing if disclosed later should be disclosed now.

Even the most open information program, however, can expect only partial success. Individuals who are uninvolved in the siting controversy will not often bother to master the information, since there is nothing they plan to do with it. Individuals who are heavily involved, on the other hand, generally know what side they are on, and read only for ammunition. This is entirely rational. If changing one's mind is neither attractive nor likely, why endure the anxiety of listening to discrepant information? When many alternatives are under consideration, as in a negotiation, information has real value and helps the parties map the road to a settlement. When the only options are victory and defeat, objective information processing is rare.

Even in a negotiation, information carries only the limited credibility of the organization that provides it. As a rule, the parties prefer to provide their own. The Siting Commission would be wise to facilitate this preference. Rather than insisting that *its* information is "objective" and berating the community for distrusting it, the Commission can guarantee that all parties have the resources to generate their own information. The information should be generated as early as possible, while positions are fluid. Finally, the Commission should make sure the community has a real opportunity to use the information it acquires—ideally in negotiation. Information without power leads only to frustration, while the power to decide leads to information-seeking and a well-informed community.

## Consider developing new communication methods.

There are a wide variety of all-purpose methodologies for developing means to facilitate interaction, communication, trust and agreement. Some are a bit trendy or "touchy-feely"; some are potentially explosive—all require careful assessment and, if appropriate at all, careful design and implementation in the hands of a skilled practitioner. The list that follows is by no means exhaustive. These are tools that are available to the Siting Commission, to a developer, to a community group, or to anyone interested in making negotiation more likely or more successful.

*1. Delphi methodology.* This is a formal technique for encouraging consensus through successive rounds of position-taking. It is appropriate only where the grounds for consensus are clear—for helping the community clarify its concerns, for example, but not for helping it reach agreement with the developer.

*2. Role-playing.* Playing out the stereotyped roles of participants in a controversy can help all sides achieve better understanding of the issues. Under some circumstances this can greatly reduce the level of tension. There are many variations. Most useful for facility siting would probably be exaggerated role-playing, in which participants burlesque their own positions. This tends to produce more moderate posturing in real interactions. Counter-attitudinal role-playing, in which participants take on each other's roles, tends to yield increased appreciation of the multi-sidedness of the issue. Both require some trust, but much can be learned even from role-playing without the "enemy" present.

*3. Gaming-simulation.* This is a variation on role-playing, in which the participants interact not just with each other but with a complex simulation of the situation they confront. Game rules control how the participants may behave and determine the results—wins, losses, or standoffs. Participants learn which behaviors are effective and which are self-defeating. As with any role-playing, the participants may play themselves or each other, and may undergo the game in homogeneous or heterogeneous groups. Massachusetts Institute of Technology has recently developed a hazardous waste facility siting gaming-simulation.

*4. Coorientation.* This is a tool to help participants come to grips with their misunderstanding of each other's positions. A series of questions is presented to all participants, individually or in groups. First they answer for themselves, then participants predict the answers of the other participants (those representing conflicting interests). Responses are then shared, so that each side learns: (a) its opponent's position; (b) the accuracy of its perception of its opponent's position; and (c) the accuracy of its opponent's perception of its position. The method assumes that positions taken will be sincere, but not that they are binding commitments.

*5. Efficacy-building.* This is a collection of techniques designed to increase a group's sense of its own power. In some cases this includes skills-training to increase the power itself. In other cases, the stress is on increasing group morale, cohesiveness, and self-esteem. To the extent that community intransigence may be due to low feelings of efficacy, then efficacy-building procedures should lead to increased flexibility.

*6. Focus groups.* A focus group is a handful of individuals selected as typical of a particular constituency. This focus group is then asked to participate in a guided discussion of a predetermined set of topics. Often the focus group is asked to respond to particular ideas or proposals, but always in interaction with each other, not in isolation as individuals. The purpose of the focus group methodology is to learn more about the values of the constituency and how it is likely to respond to certain messages—for example, a particular compensation package in a siting negotiation. Focus groups do not commit their constituency, of course, but in the hands of a skilled interviewer and interpreter they yield far better information than survey questionnaires.

*7. Fact-finding, mediation, and arbitration.* These are all third-party interventions in conflict situations. Fact-finding concentrates on helping the parties reach agreement on any facts in contention. Mediation helps the parties find a compromise. Arbitration finds a compromise for them. These approaches assume that the parties want to compromise, that each prefers agreement to deadlock or litigation. They have been used successfully in many environmental conflicts, including solid waste siting controversies. The Center for Dispute Resolution of the Public Advocate's Office offers these services, as do several specialized environmental mediation organizations.

*8. Participatory planning.* This is the label sometimes given to a collection of techniques for making public participation more useful to the decision-maker and more satisfying to the public. To a large extent the value of public participation is in the agency's hands. It depends on how early in the process participation is scheduled, now flexible agency planners are, and how much real power is given to the community. Even if these questions are resolved in ways that make participation more than mere window-dressing, the success of the enterprise still depends on technique: on how people are invited, on how the policy questions are phrased, on what speakers are allowed to talk about, what issues for how long, on who moderates the meeting, etc. Many techniques of participatory planning, in fact, do not involve a meeting at all.

*9. Feeling acceptance.* A classic misunderstanding between communities and agencies centers on their differing approaches to feeling; citizens may sometimes exaggerate their emotions while bureaucrats tend to stifle theirs. Not surprisingly, "irrational" and "uncaring" are the impressions that result. Feeling acceptance is a technique for interacting with people who feel strongly about the topic at hand. It involves identifying and acknowledging the feeling, then separating it from the issue that aroused it, and only then addressing the issue itself.

*10. School intervention.* In situations where strong feelings seem to be interfering with thoughtful consideration, it is sometimes useful to introduce the topic into the schools. Primary school pupils, in particular, are likely to approach the issue less burdened by emotion, yet they can be relied upon to carry what they are learning home to their parents. It is essential, of course, to make sure any school intervention incorporates the views—and the involvement—of all viewpoints in the community. Any effort to teach children a single "objective" agency viewpoint will bring angry charges of indoctrination. Existing curricula that are themselves multi-sided can augment the local speakers.

*11. Behavioral commitment.* People do not evolve new attitudes overnight; rather, change comes in incremental steps. The most important steps are not attitudes at all, but behaviors, preferably performed publicly so as to constitute an informal commitment. The behavioral commitment methodology, sometimes known as the "foot in the door," asks people to take small actions that will symbolize, to themselves and their associates, movement in the desired direction. Among the possible actions which can be taken: to request a booklet with more information, to urge rational discussion on the issue, to state that one is keeping an open mind, to agree to consider the final report when it is complete, to agree to serve on an advisory committee, to meet with citizens concerned about Superfund cleanup, etc.

*12. Environmental advocacy.* In a large proportion of successfully resolved siting controversies in recent years, respected environmentalists played a crucial intermediary role. Environmental organizations may need to play that role in New Jersey's hazardous waste facility siting. By counseling caution on industry assurances while agreeing that new facilities are needed and much improved, environmentalists position themselves in the credible middle.

A credible middle is badly needed on this issue, but it will take time. Now is not the time to ask *any* New Jersey community to accept a hazardous waste facility. From "no" to "yes" is far too great a jump. We should ask the community only to consider its options, to explore the possibility of a compromise. Our goal should be moderate, fair, and achievable: getting to maybe.

## Notes

[1–2. *Editors' note:* Footnotes 1 and 2 appear in the Foreword, not reprinted here.]

3. *See Superfund Strategy* (Apr. 1985) (Office of Technology Assessment).

4. BLACK'S LAW DICTIONARY (5th ed. 1979) defines "de facto" as a "phrase used to characterize a state of affairs which must be accepted for all practical purposes but is illegal or illegitimate."

5. N.J. STAT. ANN. § 13:1E–81 (West Supp. 1985) ("Eminent domain").

6. D. MORELL & C. MAGORIAN (1982).

7. Carney, *D.E.P.: The Record and the Problems*, N. Y. Times, Jan. 27, 1985, § 11 at 6.

8. BLACK'S LAW DICTIONARY (5th ed. 1979) defines "de jure" as "descriptive of a condition in which there has been total compliance with all requirements of the law." Here the term refers to the actual legal authority of the state to site a facility over the objection of a municipality, whether or not that approach will ever be taken.

9. Starr, *Social Benefit Versus Technological Risk*, 165 SCIENCE 1232–38 (1969).

10. N. FREUDENBERG (1984).

11. N.J. STAT. ANN. § 13:1E–59.d. (West Supp. 1985); *see also* N.J. STAT. ANN. § 13:1E–60.c.(4)(West Supp. 1985).

12. *See* Starr *supra* note 9.

13. Slovic, Fischoff, Layman & Coombs, *Judged Frequency of Lethal Events*, 4 JOURNAL OF EXPERIMENTAL PSYCHOLOGY: HUMAN LEARNING AND MEMORY 551–578 (1978).

14. D. MORELL & C. MAGORIAN, SITING HAZARDOUS WASTE FACILITIES: LOCAL OPPOSITION AND THE MYTH OF PREEMPTION, at 74 (1982).

15. *Id.*

16. Goldensohn, *Opponents, Officials Charge Politicizing of Waste Site Debate*, Star-Ledger (Newark, NJ), Dec. 2, 1984, at 12.

17. M. KARLINS & H. ABELSON, PERSUASION, at 62–67 (2d ed. 1970).

18. *See* Dodd, *The New Jersey Hazardous Waste Facilities Siting Process: Keeping the Debate Open* in this issue [*Seton Hall Legislative Journal*, vol. 9 (1985)].

19. *See supra* note 11.

20. *See Response to Comments on "Draft" Hazardous Waste Facilities Plan Issued September 1984* (Mar. 26, 1985) (copies available from the Siting Commission, CN-406, Trenton, NJ 08625).

21. S. ARNSTEIN, *A Ladder of Citizen Participation,* in THE POLITICS OF TECHNOLOGY, at 240–43 (1977).

22. *Id.*

23. *Id.*

24. *Id.*

25. *See* Duffy, 11 B.C. ENV. AFFAIRS L. REV. 755, 755–804 (1984).

26. *Id.*

27. *Id.*

28. D. MORELL & C. MAGORIAN, *supra* note 14, at 102.

29. N.J. STAT. ANN. § 13:1E–80.b. (West Supp. 1985).

30. N.J. STAT. ANN. § 13:1E–62 (West Supp. 1985) ("Joint and several strict liability of owners and operators").

# Questions for Thought and Discussion

1. Sandman suggests that localized hazardous waste facilities may be less practical than centralized ones but that they are more appealing psychologically. Which approach do you favor? Offer three reasons other than those given by Sandman in support of your preference.

2. The author cites a number of problems revolving around the use of experts by each side in hazardous waste siting disputes and refers to the provision in New Jersey law to fund the hiring of experts by the community. Discuss the pros and cons of this approach as opposed to one in which the developer, the government, and the community would all be required to approve in advance the selection of one team of experts to study the siting proposal.

3. It is stated in this paper that "people are less interested in risk estimation than in risk reduction." Explain what this implies about the importance of risk assessment. How does the applicability of the statement depend on the magnitude of the risk?

4. Summarize the ten most important principles of risk communication, according to Sandman, and explain how they would apply to the issue of doubling the capacity of a chemical manufacturing plant located in a small city.

# Informed Choice or Regulated Risk? Lessons from a Study in Radon Risk Communication

### F. Reed Johnson
*U.S. Environmental Protection Agency, Washington, D.C., and U.S. Naval Academy, Annapolis, Maryland*

### Ann Fisher
*U.S. Environmental Protection Agency Washington, D.C.*

### V. Kerry Smith
*North Carolina State University Raleigh, North Carolina*

### William H. Desvousges
*Research Triangle Institute Research Triangle Park, North Carolina*

The benefits of improving the public's access to information seem obvious. Better information should lead to better decisions for increasing one's welfare. Unfortunately, this simplistic view does not account for the way that people actually respond to "bad news." For example, psychologists hold that people often reject information that conflicts with their preconceptions: smokers sometimes refuse to read accounts about the health risks of smoking, and new car owners may reject negative reports about their recent purchase. On the other hand, people can be alarmed unnecessarily in the face of unfounded scare stories.

The authority or trustworthiness of the information source often affects the perceived credibility of the message. Psychologists have demonstrated that the context surrounding new information influences the way it is transferred and used. People apparently employ a variety of filters to transform and interpret information in light of certain preconceptions, attitudes, and information-processing abilities. In short, people use a complex process to convert information for use in making decisions.

Government regulatory agencies have expressed increasing interest in providing information to the public as an alternative to the often difficult or costly traditional approaches to reducing risks. The basis of such programs is the rather naive assumption that

"Informed Choice or Regulated Risk? Lessons from a Study in Radon Risk Communication" by F. Reed Johnson, Ann Fisher, V. Kerry Smith, and William H. Desvousges is reprinted from *Environment,* vol. 30, no. 4, pp. 12–15, 30–35 (May 1988) with permission of the Helen Dwight Reid Educational Foundation. Published by Heldref Publications, Washington, D.C. Copyright © 1988.

information programs will motivate people voluntarily and rationally to reduce risks. Unfortunately, the evidence on the effectiveness of risk information programs has been discouraging.[1]

In principle information programs avoid some of the ethical problems recently raised by Douglas MacLean.[2] He suggests that various types of risk taking involve a spectrum of consent. At one end of the spectrum (and typical of individual decisions), consent is actual and explicit, as in the case of deciding not to buy a smoke detector. At the other end of the spectrum (and typical of centralized decisions), consent is implicit or hypothetical, as in the case of the public's reliance on the U.S. Food and Drug Administration to decide which medicines are safe and effective. Most people would feel that it is ethically more acceptable to be at the end of the spectrum where consent is explicit. It could be concluded from MacLean's arguments that the introduction of an information program would make the consent more explicit by allowing people to choose for themselves. We argue that this conclusion is unwarranted: ethical issues cannot be avoided if the content, format, and tone of informative messages affect how people understand and use the information.

This article reports some preliminary results of a recent social experiment designed to test the sensitivity of people's responses to alternative presentations of the same facts about radon risks. Ethical issues enter the experiment in two ways. First, it was found that the way risk information is presented does matter and therefore involves ethical judgments. Second, ethical issues arose in the design of the social experiment itself. Discussed here are ways in which these two concerns limited the acceptable range and character of the experiment.

## Communicating Radon Risks

The U.S. Environmental Protection Agency (EPA) estimates that radon causes more cancer deaths per year—5,000 to 20,000—than any other pollutant under its jurisdiction.[3] Radon is a colorless, odorless gas that occurs naturally. It moves through the soil and becomes trapped in buildings. Since exposure occurs primarily in people's homes, conventional regulatory approaches are not appropriate. This situation has led EPA to turn to risk communication as a way of encouraging voluntary reductions in risk.

EPA has initiated several studies to investigate how people understand and react to new information on indoor radon risks. A general objective of these studies is to determine which approaches are most effective in communicating risk information. Radon provides a good opportunity for evaluating risk communication approaches for several reasons:

- The risk is relatively unfamiliar. Significant media coverage of radon, unlike that of smoking or seat belts, began only a few years ago. In many areas where radon is suspected to be a problem, recent surveys and other evidence indicate that people know little or nothing about the origins of the problem, the health risk, or how to reduce their exposure to it. This means a risk communication program is likely to be the main source of information for the public.

- There is usually no villain to blame. Perceptions that individuals' rights to a safe, clean environment have been violated often complicate reactions to other environmental hazards. Affected individuals tend to focus on determining who caused the problem and, therefore, who should be responsible for its solution.

- Although radon cannot be seen or smelled, testing for radon is simple and inexpensive. On the other hand, data on the cost and effectiveness of alternative mitigation techniques are limited. In some cases mitigation is complicated and expensive. This means home owners may be confronted with complex decisions that require considerable information.

- Combined with data on actual exposures and survey data on risk perceptions, home owners' mitigation choices can provide an objective measure of the impacts of a communication effort.

The prospect of testing an actual radon communication program provided an excellent opportunity for social science research, even though it also imposed ethical constraints on the research design. Because radon poses significant health risks, it was inappropriate to provide significantly more information to some people than to others. Nor was it appropriate to provide information in forms that were suspected a priori of being difficult for people to understand and use or that would alarm people unnecessarily. Although there were good research reasons to allow several months to elapse before informing home owners of their interim test results, that information was provided as quickly as possible. Despite deliberate efforts not to harm any participants, certain preliminary results have required changes in the research design because of ethical concerns.

EPA's interest in testing a radon communication program evolved from a survey of about 200 home owners in Maine. The respondents had participated in a lung cancer epidemiology study by the Maine Medical Center and received radon test results for their homes along with a pamphlet about radon. It was demonstrated that this new information did affect respondents' perceptions of the seriousness of the problem and that perceptions changed in the desired directions.[4] However, most respondents still greatly understated their risk compared with objective estimates based on their radon test results.

Despite this misunderstanding, half of the sample had undertaken some radon mitigation; about the same share of those at low and high risk had mitigated. Many home owners used simple, low-cost measures such as opening windows more often and avoiding basement areas. Even so, the results appear to suggest that the information program led some people to reduce small risks and others to do nothing about large risks.

This result is troublesome to environmental policymakers who are concerned with protecting and improving public health. However, it does not necessarily prove that people are behaving irrationally. In general people do not maximize their health to the exclusion of other considerations. For example, smokers often cite their gain in personal enjoyment in justifying their actions. The same holds true for people who eat high-cholesterol foods or ride motorcycles or do not buckle seat belts.

If people are fully informed about potential health consequences, their decisions presumably reflect a conscious assessment that the benefits of risky behavior exceed the costs. Indeed, some economists have suggested that consumers may respond to mandated safety requirements by accepting new risks in other areas. As a consequence, the overall level of public safety may not improve.[5]

If the goal of public policy is to maximize individual and social welfare, then people should be able to trade their personal health for more desired gains. This view implies that the appropriate measure of an information program's success is not whether everyone above some arbitrary exposure level takes action to reduce his or her exposure, but whether people understand the consequences of taking or not taking action. The consequences of living in a house with a certain radon concentration might be very different for a family with young children than for a single elderly person.

The Maine study was important because it showed that home owners respond systematically to risk information. However, the study did not provide sufficient data on how much respondents learned and understood about radon risks or what factors influenced home owners' responses.

## The New York Study

The New York State Energy and Research Development Authority (NYSERDA) sampled single-family homes to determine statewide exposures to radon. NYSERDA placed three radon monitors in each of about 2,300 homes. The first of these (placed in the living area) was to be sent back for analysis after two to three months; the other monitors were to be returned and analyzed after a year. This protocol would enable NYSERDA to judge whether the two-to-three-month readings were acceptable approximations of annual averages. The homes were selected randomly within seven areas representing major geological formations across the state.

When home owners agreed to participate in the study, they were promised the radon readings for their own homes, but it was not clear what information they would receive to interpret those readings. State officials were concerned about motivating households to take appropriate remedial actions without creating undue anxiety. This situation provided an opportunity to evaluate the effectiveness of alternative designs for the information materials used in communicating the risks from radon.

EPA provided the resources to inform the 2,300 home owners about the potential health risks and ways of reducing those risks. Several alternative information "treatments" were designed to test the effectiveness of different formats. Telephone surveys provided data on changes in knowledge, perceptions, and intentions. The surveys were designed to answer three questions:

- How much did people learn about radon and its associated risks?
- To what extent are perceived risks consistent with objective measures of risk?
- How much more mitigation is undertaken by those at higher risk (controlling for other factors that might influence the perceived benefits and costs of mitigation)?

The original plan was to send the radon readings and interpretive materials to home owners after the one-year monitors were returned. NYSERDA staff and the research team became concerned about the implications of not informing participants immediately after the two-to-three-month reading became available. Some homes had radon levels at which EPA recommends taking remedial action "within a few months." These public health concerns changed the research design so that people received the interim readings and interpretive materials as quickly as possible. Another set of information accompanied the annual readings that were distributed in fall 1987. Having to deal with two separate information distributions instead of one greatly complicated the research design but was one of the areas in which ethical considerations affected the social experiment.

The results reported here are based on home owners' reactions to the interim readings and information materials. Home owners were discouraged from taking action until the annual results were available, so the third question of who mitigated and why cannot be addressed. Although the study is still under way—final data on mitigation will be collected in fall 1988—the preliminary results have some important implications for risk communication.

## The Base-Line Surveys

Before sending the radon readings and interpretive information, it was necessary to measure people's base-line knowledge about radon and their attitudes and perceptions about risks. The 2,300 NYSERDA home owners were interviewed during summer 1986; 97 percent of them (referred to hereafter as the monitored group) responded. Because their partici-

pation in the monitoring study might have changed their awareness of radon and its risks, similar data were collected from approximately 250 home owners who were not part of the NYSERDA monitoring study but were randomly selected in the same way as those in the study (referred to hereafter as the comparison group).

Despite widespread media coverage of radon, the base-line surveys indicated that only half of the monitored group and only a quarter of the comparison group remembered reading or hearing about the problem in the previous three months. Those in the monitored group knew little about radon (answering half of a set of simple quiz questions correctly), but they still knew more than those not having their homes monitored (who answered a third of the questions correctly). Quiz scores were higher for those who had heard about radon recently through the media and lower for older people. Other household characteristics had much smaller effects on the number of correct answers.

Respondents in the monitored group were somewhat more likely to agree that radon is a serious concern in their neighborhood, town, or city, but many in both groups said they just did not know how serious a risk radon might be to themselves or to the population in general. Even so, respondents perceived their own risk from radon to be only slightly less than that from auto accidents, home accidents, and hazardous wastes.

The effectiveness of a program designed to induce households to take action to reduce their risk is very likely to depend partly on whether they are willing to assume responsibility for the risk. If people believe something else is imposing the risk on them, they will be less likely to reduce the risk themselves. The base-line survey indicated that less than 5 percent of both the monitored group and the comparison group thought the government is responsible for radon in homes; 27 percent of the monitored group (12 percent of the comparison group) blamed radon on nature; but 46 percent (62 percent of the comparison group) did not know who or what is responsible for high radon levels.

The expected effectiveness of a risk communication program also depends on whether people can get more information about the risk. The New York State Health Department has primary responsibility for providing radon information, but less than 5 percent of the home owners mentioned this agency when asked where they would go for more information on radon. About 20 percent of each group (monitored and comparison) said they would turn to the state EPA, with a somewhat smaller share turning to the federal EPA. However, over 40 percent reported that they did not know where to get more information. This figure probably includes those who think that they do not need additional information. Even so, these results suggest a need to let people know how to acquire more radon information.

The mechanism used to provide information will affect whether the target audience will use it. About 90 percent of both groups said that they would be likely to use a booklet mailed to their homes or to view a TV special on radon. About half said they would attend a town meeting with a panel of experts, and one-third said they would attend a neighborhood meeting or use a library videocassette.

## Designing Information Treatments

The literature on risk communication clearly indicates that the way information is presented affects the way it is received.[6] Baruch Fischhoff has observed that "presenting information in a usable form may require a fairly deep understanding of the cognitive processes of the intended audience."[7] Yet despite agreement that the way information is presented matters, there is no clear consensus in the literature about what specific features communicate risk concepts well. It was therefore necessary to use other guides to structure the features of the radon information materials for the New York study. An advisory panel of experts in marketing research, cancer risk communication, psychology, and environmental journalism helped narrow the range of design alternatives for presenting the radon risk information. The members agreed that a booklet should be relatively short, personal, and include color.

They stressed the importance of testing two issues as part of the research design. First, do people respond better to risk information that is quantitative rather than qualitative? Second, do people respond better to a directive format that gives explicit instructions about what they should do under given circumstances or to a format encouraging judgment and

evaluation in what might be considered a *Consumer Reports* framework? The latter distinction corresponds to the difference between imposing a public health view of risk management and an individual decision view.

These two formats were tested with a series of focus groups. (In a focus group, a trained moderator guides the discussion among six to ten people to yield qualitative information about their knowledge, perceptions, and reactions to the topic being explored.) Participants were quite clearly divided between those who preferred to have a trusted authority tell them how they should deal with the problem and those who preferred to evaluate the facts and make a decision themselves. Since there was no means of identifying people's orientation in advance, it was desirable to test the effects of each approach on random samples of the population.

These considerations yielded four different information booklets on radon risk. They varied according to the dimensions the advisory panel agreed were most important: a quantitative versus qualitative approach and a directive versus *Consumer Reports* (command versus cajole) tone. Focus group sessions helped refine the distinctions among the booklets and suggested changes in language, format, and examples that readers could understand easily.[8] The experiment also incorporated an August 1986 EPA publication, *A Citizen's Guide to Radon*, a middle-case treatment of information that includes some quantitative and some qualitative information and is partly directive and partly suggestive.[9]

Because this social experiment involved real people facing real risks, all participants received the same basic information about radon risks. The core risk information is the same for the five booklets, and all are consistent with EPA's action guidelines. According to these guidelines, home owners should take mitigative action within a few years if their radon readings are between 4 picocuries of radon per liter (pCi/l) of air and 20 pCi/l; within several months if readings range from 20 to 200 pCi/l; and within several weeks if they exceed this range.

The primary differences between the quantitative and qualitative versions are in the risk chart and its explanations. Both versions link radon concentrations to activities with comparable lifetime risks (see figure 1). The actual booklets use a color scale to

**Figure 1. Comparison of approaches to conveying radon risk.**

| QUANTITATIVE<br>Radon Risk Chart | | | QUALITATIVE<br>Radon Risk Chart | |
|---|---|---|---|---|
| Lifetime exposure<br>(picocuries per liter) | Lifetime risk of dying from radon<br>(out of 1,000) | Comparable risks of fatal lung cancer<br>(lifetime or entire working life) | Lifetime exposure<br>(picocuries per liter) | Comparable risks of fatal lung cancer<br>(lifetime or entire working life) |
| 75 | 214−554 | | 75 | |
| 40 | 120−380 | | 40 | |
| 20 | 60−210 | Working with asbestos | 20 | Working with asbestos |
| 10 | 30−120 | Smoking 1 pack cigarettes/day | 10 | Smoking 1 pack cigarettes/day |
| 4 | 13−50 | | 4 | |
| 2 | 7−30 | Having 200 chest x-rays per year | 2 | Having 200 chest x-rays per year |
| 1 | 3−13 | | 1 | |
| 0.2 | 1−3 | | 0.2 | |

*Source:* V. Kerry Smith, William H. Desvousges, Ann Fisher, and F. Reed Johnson, "Communicating Radon Risk Effectively: Interim Report," Research Triangle Institute report (U.S. Environmental Protection Agency, Washington, D.C., July 1987).

distinguish low increments to lifetime risks (green); moderate increments (yellow and orange); and high increments (red). The quantitative version also includes the range of lifetime fatality risk estimates at each radon level and has an example of how to interpret the ranges: "If you have a lifetime exposure to 10 picocuries per liter, your risk of dying from lung cancer caused by radon could be as low as 30 out of 1,000 or as high as 120 out of 1,000."

The differences between the command and cajole versions are in tone (see figure 2). The command version uses only the EPA action guidelines for radon concentrations, while the cajole version includes guidelines from the National Council on Radiation Protection and the Canadian government. Similar subtle differences in tone appear throughout the brochure, with the command version emphasizing what the reader should do and the cajole version emphasizing what the reader may want to consider in reaching a decision.

New York State also developed a sixth information treatment, a one-page fact sheet typical of the information that households usually receive from state agencies and testing companies. The fact sheet includes some background information and two paragraphs about radon risks but contains somewhat less information than the other information treatments. Consequently, NYSERDA and the research team felt that this treatment should be limited to people at very low risk, and only home owners with radon levels below 1 pCi/l were sent the fact sheet.

(One pCi/l is one-fourth of the minimum EPA action level; the national average exposure is 0.8 pCi/l.)

Explicit ethical considerations motivated this decision. In our judgment, the decision satisfied MacLean's requirement for hypothetical consent. The requirement suggests that rational home owners would not have wanted any additional information at such low risk levels.

The final experiment design included all six types of information treatments. Half of those participants who had readings below 1 pCi/l received the fact sheet. One of the five other booklets was randomly assigned to everyone else in the study. Those with readings above 1 pCi/l also received EPA's *Radon Reduction Methods: A Home Owner's Guide.*[10] In December 1986 the interim readings and information materials were mailed to the households that had returned the two-to-three-month monitors. Shortly thereafter participating home owners were interviewed a second time to find out what they had learned and how they had reacted to the information they received.

## Preliminary Results

The analysis is keyed to the three major experimental questions: the extent of learning about radon, the degree of convergence of perceived and measured risk, and whether mitigation is tied to radon levels.[11] The only data on the mitigation question at this time relate to plans for future actions and a hypothetical

**Figure 2.  Major differences in tone of radon brochures.**

| COMMAND | CAJOLE |

**Action Guidelines** (issued by the U.S. Environmental Protection Agency)

**Red:** These levels are very high risks. You should act to reduce these levels, preferably within several months.

**Orange:** Living in these levels for many years presents a high risk. You should act within the next few years to reduce these levels.

**Yellow:** Living in these levels for many years still has some risk. You should see if it is feasible to reduce these levels.

**Green:** These are low levels and have lower risk. The average outdoor level is about 0.2 picocuries per liter. The average indoor level is about 0.8 picocuries per liter.

---

**Because radon risk is cumulative,** it usually is given as lifetime risk. This risk is based on two factors:

- **How long** you are exposed to your radon level: Lifetime risk calculations assume an average "lifetime" of 74 years in a house with a particular radon level.
- **Hours at home** each day: Lifetime risk calculations usually assume you spend about three-quarters of your time, or 18 hours, at home each day.

These assumptions will not fit you exactly, but you should use lifetime risk as a benchmark in making any decisions.

---

**Should I have additional radon tests?**

The monitors still in your home will measure the average amount of radon in your living area for an entire year. You will also get a reading for your basement, where radon levels are likely to be highest. Even if your risks are in the red or orange areas of the colored chart, you should have more than one test before spending any money to fix your home.

---

**Are there any guidelines for radon levels?**

Several government agencies and scientific groups have recommended that actions be taken at various levels.

| Agency or Organization | Radon Level (picocuries) per liter | Action Guidelines |
|---|---|---|
| U.S. Environmental Protection Agency | 20 | Remedial action, preferably within several months |
|  | 4 | Remedial action within next few years |
| National Council on Radiation Protection | 8 | Remedial action |
| Canadian Government | 30 | Prompt action |
|  | 4 | Remedial action |

**What is a lifetime risk?**

Because radon risk is cumulative, it usually is given as lifetime risk. This risk is based on two factors:

- **How long** you are exposed to your radon level: Lifetime risk calculations assume an average "lifetime" of 74 years in a house with a particular radon level.
- **Hours at home** each day: Lifetime risk calculations usually assume you spend about three-quarters of your time, or 18 hours, at home each day.

**Because every household is different,** you may want to adjust the typical risks to fit your circumstances. For example, if you had a reading of 10 picocuries per liter but spend only 9 hours inside your home on a typical day, you would multiply your risk from the colored risk chart on page 4 by one-half or .50. In this case, your risk would now range from as low as 15 out of 1,000 to as high as 60 out of 1,000. Your risk would now be in the beginning of the orange area of the risk chart. If you think lifetime risks are not appropriate for your situation, the next page shows a chart with risks for different exposure periods.

---

**Should I have additional radon tests?**

The monitors still in your home will measure the average amount of radon in your living area for an entire year. You will also get a reading for your basement, where radon levels are likely to be highest. In any case, it is a good idea to check the accuracy of a single test by having more tests before spending any money to fix your home.

*Source*: V. Kerry Smith, William H. Desvousges, Ann Fisher, and F. Reed Johnson, "Communicating Radon Risk Effectively: Interim Report, " Research Triangle Institute report (U.S. Environmental Protection Agency, Washington, D.C., July 1987).

question about whether the respondent would urge a neighbor to act if the neighbor had a specified radon reading.

## Learning about Radon

One simple indicator of the effectiveness of an information program is whether it successfully transfers new information to the target audience. The monitored group was quizzed both before and after it received the readings and materials. The comparison group also was quizzed, but it did not receive any readings or information materials. The quizzes had multiple-choice questions about the risk from radon and how to measure and mitigate it. There was significant learning for all three question categories, although less than half of the respondents could answer the mitigation question correctly, even after reading the materials sent to them.

Quiz performance differs when the responses are separated by type of information materials received. Home owners who received the single-page fact sheet did not improve their scores on the risk questions in the follow-up survey. The comparison group—without receiving any information materials—actually had improved scores by about as much as the average monitored home owner. (The comparison group's learning may reflect increased media coverage of radon, more attention paid to the issue after being contacted for the base-line survey, or the effects of selection bias.) Respondents with higher radon readings generally scored better on the follow-up quiz. Perhaps the higher readings motivated them to pay more attention to the material and retain it.

Socioeconomic characteristics also affected the number of correct responses. Older people had fewer correct answers in the follow-up quiz, as was the case in the base-line quiz. People with more education and prior awareness of radon performed better.

Analysis of each person's responses to questions that were in the base-line survey and were repeated in the follow-up survey gives another indication of the degree of learning that took place. For example, the command-qualitative version increased learning about the health effects of radon. The health information was the same across all five booklets, so something about the command-qualitative booklet must have motivated people to learn more effectively.

This is an example of how format can influence the transfer of information, since there was differential learning about material that was the same in all of the booklets.

Although these variables are fairly crude measures of learning, the results indicate that fairly subtle differences in the information materials can make a difference. All the booklets performed better than the fact sheet, but no single format—for example, command-quantitative versus cajole-qualitative—appeared to be best for all categories of test questions.

## Changes in Risk Perceptions

An important policy concern is whether the materials lead to greater consistency between the risks people perceive and objective measures of those risks based on actual radon readings (and personal characteristics and living conditions). In short, do the "right" people start worrying and the others stop? One indication of changes in perceptions is the perceived seriousness of radon risk compared with other risks.

The personal radon risk question revealed two dramatic changes. The share of people who said they just did not known how serious their risk might be fell from about 25 percent in the base-line survey to less than 5 percent in the follow-up survey. This indicates that the materials helped them form a judgment about their risk. The share who thought their risk was low grew from less than a quarter to nearly half of the respondents in the follow-up survey. Less than 10 percent of home owners actually had radon readings above the EPA guidelines' lowest action level of 4 pCi/l. The follow-up group with low risk perceptions probably includes some people who previously felt they could not judge how serious their risk was. Even so, the information apparently reduced the anxiety of some people at low risk.

About half of the respondents scored radon in the top half of a scale of seriousness that compared risks people face, but only a quarter thought their own risk from radon would be that high. This is consistent with research that shows people believe themselves to be at lower risk from a given cause than most others. For example, nearly all people consider themselves to be safer drivers than average, or less likely to get cancer than average.

There is considerable evidence that people's subjective perceptions of risk differ substantially from the technical estimates developed by experts.[12] For radon, participants in the Maine study seriously underestimated their risk. Other researchers have found little relation between perceived seriousness of risk and radon readings in New Jersey.[13] In the New York study, the reported subjective risks overstated the estimates of the objective risks from radon exposures, especially for people with higher radon readings.

The Maine data were reanalyzed, using a model that allowed people to update their risk perceptions in response to new information and that accounted for differences in individual characteristics. People did revise their estimates of risk in the correct direction after receiving radon information, but the relationship between perceived risk and radon readings was complicated.[14]

Based on a model analysis similar to that used for the Maine data, there is even stronger evidence that the monitored group in the New York study updated risk perceptions in a systematic manner. Those with high radon readings had higher perceived risk in the follow-up survey. The perceptions are scaled with radon readings better than in the Maine data, a difference that could reflect limitations in the Maine survey data or deficiencies in the Maine brochure. The information treatments did significantly affect risk perceptions, with the quantitative NYSERDA booklets leading to greater consistency between perceived and objective risk than did the fact sheet. However, those receiving the qualitative versions or EPA's *Citizen's Guide* had no better convergence between perceived and objective risk than did those receiving the fact sheet.

The exposure assumptions behind the risk estimates (see figure 2) probably overstate exposures for many people, so the cajole versions (in which people were encouraged to adjust their risk estimates for their own circumstances) might have been expected to lead to smaller overstatements. However, this hypothesis is not supported by the analysis so far. On the other hand, having more education, having less interest in health issues, and being white were factors that did tend to reduce the overstatement of objective risk measures.

Responses to a question about which portion of the risk chart contained their scores provide another clue as to how effectively home owners understood risk information. Less than a third of those receiving the *Citizen's Guide* answered this question correctly, compared with 49 to 56 percent of those receiving one of the other booklets. (These results should be viewed tentatively because of the assumptions that had to be made about where to divide the *Citizen's Guide* chart to judge correct responses. Nonetheless, the difference still appears to be substantial.)

People who received the cajole versions did not identify their location on the colored risk chart as well as those who received the command versions of the booklets designed for the experiment. This result may also reflect that readers of the cajole versions were encouraged to adjust their radon readings if their own circumstances differed from the assumptions used to calculate risk in the colored chart. People may be more likely to remember the color matching their adjusted reading than the color matching the reading originally sent to them. In contrast, readers of the command version were instructed to use the population risks on the chart regardless of personal circumstances.

## Advising Neighbors About Radon Risk

There were several reasons to include some questions about what advice a person would give to a neighbor with a specified radon reading. First, this serves to check whether a person fully understands his or her own risk. If people processed the information correctly, they should be able to inform others about how serious the risk would be from a particular level of radon exposure. Second, the distribution of radon readings in the NYSERDA study was skewed toward the low end of the scale. Of course, this distribution was desirable from NYSERDA's perspective, but it narrowed the conclusions to be drawn about responses to a wide range of radon readings and their corresponding risks. Examining the advice one would give a neighbor allowed us to draw from a distribution of higher readings. Third, since people would be unlikely to take action on the basis of a single reading, the questions about advising neighbors would yield some information on whether mitigation might be undertaken at various radon levels.

The main hypothesis is that if people understand the risks from radon, the likelihood of advising neighbors to act should increase with the level of radon. A secondary hypothesis is that they should be able to give their neighbors the correct advice on how soon they should mitigate. The results strongly support our hypotheses: the higher the level of radon depicted in the neighbor's home, the more likely respondents were to recommend that action be taken, and the more likely they were to advise that the action be taken in the time frame recommended by EPA's action guidelines. Those who were more educated and thought lifetime risk was a useful way both to understand risk and plan mitigation were more likely to advise action.

The cajole-qualitative version had a positive effect on the likelihood of one's making an appropriate recommendation about timing the neighbor's action. Home owners who received this version were most likely to recommend the correct timing for their neighbor's mitigation. This version uses the three colored scales to illustrate the difference between annual risk and lifetime risk. It was more effective in helping people make fundamental distinctions among key risk concepts. In addition to the neighbor's radon level, education was important in the recommendation of when to take action.

## Affecting Future Behavior

Too often, risk communication programs have been simple public relations exercises to justify a regulatory agency's decisions. When agencies have tried to provide objective information, they generally have turned to technical experts, who are concerned primarily with the scientific accuracy of the information. Rarely have the agencies considered or evaluated how the program design actually affects behavior and risk exposures.

Abundant evidence from clinical experiments indicates that perceptions are influenced by how risks and related choices are described. However, this study is one of the first attempts to test alternative formats in a field setting with people facing real risks. The results provide unequivocal evidence that differences in information treatment do influence learning, formation of risk perceptions, and intended or recommended behavior:

> The framing of an action sometimes affects the actual experience of its outcomes. For example, framing outcomes [in particular ways] . . . may attenuate one's emotional response to an occasional loss. . . . The framing of acts and outcomes can also reflect the acceptance or rejection of responsibility for particular consequences. . . . When framing influences the experience of consequences, the adoption of a decision frame is an ethically significant act.[15]

If attempts to inform people in objectively identical circumstances lead to different behavioral responses, then public agencies' information programs involve significant ethical responsibilities. In the past, concerns about implicit or hypothetical consent related primarily to mandatory safety regulations. The results here indicate that such concerns are equally relevant for information programs.

Our finding that different groups (for example, older people) respond to the same message differently may indicate the need for specially targeted materials and delivery vehicles. The apparently reasonable strategy of providing less information to groups at low risk generated anxiety and increased the likelihood that such people would overinvest in protective measures. These effects occurred even though most of those receiving the fact sheet indicated that it was useful and understandable. The next stage of the study will eliminate this information treatment, and complete information will be provided to all participants.

The policy implication is that too little risk information can actually do harm even when risks are negligible. Commercial testing companies routinely provide minimal interpretive information when reporting test results. And since people have very little notion of where to turn for reliable help, according to this study, EPA should require appropriate information practices in certifying testing companies.

## Acknowledgment

This research was funded in part by the U.S. Environmental Protection Agency under Cooperative Agreement no. CR-811075.

## Notes

1. U.S. Environmental Protection Agency, "Evaluating and Improving EPA's Risk Advisory Programs" (Program

Evaluation Division, Washington, D.C., May 1987); Robert S. Adler and R. David Pittle, "Cajolery or Command: Are Education Campaigns an Adequate Substitute for Regulation?" *Yale Journal on Regulation* 1(1984):159–93; Abt Associates, "Evaluation of the Effectiveness of Outdoor Power Equipment Information and Education Programs" (U.S. Consumer Product Safety Commission, Washington, D.C., March 1978).

2. Douglas MacLean, "Risk and Consent: Philosophical Issues for Centralized Decisions," in Douglas MacLean, ed., *Values at Risk* (Totowa, N.J.: Rowman and Allanheld, 1986), 17–30.

3. U.S. Environmental Protection Agency, "Unfinished Business: A Comparative Assessment of Environmental Problems" (Washington, D.C., February 1987).

4. F. Reed Johnson and Ralph A. Luken, "Radon Risk Information and Voluntary Protection: Evidence from a Natural Experiment," *Risk Analysis* 7, no. 1 (1987):97–107.

5. Sam Peltzman, "The Effects of Automobile Safety Regulation," *Journal of Political Economy* 83(August-December 1975):677–725.

6. David E. Konouse and Barbara Hayes-Roth, "Cognitive Considerations in the Design of Product Warnings," in L. A. Morris, M. B. Mayis, and I. Borofsky, eds., *Product Labeling and Health Risks*, Banbury Report 6 (Cold Spring Harbor, N.Y.: Cold Spring Harbor Laboratory, 1980); Paul Slovic, Baruch Fischhoff, and Sarah Lichtenstein, "Regulation or Risk: A Psychological Perspective," in Roger Noll, ed., *Regulatory Policy and the Social Sciences* (Berkeley: University of California Press, 1985); Ola Svenson and Baruch Fischhoff, "Levels of Environment Decisions," *Journal of Environmental Psychology* 5(1985):55–67; Vincent T. Covello, Paul Slovic, and Detlof von Winterfeld, *Risk Communication: A Review of the Literature* (New York: Cambridge University Press, forthcoming).

7. Baruch Fischhoff, "Cost Benefit Analysis and the Art of Motorcycle Maintenance," *Policy Sciences* 8(1977):177.

8. William H. Desvousges and Mel Kollander, "Radon Focus Groups: A Summary," Research Triangle Institute report (Office of Policy, Planning, and Evaluation, U.S. Environmental Protection Agency, Washington, D.C., January 1986).

9. U.S. Environmental Protection Agency and Centers for Disease Control, *A Citizen's Guide to Radon: What It Is and What to Do About It*, OPA-86-004 (Washington, D.C.: EPA, August 1986).

10. U.S. Environmental Protection Agency, *Radon Reduction Methods: A Home Owner's Guide*, OPA-86-005 (Washington, D.C.: EPA, August 1986).

11. V. Kerry Smith, William H. Desvousges, Ann Fisher, and F. Reed Johnson, "Communicating Radon Risk Effectively: Interim Report," Research Triangle Institute report (U.S. Environmental Protection Agency, Washington, D.C., July 1987).

12. Paul Slovic, "Perception of Risk," *Science* 236(1987):236–85; Slovic, Fischhoff, and Lichtenstein, note 6 above; W. Kip Viscusi and Charles J. O'Connor, "Adaptive Responses to Chemical Labeling: Are Workers Bayesian Decision Makers?" *American Economic Review* 74(December 1984):942–56.

13. Neil D. Weinstein, Peter M. Sandman, and M. L. Klotz, "Public Response to the Risk from Radon, 1986," Rutgers University Final Report (New Jersey Department of Environmental Protection, Trenton, January 1987).

14. V. Kerry Smith and F. Reed Johnson, "How Do Risk Perceptions Respond to Information? The Case of Radon," *Review of Economics and Statistics* 70, no. 1(1988):1–8.

15. Amos Tversky and Daniel Kahneman, "The Framing of Decisions and the Psychology of Choice," *Science* 211(1981):458.

# Questions for Thought and Discussion

1. The authors refer to the paper by Douglas MacLean dealing with consent for risk taking, which makes the point that social decisions are made by centralized authorities when any of the following conditions exists:

a. when an individual's actions can reduce a risk only if everyone acts similarly, but there is no assurance that others will cooperate;

b. when individual risk-reduction solutions are prohibited because many different actors are each imposing slight risks that are intolerable in the aggregate;

c. when a social consensus about the safety of a large-scale project may be difficult or impossible to obtain.

In each case, describe a real-world situation that is characterized by the given set of conditions.

2. Radon risks have characteristics that are different from risks related to hazardous waste facilities. In what ways do these differences influence public risk perceptions? What implications do these differences have for risk communication? What constitutes effective risk communication in each case?

3. Suppose that someone perceives the magnitude of a man-made risk (e.g., the contamination of drinking water by a leaking underground storage tank) to be far greater than any objective measure of that risk. Explain the circumstances under which you would consider this a misunderstanding and those under which you would consider it a legitimate difference in point of view. What if the risk is not man-made but natural (e.g., the inhalation of radon in the home) and the perception is far *less* than the best technical estimate?

4. EPA's guidelines state that homeowners should take mitigative action within a few years if radon readings are between 4 and 20 picocuries of radon per liter (pCi/l) of air, within several months if readings are between 20 and 200 pCi/l, or within several weeks if the readings are higher. What are the cancer risks associated with lifetime exposure to those levels? In general, EPA regulates chemical carcinogens so that the lifetime risk of dying from exposure to a regulated carcinogen is on the order of $10^{-6}$ to $10^{-5}$. Why do you think EPA recommends mitigation of indoor radon at risk levels different from the levels at which it regulates environmental chemicals?

5. In analyzing the results of their experiments, the authors found that "about half of the respondents scored radon in the top half of a scale of seriousness that compared risks people face, but only a quarter thought their own risk from radon would be that high." They say that this is consistent with research on personal attitudes about driving and about cancer. Why do you suppose people tend to think they are better-than-average drivers? Why do you suppose they think their chance of getting cancer is lower than average? In your opinion, do these reasons help explain the radon results? Why or why not?

# Suggestions for Further Reading

The Conservation Foundation. 1985. *Risk Assessment and Risk Control.* Washington, D.C.: The Conservation Foundation.

A brief but informative report on the basics of the practice of risk assessment as it relates to environmental problems.

Douglas, Mary, and Aaron Wildavsky. 1982. *Risk and Culture: An Essay on the Selection of Technologies and Environmental Dangers.* Berkeley: University of California Press.

An essay about how cultures select the risks to worry about.

Fischhoff, Baruch, Sarah Lichtenstein, Paul Slovic, Stephen L. Derby, and Ralph L. Keeney. 1981. *Acceptable Risk.* New York: Cambridge University Press.

Expert views on risk perception and related topics.

Graham, John D., Laura C. Green, and Marc J. Roberts. 1988. *In Search of Safety.* Cambridge, Mass.: Harvard University Press.

An analysis of how science and regulatory policy interact when chemicals are thought to cause cancer.

Kates, Robert W., Christoph Hohenemser, and Jeanne X. Kasperson, eds. 1985. *Perilous Progress: Managing the Hazards of Technology.* Boulder, Colo.: Westview Press.

Readings on conceptualizing hazards, measuring consequences, and assessing and managing risks of various kinds.

Lowrance, William W. 1976. *Of Acceptable Risk: Science and the Determination of Safety.* Los Altos, Calif.: William Kaufmann.

A landmark discussion of risk, acceptability, and perception.

McCormick, Norman J. 1981. *Reliability and Risk Analysis: Methods and Nuclear Power Applications.* London: Academic Press.

A technical textbook on the analysis of risks from nuclear power and other technologies.

National Research Council. 1983. *Risk Assessment in the Federal Government: Managing the Process*. Washington, D.C.: National Academy Press.

The "red book" that codifies the separation of risk assessment from risk management in the federal government.

National Research Council. 1989. *Improving Risk Communication*. Washington, D.C.: National Academy Press.

Report of a committee on risk perception and risk communication.

Paustenbach, Dennis J., ed. 1989. *The Risk Assessment of Environmental Hazards*. New York: John Wiley & Sons.

A collection of risk assessments from diverse fields.

Petak, William J., and Arthur A. Atkisson. 1982. *Natural Hazard Risk Assessment and Public Policy: Anticipating the Unexpected*. New York: Springer-Verlag.

Facts about reducing the adverse effects of natural hazards on people and property.

Schwing, Richard C., and Walter A. Albers, Jr., eds. 1980. *Societal Risk Assessment: How Safe Is Safe Enough?* New York: Plenum.

A wide-ranging collection of papers about risk assessment, risk management, and risk perception.

Shrader-Frechette, Kristin S. 1985. *Risk Analysis and Scientific Method*. Dordrecht/Boston/Lancaster: D. Reidel.

A philosophical examination of the underpinnings of risk analysis with special emphasis on epistemology, logic, and ethics.

## Also from Resources for the Future

**Worst Things First?  The Debate over Risk-Based National Environmental Priorities**

*Edited by Adam M. Finkel and Dominic Golding*

This book presents findings from a forum convened to explore the controversy over EPA's risk-based approach for setting the nation's environmental priorities. Agreeing that alternative ways exist to target the nation's resources for environmental protection, participants differ sharply as to whether these varied approaches complement each other or would disrupt environmental policy-making. 1994 • 346 pages • ISBN 0-915707-74-8